ADVANCES IN EXPERIMENTAL MEDICINE AND BIOLOGY

Recent Volumes in this Series

Kyung A. Kang · David K. Harrison ·
Duane F. Bruley
Editors

Oxygen Transport to Tissue XXIX

Springer

Editors
Kyung A. Kang
Department of Chemical Engineering
University of Louisville, Louisville
KY 40292, USA
kyung.kang@louisville.edu

David K. Harrison
Durham Unit, Regional Medical Physics
Department, University Hospital of
North Durham, Durham, UK
d.k.harrison@ncl.ac.uk

Duane F. Bruley
Synthesizer, Inc., Ellicott City, MD
USA, UMBC, Baltimore, MD, USA
bruley@umbc2.umbc.edu

ISBN: 978-0-387-74910-5 e-ISBN: 978-0-387-74911-2

Library of Congress Control Number: 2007936187

Printed on acid-free paper

9 8 7 6 5 4 3 2 1

springer.com

Dr. Duane Frederick Bruley
During the Annual 34th ISOTT Conference, August 12–17, 2006
Louisville, Kentucky, USA

The 34th ISOTT Conference President, Kyung A. Kang, would like to dedicate this volume to one of her mentors, Dr. Duane F. Bruley, one of the two founders (the other one being Dr. Haim Bicher) of ISOTT. He has been continuously supportive and helped the conference behind the scenes. Dr. Bruley's studies on oxygen transport to tissue started in 1962 with a prominent medical doctor, Dr. Melvin H. Knisely (Medical School of South Carolina), who was interested in the influence of blood agglutination (coagulation) on oxygen transport to tissue, which stimulated Dr. Bruley to form the Society in 1973. Drs. Bruley and Bicher organized the first ISOTT conference both in Clemson and Charleston, SC, USA, with Dr. Knisely as honorary President. Since then, Dr. Bruley has been a major researcher on oxygen transfer in the human (especially brain) tissue, as well as a pioneer in the mathematical modeling and computer simulation of the human microcirculation system. His mathematical modeling related to oxygen transport led to a unique computational strategy (BWK technique) that performs three-dimensional, time-dependent, heterogeneous, convection, diffusion, conduction, and reaction simulations. He organized another ISOTT conference in 1983, in Ruston, LA, USA. His current research focuses on Protein C, an anticoagulant, antithrombotic, anti-inflammatory and anti-apoptotic protein in blood plasma.

He is an elected fellow of the American Institute of Chemical Engineers, American Society of Mechanical Engineers, and Biomedical Engineering Society, and also an elected founding fellow of the American Institute of Medical and Biological Engineering.

Dr. Britton Chance
During the Annual 34th ISOTT Conference, August 12–17, 2006
Louisville, Kentucky, USA

The 34th ISOTT Conference President, Kyung A. Kang, would also like to dedicate this volume to another mentor of hers, Dr. Britton Chance. He has been a long time ISOTT member and was President of the 4th ISOTT conference in Philadelphia, PA, USA.

Dr. Chance is Eldridge Reeves Johnson University Professor Emeritus of Biophysics at the University of Pennsylvania. His studies on the control of metabolism, especially as it is related to mitochondria, have been just one of the ties to ISOTT. One of his most current research interests is the use of infrared light to characterize the properties of various tissues and cancer. He joined the United States National Academy of Sciences in 1952 and received the National Medal of Science in 1974. He is the inventor of both fNIR and an LED breast cancer screening device using the technology. He also won a gold medal for the United States at the 1952 Summer Olympics in the 5½ Meter Class. He has published more than 700 peer reviewed journal articles.

Kyung A. Kang especially appreciates Dr. Chance's support and his attendance at ISOTT-2006, particularly because he was not in the best of health and was in the middle of preparing to move to Singapore for a long-term research project immediately after the conference.

Group Photo of ISOTT-2006
Kentucky Derby Museum, Churchill Downs, Louisville, Kentucky, USA

This volume is also dedicated to all ISOTT members, participants of ISOTT-2006, and to the contributors of manuscripts for this volume.

Preface

The 34th Annual Conference of the International Society on Oxygen Transport to Tissue (ISOTT) was held during August 12–17, 2006 in Louisville, Kentucky, USA. The emphasis of ISOTT-2006 was on "Expanding our Horizon." In terms of research topics, we added some newer ones – Translational Studies, Tissue Engineering, and Nanobiotechnology. In terms of participants, we put extra effort into including more junior researchers because we felt that they were the future of our society and for the first time in ISOTT history we had presentations made by high school students. In terms of organization, it was truly local, national, and international.

The support in organizing ISOTT-2006 came from various directions: We would like to express our special thanks to the University of Louisville (UofL), especially to the UofL Provost Office, Speed School of Engineering, School of Medicine, and the office of the Vice President of Research. UofL's financial support allowed us to supplement the student's registration fees substantially and to invite so many excellent distinguished lecturers. It should be noted that none of our invited speakers requested an honorarium – we thank them immensely. I would personally like to thank the Chemical Engineering Department of UofL for allowing me to take time off from my teaching responsibility. We appreciate UofL President Ramsey's visit to our dinner held at the Derby Museum. We thank the Case Western Reserve University for having their MIMS Center symposium with us. Their financial support through NIH and participation enable us to have more diverse scientific sessions and more attractive social events. We truly appreciate the participating industries and the many local companies and friends who provided us with monetary support and valuable gifts. We also appreciate the encouragement received from the offices of the Kentucky State Governor and Louisville City Mayor. The international, USA, and local ISOTT-2006 organizing committee members are acknowledged for their constant help, suggestions, and valuable criticisms.

Three ladies, without whom, ISOTT-2006 may not have been possible are: Barbara Johnson, Trinia S. Hill and Carmel F. Mackin. They mysteriously appeared just to help ISOTT-2006 and then quietly disappeared.

ISOTT-2006 had 100 participants and 77 presentations. In total, 42 papers were submitted, reviewed, and accepted for publication. We are very proud of the quality and quantity of the scientific content that we have exchanged/discussed during the conference and published in this volume.

Kyung A. Kang, President of the 34th ISOTT Conference
Ph.D. and Professor of the Chemical Engineering Department
University of Louisville

Organization of ISOTT-2006

The International Society on Oxygen Transport to Tissue (ISOTT) is an interdisciplinary society comprising about 250 members worldwide. Its purpose is to further the understanding of all aspects of the processes involved in the transport of oxygen from the air to its ultimate consumption in the cells of the various organs of the body.

Founded in 1973 by Drs. Duane F. Bruley and Haim Bicher, the society has been the leading platform for the presentation of many of the technological and conceptual developments within the field both at the meetings themselves and in the proceedings of the society.

The annual meeting brings together scientists, engineers, clinicians and mathematicians in a unique international forum for the exchange of information and knowledge, the updating of participants on latest developments and techniques, and the discussion of controversial issues within the field of oxygen transport to tissue.

ISOTT-2006 Officers

Kyung A. Kang, USA	President
David J. Maguire, Australia	Past President
Per Liss, Sweden	President-Elect
Oliver Thews, Germany	Secretary
Peter E. Keipert, USA	Treasurer
Duane F. Bruley, USA	Chairman, Knisely Award Committee

Executive Committee

Chris Cooper, UK	Jerry D. Glickson, USA
Fahmeed Hyder, UK	Paul Okunieff, USA
Valentina Quaresima, Italy	Akitoshi Seiyama, Japan
Peter Vaupel, Germany	Christopher B. Wolff, UK

USA Committee

Joseph LaManna
Chia-Chi Ho
Anthony Hudetz
H. Fred Downey
Fahmeed Hyder
Peter Keipert
Sergei Vinogradov

International Committee

Duane F. Bruley, USA
Chis Cooper, UK
Fredrik Palm, Sweden
Oliver Thews, Germany
Giuseppe Cicco, Italy
Britton Chance, USA
Louis Hoofd, Netherlands
Eiji Takahashi, Japan
Martin P. Wolf, Switzerland

Local Organizing Committee

Kyung A. Kang, President

Meeting Administration Committee

Matt Becker
Cassandra Carmichael
Trinia Simmons Hill
Barbara Johnson
Patricia Lumley
Carmel F. Mackin
Kurt Bendl
A. Maria Utley
Greater Louisville Convention and Visitor's Bureau

Scientific Program Committee

John Barker
Richard L. Benton
Eric Berson
Douglas B. Borchman
Sham S. Kakar
David Magnuson
Rosalie Mainous
David A. Scott
Yang Wang

Student Assistants

James J. Lee
Yongjie Ren
Samin Rezania
Bin Hong
Hanzhu Jin
Karen Boone
Rebecca Vitale

Awards

The Melvin H. Knisely Award was first presented by ISOTT at the 1983 annual banquet to acknowledge a young investigator (35 years of age or younger) for outstanding achievements in research related to oxygen transport to tissue. This award acknowledges the pleasure that Dr. Knisely derived from assisting and encouraging young scientists and engineers to contribute to the study of the transport of anabolites and metabolites in the microcirculation. His many accomplishments in the field have inspired developing investigators to follow in his footsteps. The continuation of this award aims to encourage young scientists and engineers to join ISOTT and aspire to generate high quality research in the area of oxygen transport to tissue. Members of the society are invited to nominate eligible candidates for this award. The award usually includes a Melvin H. Knisely plaque and a cash prize.

Melvin H. Knisely Award Recipients:

1983	Antal G. Hudetz, Hungary	1984	Andras Eke, Hungary
1985	Nathan A. Bush, USA	1986	Karlfried Groebe, Germany
1987	Isumi Shibuya, Japan	1988	Kyung A. Kang, Korea/USA
1989	Sanja Batra, Canada	1990	Stephen J. Cringle, Australia
1991	Paul Okunieff, USA	1992	Hans Degens, Netherlands
1993	David A. Benaron, USA	1994	Koen van Rossem, Belgium
1995	Clare E Elwell, UK	1996	Sergei A. Vinogradov, USA
1997	Chris Cooper, UK	1998	Martin Wolf, Switzerland
1999	Huiping Wu, USA	2000	Valentina Quaresima, Italy
2001	Fahmeed Hyder, Bangladesh	2002	Geofrey De Visscher, Belgium
2003	Mohammad N. Khan, USA	2004	Fredrick Palm, Sweden
2005	Nicholas Lintell, Australia	***2006***	***No award was made***

The Dietrich W. Lübbers Award was established in honor of Professor Lübbers's long-standing commitment, interest, and contributions to the problems of oxygen transport to tissue and to the society. The Lübbers Award is made to a young investigator 30 years of age or younter (with the nomination and sponsorship of an ISOTT member) and will consist of travel support to the meeting at which the award is made. The selection will be based on the scientific excellence of the individual's first authored manuscript on the topic of oxygen transport as judged by the members of the organizing committee of the annual meeting.

Dietrich W. Lübbers Award Recipients:

1994 Michael Dubina, Russia
1995 Philip E. James, UK/USA
1996 Resit Demit, Germany
1997 Juan Carlos Chavez, Peru
1998 Nathan A. Davis, UK
1999 Paola Pichiule, USA
2000 Ian Balcer, USA
2001 Theresa M. Busch, USA
2002 Link K. Korah, USA
2003 James J. Lee, USA
2004 Richard Olson, Sweden
2005 Charlotte Ives, UK
2006 Bin Hong, China/USA

The Britton Chance Award was established in honor of Professor Chance's long-standing commitment, interest and contributions to many aspects of oxygen transport to tissue and to the society. The Chance Award is made to a young investigator 30 years of age or less (with the nomination and sponsorship of an ISOTT member) and will consist of travel support to the meeting at which the award is made. The selection will be based on the scientific excellence of the individual's first authored manuscript on the topic of oxygen transport as judged by the members of the organizing committee of the annual meeting.

Britton Chance Award Recipients:

2004 Derek Brown, Switzerland
2005 James Lee, USA
2006 Hanzhu Jin, China/USA

The Duane F. Bruley Awards were first presented by ISOTT at the 2004 annual meeting in Bari, Italy. They were established to support travel funds for student researchers in all areas of oxygen transport to tissue. The Awards signify Dr. Bruley's interest in seeking young scientists and engineers to maintain the image and quality of research associated with the society. As a co-founder of ISOTT in 1973, Dr. Bruley emphasizes cross-disciplinary research among basic scientists, engineers, medical scientists, and clinicians. It is hoped that receiving the Duane F. Bruley Award will inspire students to excel in their research and will assist in securing future leadership for ISOTT.

Duane F. Bruley Award reciptents:

2004 Helga Blocks (Belgium); Jennifer Caddick (UK); Charlotte Ives (UK); Nicholas Lintell (Australia); Leonardo Mottola (Italy); Samin Rezania (Iran/USA); Ilias Tachtsidis (UK); Liang Tang (China/USA); Iyichi Sonoro (Japan); Antonio Franco (Italy)

2005 Robert Bradley (UK) Harald Oey (Australia) Kathy Hsieh (Australia); Jan Shah (Australia)

2006 Benn S. Gooch (UK); Ulf R. Jensen (Germany); Smruta S. Koppaka (USA); Daya Singh (UK); Martin Tisdall (UK); Bin Wang (China/USA); Kui Xu (China/USA)

Sponsorship

University of Louisville, Office of the University Provost
University of Louisville, Office of Senior Vice President for Research
University of Louisville, Speed School of Engineering
University of Louisville, School of Medicine and the Health Sciences Center
University of Louisville, Chemical Engineering Department
Center for Modeling Integrated Metabolic Systems (MIMS), Supported by a grant (GM66309) from the National Institute of General Medical Sciences, NIH, to the Case Western Reserve University, Cleveland, OH

LOUISVILLE FRIENDS WITH GIFTS

BROWN-FORMAN

ROHM AND HAAS

BEAM GLOBAL SPIRITS & WINE, INC

Samiteon, LLC

Maker's Mark

U of L Print

Metro Louisville Office for International Affairs

Louisville — The Greater Louisville Convention and Visitor's Bureau

INSTITUTIONAL/ INDUSTRIAL PARTICIPATIONS

SOMANETICS

BTI Discovery Technology International

UofL - The Office of Technology Transfer

abcam

ISOTT-2006 editors would like to thank the following reviewers

Panel of Scientific Review:

James Bassingthwaighte, University of Washington, USA
Eric Berson, University of Louisville, USA
Duane Bruley, University of Maryland Baltimore County, USA
Simon Faithfull, Fidelis Consulting, USA
Arthur Fournell, University of Düsseldorf, Germany
Peter Hansell, University of Uppsala, Sweden
David Harrison, University Hospital of North Durham, UK
Fahmeed Hyder, Yale University, USA
Fahmeed Hyder, Yale University, USA
Sham Kakar, University of Louisville, USA
Kyung A. Kang, University of Louisville, USA
Peter Keipert, Sangart Inc., USA
Joe LaManna, Case Western Reserve University, USA
Edwin Nemoto, University of Pittsburgh, USA
Paul Okunieff, University of Rochester, USA
Gerald Saidel, Case Western Reserve University, USA
Oliver Thews, University of Mainz, Germany
Peter Vaupel, University of Mainz, Germany
David Wilson, University of Pennsylvania, USA
Martin Wolf, University Hospital Zurich, Switzerland
Christopher Wolff, St. Bartholomew's and the Royal London Hospital, UK

Technical Review:

Laraine Visser-Isles, Rotterdam, the Netherlands
Eileen Harrison, Durham, UK

Contents

Chapter 1
ISOTT: Roots, Founding and Beyond

Duane Frederick Bruley[1]

Abstract The International Society on Oxygen Transport to Tissue (ISOTT) was founded in April, 1973 by Drs. Duane F. Bruley and Haim I. Bicher. However, the roots of ISOTT go back to Drs. Christian Bohr and August Krogh. Dr. Bruley first wanted to sponsor an international symposium on oxygen transport to tissue to highlight the research activity between his group at Clemson University and Dr. Melvin H. Knisely's group at the Medical College of South Carolina. It was also intended to honor Dr. Knisely for his ingenious development of the Quartz Rod Crystal technique for observing blood flow in-vivo. Later Dr. Bicher was selected to organize the program from the Medical College of South Carolina

With an overwhelming response to the initial call for papers, Drs. Bruley and Bicher made the decision to found an International Society. They then decided on a name, developed the society logo, assigned a mission, developed a charter, sketched the by-laws, and contracted a publisher for the proceedings. The new society was to include a focus on inter and cross-disciplinary research involving theoretical and experimental investigations of oxygen transport to tissue in a single session format. The society meets annually at different venues throughout the world.

1.1 Body

This paper represents an extension of the presentation and paper prepared for the twenty fifth (25th) anniversary of the *International Society on Oxygen Transport to Tissue* on the founding of ISOTT [1]. Similar to most successful research projects it was an exciting serendipitous process, therefore it is important to step back and record the sequence of events that took place before this *special* society (ISOTT) was born.

The roots of ISOTT date back as far as Dr. Christian Bohr (1855–1911), for his pioneering work in respiratory physiology and to August Krogh (1847–1949), when his work conceptualizing the capillary-tissue cylinder for

[1]Synthesizer, Inc., Ellicott City, MD, USA; UMBC, Baltimore, MD, USA.

K.A. Kang et al. (eds.), *Oxygen Transport to Tissue XXIX*

oxygen transport was awarded the Nobel Prize (1920). In my opinion Krogh was the first Tissue Engineer because he quantified the physical system using mathematical modeling and calculated molecular transport in and around the microcirculation. Dr. Krogh studied under Dr. Bohr as his teaching assistant and continued his studies throughout his professional career. Dr. Melvin H. Knisely (1900–1975) served as Dr. Krogh's Post Doctoral Fellow which propelled him on a career of studies related to blood agglutination in the capillaries and experimental research on the resulting tissue destruction and disease. Dr. Knisely has been cited as the first person to observe the pathological clumping of red and white cells, in vivo, at the capillary level. He identified these phenomena as "blood sludging" and pointed out its negative impact on oxygen transport to the tissue and to the removal of toxic metabolic byproducts.

Colleagues, on occasion, have said that there were discussions regarding the possible establishment of a society on oxygen transport. If that is true I was never a part of any of the discussions nor was I even contacted by a colleague to be part of such an effort. Also, I have been asked whether or not Dr. Knisely approached me to create a symposium in his honor or to form a society. Just for clarification, I never had any discussions related to the development of a symposium or a society with Dr. Knisely prior to gaining permission from Clemson to host a meeting at Clemson University, Clemson, SC, USA.

My formal education was in traditional chemical, mechanical, and nuclear engineering. In the fall of 1962 I accepted a position as Assistant Professor of Chemical Engineering and Head Varsity Tennis coach at Clemson University in Clemson, South Carolina, USA. That fall a colleague, Dr. William Barlage, and I were discussing possible new research areas; thus, we decided to take a five hundred mile round trip to the Medical College of South Carolina in Charleston, SC to see if there were problems involving "living systems" that we could apply our engineering skills to. Being traditional engineers neither of us had a formal education in the biological or life sciences and had studied only non-living systems. To clarify, even though traditional engineers can make significant contributions to the engineering of living systems a new breed has evolved, the Bioengineer, which represents the fifth traditional discipline of engineering [2]. A definition that I have frequently used for bioengineering is as follows: "Bioengineering is The Application of Engineering Principles and Fundamentals to Engineering Problems that *Require* Basic Understanding of the Biological and/or Life Sciences." This definition states that modern Bioengineers must have a formal education that includes the biological and/or life sciences thus giving them insight into processes involved in living systems that would not be obvious to traditional engineers. This concept has a **foundation** in the principles upon which ISOTT was founded.

On the second day of our visit to the Medical College and after several meetings, without success, we were standing outside of the Anatomy Department when Dr. Melvin H. Knisely (Head of the Department) appeared and introduced himself. After a brief discussion he invited us to lunch where he stated his interest in mathematical modeling and computer simulation of oxygen transport in the grey matter of brain. He was concerned about the viability of neurons under different pathological conditions and he thought that computer predication could be valuable.

This problem was of interest to me since I had recently completed my Ph.D. Dissertation that consisted of experimental and theoretical work on the thermal dynamics of a wetted-wall-column [3]. My theoretical model consisted of a computer simulation of a coupled set of partial differential equations describing simultaneous heat and mass transfer in cylindrical coordinates. The equations contained terms for convection and conduction in two space dimensions and time and were solved using finite difference techniques via Fortran programming. This research fit perfectly with the description of the Krogh Capillary Tissue model and the problems associated with the solution of representative models that scientists and engineers around the world were then exploring to quantify the microcirculation. After a year of study to learn the necessary physiology and anatomy and the translation of two German articles, one by Opitz and Schneider [4] and the other by Thews [5] (help in translation was provided by Isebel Lockard and Elsie Tabor in Dr. Knisely's Laboratory) I derived a mathematical model, from basic principles (the Bruley Model), that was solved by various graduate students on digital, analog and hybrid computers, for different anatomical and physiological conditions. This research represented the first computer simulations of the microcirculation, and a major step forward in quantitative analysis because computer simulation allowed investigation of the dynamic and non-linear characteristics of the system.

These studies started in 1962 and we worked together until Dr Knisely's death in 1975. During that period we published about 35 papers together regarding theoretical and experimental investigations of oxygen transport to tissue.

In 1968 Dr. Haim I. Bicher was recruited to our team because of his knowledge of blood agglutination and his expertise in the construction and use of oxygen micro electrodes. His contribution to our research effort allowed us to work back and forth between theory and experiment thus giving us the best possible research environment. We presented our work primarily at the European Microcirculation meetings and published in a variety of journals. It was then that we started to examine anti-adhesive drugs in an attempt to prevent clotting and to reverse the consequences of blood agglutination [6]. This initial work has led to my current studies of Protein C, a blood factor that might be the ultimate anticoagulant/antithrombotic/anti-inflammatory/anti-apoptotic for Protein C deficient patients, because there are little or no known side effects such as, bleeding complications with the zymogen [7].

In 1971 our team attended a workshop on oxygen supply at The Max-Planck Institute in Dortmund, Germany. It was then that I decided to inquire about sponsoring a symposium at Clemson University to highlight our team work with Dr. Knisely's group at The Medical School of South Carolina. Immediately after I returned to the United States I asked Dr. Edwards, the President of Clemson University, for permission to host an oxygen transport to tissue symposium at Clemson University and with it honor Dr. Melvin H. Knisely for his many contributions to the field of microcirculation. Particularly for his development of the quartz rod crystal illumination technique that allowed him to visualize the sticking together of blood components, in vivo, in many disease states [8]. Permission was granted so I called Dr. Knisely's wife, Verona, to find out what she thought about it. After a short time Verona called back and said it was a good idea but she thought it would be better to have the symposium at The Medical College of South Carolina. With further discussion we decided to have a symposium at both campuses, with bus transportation in between. Both Dr. Edwards, President at Clemson University and Dr. McCord, President of The Medical School of South Carolina agreed to help fund the symposium.

When Dr. Bicher returned from an extended trip to Israel, I asked him if he would like to participate in setting up the symposium. He was anxious to do so and he then took responsibility for further arrangements at the Medical School while I handled all arrangements at Clemson University and the combined meeting. Together we obtained additional support from other companies and agencies to fund the meeting.

The intended purpose of the symposium was to promote interdisciplinary and cross-disciplinary research involving theoretical and experimental investigations for oxygen transport in tissue. It was to bring together life scientists and engineers in a single session format to examine the many complex phenomena of normal tissue growth and maintenance, and tissue survival and repair under pathological conditions. This has remained the mission for ISOTT since its birth and is probably the precursor to what is defined as "Tissue Engineering," today.

After an intensive period of planning and preparation an initial meeting announcement was sent out to sample community interest. The results demonstrated enthusiasm far beyond projections and triggered Drs. Bruley and Bicher to consider the meeting as a launching pad for a very focused international society regarding oxygen transport to tissue. We presented our idea to several other investigators and then we decided that a formal society would be in the best interest of groups around the world to achieve research goals related to oxygen transport in tissue and that the Charleston/Clemson meeting would be an appropriate forum to formalize and begin an international society. We then decided on the name "International Society on Oxygen Transport to Tissue," designed a society "logo," assigned a mission, developed a charter, sketched the by-laws, contracted with Plenum Publishers to publish the meeting proceedings, and selected members to comprise an International Committee for the Clemson/Charleston meeting. The membership consisted of the following scientists and engineers:

Dr. Melvin H. Knisely, Charleston, USA
Dr. Haim I. Bicher, Charleston, USA
Dr. Ian A. Silver, Bristol, England
Dr. Britton Chance, Philadelphia, USA
Dr. Lars-Erik Gelin, Goteborg, Sweden
Dr. Manfred Kessler, Dortmund, Germany
Dr. William J. Whalen, Cleveland, USA
Dr. Duane F. Bruley, Clemson, USA
Dr. Gerhard Thews, Mainz, West Germany
Dr. Herbert J. Berman, Boston, USA
Dr. Leland C. Clark, Jr., Cincinnati, USA
Dr. Jurgen Grote, Mainz, West Germany
Dr. Jose Strauss, Miami, USA
Dr. Daniel D. Reneau, Ruston, USA

Drs. Bruley and Bicher solicited Dr. Melvin H. Knisely to serve as an Honorary President of the Society for the initial symposium. At the Clemson/Charleston meeting, ISOTT was founded, and the following slate of officers were elected:

President-Elect- Dr. Gerhard Thews, Mainz, West Germany
Secretary- Dr. Haim I. Bicher, Charleston, USA
Treasurer- Dr. Ian A. Silver, Bristol, England

The first symposium of ISOTT surpassed all expectations and established a society that has continued to meet annually at various locations around the world. The registered participants numbered 267 and two proceedings volumes consisting of 133 papers were published by Plenum Press in their "Advances in Experimental Medicine and Biology" series [9, 10].

Society meetings have been held at the following locations under the leadership of the listed presidents:

1973	Charleston, SC, USA (Honorary)	M.H. Knisely, Founding Meeting
1974		Group Meeting
1975	Mainz, Germany	G. Thews (First Elected President)
1976	Anaheim, CA, USA	B. Chance
1977	Cambridge, U.K.	I. A. Silver
1978	Atlantic City, NJ, USA	J. Strauss
1979	La Jolla, CA, USA	J. Strauss
1980	Budapest, Hungary	A. Kovach
1981	Detroit, MI, USA	H. Bicher
1982	Dortmund, Germany	D. Lubbers
1983	Ruston, LA, USA	D. F. Bruley
1984	Nijmegen, The Netherlands	F. Kreuzer
1985	Raleigh, NC, USA	I.S. Longmuir
1986	Cambridge, UK	I.A. Silver
1987	Sapporo, Japan	M. Mochizuki (Carl Honig)
1988	Ottawa, Canada	K. Rakusan
1989	Gottingen, Germany	J. Piiper
1990	Townsville, Australia	M. McCabe
1991	Curacao, Dutch Antilles	W. Erdmann
1992	Mainz, Germany	P. Vaupel
1993	San Diego, CA, USA	P.D. Wagner
1994	Istanbul, Turkey	C. Ince (K. Akpir)
1995	Pittsburgh, PA, USA	E. M. Nemoto
1996	Dundee, Scotland	D.K. Harrison
1997	Milwaukee, WI, USA	A.G. Hudetz (25th Anniversary)
1998	Budapest, Hungary	A. Eke
1999	Hanover, NH, USA	H. Swartz
2000	Nijmegen, The Netherlands	B. Oeseburg
2001	Philadelphia, USA	D.F. Wilson
2002	Manchester, UK	M.S. Thorniley
2003	Rochester, USA	P. Okunieff
2004	Bari, Italy	G. Cicco
2005	Brisbane, Australia	D. Maguire
2006	Louisville, USA	K. Kang

The 2007 meeting will be held in Uppsala, Sweden where Dr. Per Liss will serve as President.

In 1983 at the Ruston, Louisiana meeting Dr. Bruley initiated the first Melvin H. Knisely Award to a promising young investigator. This award was then approved and established by the Executive Committee to express the spirit and willingness of Dr. Knisely to work with and contribute to the growth of beginning scientists and engineers addressing the problems of oxygen transport to tissue. Dr. Bruley was then elected as the Chairman of the "Melvin H. Knisely Award" selection committee and nominees have been reviewed each year with those selected being honored at the annual banquet.

The recipients, through the 2006 meeting in Louisville, USA are as follows:

1983 Antal G. Hudetz (Hungary)
1984 Andras Eke (Hungary)
1985 Nathan A. Bush (USA)
1986 Karlfried Groebe (Germany)
1987 Isumi Shibuya (Japan)
1988 Kyung A. Kang (Korea/USA)
1989 Sanjay Batra (Canada)
1990 Stephen J. Cringle (Australia)
1991 Paul Okunieff (USA)
1992 Hans Degens (The Netherlands)
1993 David A. Benaron (USA)
1994 Koen van Rossem (Belgium)
1995 Clare E Elwell (UK)
1996 Sergei A. Vinogradov (USA)
1997 Chris Cooper (UK)
1998 Martin Wolf (Switzerland)
1999 Huiping Wu (USA)
2000 Valentina Quaresima (Italy)
2001 Fahmeed Hyder (Bangladesh)
2002 Geoffrey De Visscher (Belgium)
2003 Mohammad Nadeem Khan (USA)
2004 Frederick Palm (Sweden)
2005 Nicholas Lintell (Australia)
2006 No Awardee Selected

In 1994 a second Award to support travel for a young investigator was approved by the Executive Committee. The recipients of the "Dietrich W. Lubbers Award" are as follows:

1994 Michael Dubina (Russia)
1995 Philip E. James (UK/USA)
1996 Resit Demir (Germany)
1997 Juan Carlos Chavez (USA)
1998 Nathan A. Davis (UK)
1999 Paolo Pichiule (USA)
2000 Ian Balcer (USA)
2001 Theresa M. Busch (USA)
2002 Lino K. Korah (USA)
2003 James J. Lee (USA)
2004 Richard Olson (Sweden)
2005 Charlotte Ives (UK)
2006 Bin Hong (China/USA)

The Britton Chance Award was established in 2003 in honor of Professor Chance's long-standing commitment, interest and contributions to many aspects of oxygen transport to tissue and to the society. The award is to recognize outstanding contributions to research by a young investigator to help support travel to the ISOTT meeting. The Britton Chance Awardees are as follows:

2004 Derek Brown (Switzerland)
2005 James Lee (USA)
2006 Hanzhu Jin (China/USA)

The Duane F. Bruley Awards were established and were first presented by ISOTT at the 2004 annual meeting in Bari, Italy. They were established to support travel funds for student researchers in all areas of oxygen transport to tissue. The Awards signify Dr. Bruley's interest in seeking young scientists and engineers to maintain the image and quality of research associated with the society. As a co-founder of ISOTT in 1973, Dr. Bruley emphasizes cross-disciplinary research among basic scientists, engineers, medical scientists, and clinicians. His pioneering work constructing mathematical models for oxygen and other anabolite/metabolite transport in the microcirculation, employing computer solutions, were the first to consider system non-linearities, time dependence, including multi-dimensional diffusion, convection, and reaction kinetics. It is hoped that receiving the Duane F. Bruley Award will inspire students to excel in their research and will assist in securing future leadership for ISOTT. The Duane F. Bruley Awardees are as follows:

2004	2005	2006
Helga Blocks (Belgium)	Robert Bradley (UK)	Ben Gooch (UK)
Jennifer Caddick (UK)	Harald Oey (Australia)	Ulf Jensen (Germany)
Charlotte Ives (UK)	Kathy Hsieh (Australia)	Smruta Koppaka (USA)
Nicholas Lintell (Australia)	Jan Shah (Australia)	Daya Singh (UK)
Leonardo Mottola (Italy)		Martin Tisdall (UK)
Samin Rezania (USA/Iran)		Bin Wong (USA)
Ilias Tachtsidis (UK)		Kui Xu (USA)
Liang Tang (USA/China)		
Iyichi Sonoro (Japan)		
Antonio Franco (Italy)		

As pointed out earlier the first society proceedings were published by Plenum Press [9, 10]. However, the number of total proceedings published has been confused by the mixing of two different publisher's "mistaken" use of two different names. Some of the first meeting proceedings were published under the Library of Congress Cataloging title of "International *Symposium* on Oxygen Transport to Tissue" rather than the official title of "International *Society* on Oxygen Transport to Tissue." Since the two titles are listed separately the uninformed might not be aware of both sets of proceedings and some libraries do not have all of the volumes.

At the 25th Anniversary it was approved by the Executive Committee and the membership-at-large to proceed with arrangements to establish a

Journal for ISOTT with Plenum Press. The publications committee now consists of:

Duane F. Bruley, Chairman	Chris Cooper
Antal G. Hudetz	Joe C. LaManna
Kyung A. Kang	Hal Schwartz
David Harrison	Britton Chance

Many attempts to start a journal have failed for various reasons. However, we are still active and working with several publishers to develop a society journal. Because ISOTT remains small in numbers, by choice, most publishers do not feel a journal would be profitable.

The future of ISOTT will be determined by our young and new members, with the dedicated mentoring of our old time membership. It will be important to stay current with new technology and be flexible enough to embrace new directions in the area of oxygen transport to tissue. The vision of ISOTT members will be critical in guiding this very special international scientific and engineering society through the troubled waters created by politics and religion.

References

1. Bruley, D.F., The Genesis of ISOTT, *Oxygen Transport to Tissue XX,* edited by A. G. Hudetz and D. F. Bruley, Plenum Press, New York, 1998.
2. Bruley, D. F., "Bioengineering: The Fifth Traditional Engineering Discipline," edited by W. Erdmann and D. F. Bruley, Advances in Experimental Medicine and Biology, Plenum Press, Vol. 317:3–6, 1992.
3. Bruley, D. F., and J. W. Prados, "The Frequency Response Analysis of a Wetted Wall Adiabatic Humidifier," *AlChE Journal*, 11,612, Septmeber, 1964.
4. Opitz, E., and M. Schneider, "The oxygen Supply of the Brain and the Mechanism of Deficiency Effects," *Ergebnisse der Physiologie, Biologischem Chemic, und Experimentellen Pharmakologic*, 46:126–260, 1950.
5. Thews, G., "Oxygen Diffusion in the Brain. A Contribution to the Question of the Oxygen Supply of the Organs," *Pflugers Archiv*., 271:197–226, 1960.
6. Bicher, H. I., Bruley, D. F., and M. H. Knisely, "Anti-Adhesive Drugs and Tissue Oxygenation," edited by D. F. Bruley and H. I. Bicher, Advances in Experimental Medicine and Biology, Plenum, Press, Vol. 37B657–667, 1973.
7. Bruley, D. F., and W. N. Drohan, "Protein C and Related Anticoagulants," Advances in Applied Biotechnology Series, Vol. 11, Gulf Publishing Company (Portfolio Publishing Company), 1990.
8. Goro, F. W., "Blood Sludge," *Life magazine*, Vol. 24, No. 22:49–59, May 31, 1948.
9. *Oxygen Transport to Tissue- Instrumentation, methods, and physiology*, edited by H. I. Bicher and D. F. Bruley, Advances in Experimental Medicine and Biology, Vol. 37A, Plenum Press, 1973.
10. *Oxygen Transport to Tissue- Pharmacology, mathematical studies, and nematology*, edited by H. I. Bicher and D. F. Bruley, Advances in Experimental Medicine and Biology, Vol.37B, Plenum Press, 1973.

Chapter 2
Dietrich W. Lübbers

Celebration of a Life Dedicated to Research into Oxygen Transport to Tissue

David K. Harrison[1]

Dietrich Werner Lübbers: 1917–2005.

2.1 Biography

It was with great sadness that members of the International Society on Oxygen Transport to Tissue heard of the death on 15th November 2005 of Dietrich Werner Lübbers, one of its most distinguished and long-standing members.

He was born on 12th May 1917 in the Harburg district of Hamburg. He attended the Landesschule Pforta, a celebrated German public boarding school

[1]Durham Unit, Regional Medical Physics Department, University Hospital of North Durham, DH1 5TW, UK.

near Naumburg on the Saale river, where he completed his "Abitur" (the school qualification for entry to university) in 1935. He completed his 6 months compulsory labour service before being called up for compulsory military service.

As early as his last years at high school, and throughout his medical course, he developed the ambition to pursue a career in scientific research applied to medicine. From 1937 to 1939 he studied medicine at the University of Heidelberg (5 semesters). However, in order to gain a basic scientific education that had been missing from the curriculum at school, in addition to his medical course, he studied chemistry for 4 semesters. At the outbreak of the second world war he was conscripted for active service, but was able to continue his study of medicine, but not chemistry (with interruptions for active service) in Halle, Leipzig and Berlin.

Dietrich's scientific career began in 1941 with the research project for his medical dissertation in the Institute of Physiology at the University of Berlin under Professor Kurt Kramer. Kramer had demonstrated that near infrared light penetrated deep into tissue so that spectral changes could be detected from the outside. Dietrich's project was to investigate the oxygen supply in the frog heart and, to this end, using haemoglobin as the indicator, applied near infrared spectroscopy to measure oxygen. The first problem he encountered was that the frequency response of the existing manometers was too low to measure the pulsatile pressure. In order to overcome this problem he began the first of his technical developments. Together with Professor Gerlach of the Institute of Physics, University of Berlin, he built a glass plate manometer, which used the capacitance principle, and this enabled him to successfully complete his dissertation in 1944. His thesis "A method for measurement of the O_2 consumption and dynamics of the isolated cold blooded animal heart" clearly set the theme for much of his future research. In the meantime he had passed his final medical examinations in 1943 and was working in the army medical corps.

In December 1944 he found himself a prisoner of war in France where, from time to time, he acted as the camp doctor. He was not released until June 1948. From 1948 to 1950 he held a clinical post at the Borstel Tuberculosis Research Institute near Hamburg

On the basis of the experimental experience gained during his Dr.med. studies he was awarded a post with Professor Erich Opitz in Kiel in 1950. Opitz's field of research was the exchange processes between oxygen in capillary blood and mitochondria. Dietrich's research project was to investigate the time course of the oxygen supply and oxygen consumption in the beating mammalian heart. Since his study would involve measurements of the oxygenation of haemoglobin, myoglobin and the redox state of cytochromes, he decided on a spectrophotometric approach to the problem. However, this could clearly only be achieved with a very fast measuring instrument that scanned a wide range of wavelengths, and multi-component analysis of the absorption spectra. After intensive discussions with Dr Köhler, a physicist, he set about with Walter Niesel to develop the so-called short-time spectral analyser, which was completed in 1957. This was further developed by the Howaldtswerke in Kiel as the "Rapidspektroskop" (see below). He completed his Habilitation whilst at Kiel and in 1956/57 was the guest of Briton Chance at The Johnson Foundation, University of Pennsylvania, where they were investigating the redox state of cytochrome c using spectrophotometry. At Kiel, Dietrich also started other research in the field of electrodes and blood gas measurement (see below).

From 1959 to 1961 he was a supernumerary assistant professor, at the Institue of Physiology, University of Cologne where the Director was Max Schneider.

In 1961 he was appointed to a Personal Chair at the Institute for Applied Physiology and Occupational Physiology, University of Marburg and in 1965 was appointed Professor of Applied Physiology and Director of the Institute in Marburg after turning down a chair in Hanover. In Marburg Dietrich continued his development of oxygen electrodes and, together with Albert and Renate Huch, started to develop the concept of the transcutaneous pO_2 for monitoring neonates. Horst Baumgärtl joined Dietrich in Marburg and together they went on to build the multiwire surface pO_2 electrode and the finest of needle electrodes for quantitative measurements of pO_2 in tissue. Manfred Kessler completed his Habilitation with Dietrich in Marburg.

In 1968 he was appointed Director of the Max Planck Institute for Occupational Physiology in Dortmund. In 1973 the Institute was renamed the Max Planck Institute for Systems Physiology reflecting Dietrich's approach to the investigation of biological systems. The renaming of the Institute caused great local controversy as the original institute was seen as one that carried out scientific research for the benefit of the ordinary worker. Dietrich had to weather a fierce barrage of criticism in the local press [1]. This was quite unfair as he had been instrumental, along with the state of North Rhein Westphalia, for the founding of an Institute for Occupational Physiology at the University of Dortmund, and of which he was the acting Director initially [2]. In 1985 he "Retired" and became Emeritus director of the Institute, which moved to a new building and was renamed the Institute for Molecular Physiology in 1994. Dietrich retained a laboratory in the Institute until 2003.

Amongst those who went with Dietrich to Dortmund were Manfred Kessler and Horst Baumgärtl. There, of course, with scientists such as Elfriede Leninger-Follert, Helmut Acker, Wolfgang Grünewald, Renate Huch, Sebastian Schuchhardt and Reinhard Wodick – to mention but a few – his institute was enormously productive and unravelled many of the mysteries of local regulation of oxygen supply to tissue.

It was in 1977 that I first met Dietrich – and not through oxygen. My PhD project was the development of a pH electrode for use in human skin and he had organized a symposium on the Theory and Application of Ion-Selective Electrodes in Physiology and Medicine at the Dortmund Institute. That was the first time, too, that I met my good friend and colleague, the late Jens Höper and of course Manfred Kessler whom I worked for in Erlangen from 1981 to 1990. I had, of course, come to know of Dietrich's work as soon as I started my project in 1974. My colleague, Vance Spence had started two years earlier on a project to develop a skin pO_2 electrode and had been the guest of Dietrich who introduced

Vance to the art of making needle electrodes. Enough of my biography. Suffice it to say that it was at this time that Dietrich's philosophy of the systems approach to physiology had a huge influence on me and has remained with me throughout my scientific career.

2.2 The Inventions

"The biological problem was always the basic drive for him. If it turned out that the known methods were not good enough to enable him to solve it, he undertook the laborious task of developing the necessary tools himself", Gerhard Thews [3].

Below is the ranking Dietrich himself put to his inventions when he was awarded the Diesel Gold Medal in 1997 of the German Institute for Inventions[4]:

- Photometry at the surfaces of scattering media such as on live organs (6 Patents)
- Blood gas analysis (electrochemical sensors) (3 Patents)
- pO_2 and pCO_2 measurements in situ (6 Patents)
- Optical sensors with absorbent and fluorescent optical indicators (optodes) (22 Patents).

As mentioned earlier, by 1957 Dietrich Lübbers, along with Walter Niesel, had built their first fast spectrophotometer in the Institute workshop in Kiel. In collaboration with the Instrumentation Department of the Hohwaldtwerke shipyard the spectrometer was further developed and marketed as the T 13/3 Rapidspektroskop [5]. This instrument could record 100 spectra (25,000 measurements) per second, and was the fastest commercially available instrument of its kind in 1964. It was a dual beam spectrometer which relied, of course, at that time on analogue electronics. Depending on the diffraction grating used, it had a wavelength range of 230–600 nm or 350–700 nm. Use of the full spectral range allowed the use of multicomponent analysis to deconvolute the spectra of individual pigments such as oxy- and deoxyhaemoglobin, myoglobin and cytochromes. This was further facilitated after he moved to Dortmund by the rapid evolution of the digital computer and software for the evaluation of the spectra which was developed by Hoffmann. By now the spectrophotometer and dedicated computer hardware was so large that for clinical research, the patients had to go to the laboratory at Max Planck Institute for the measurements. In order to get round this, together with Wodick and Pieroth, he developed a "portable" lightguide spectrophotometer that later was marketed by Sigma as the Oxyscan (see, for example, Merschbrock et al. [6]).

The aim of Dietrich Lübbers' research was to understand the entire pathway and regulation of oxygen transport from the blood into the mitochondria. To do this he needed to be able to measure the pO_2 in the tissue itself. Until the late 1950s polarographic measurements in tissue were fraught with difficulties and interferences. However, the invention of the fully integrated pO_2 electrode by

Leland Clark [7] changed all that and opened up a whole new realm of inventiveness for Dietrich. One of the earliest electrodes Dietrich developed was for in vitro measurements. However, his development of electrodes was not limited to oxygen. His team also constructed pH and pCO_2 electrodes so that they could carry out blood gas analysis during their physiological experiments. Later on he developed electrodes for other ions. As a result of these developments, the company Eschweiler, also based in Kiel, produced one of the first blood gas analysers to appear on the market – the Combi-Analyser U in 1961. Its successor is available nowadays as the Combi Line.

After his move to Marburg, together with Horst Baumgärtl and Manfred Kessler, Dietrich continued to develop electrodes for measurements in tissue. The multiwire surface electrode (MDO) was one of the trusty tools of the physiological investigation of oxygen transport to tissue. He also applied the polarographic principal to the measurement of blood flow using hydrogen clearance – a technique that I also became very involved in during the 1980s. The advantages of the MDO were that it was non-invasive and had a high resolution: 98% catchment depth of each wire approx. 60 μm. The random distribution of its 8 wires meant that 13 small rotations of the electrode gave a statistical distribution of pO_2 consisting of more than 100 values – a process that took only 3 or 4 minutes. Assessment of pO_2 histograms on most organs, revealed a remarkable similarity under physiological conditions: a Gaussian distribution always with less than 5% of values less than 5mmHg. The histogram brought life to the Krogh model of oxygen supply from the capillaries to the tissue and demonstrated that, again under physiological conditions, a highly efficient regulation of blood flow prevents anoxia occurring in the so-called lethal corner – the cells at the venous end of the capillaries.

Dietrich always questioned his own methods, and he wanted to test how representative the pO_2 histograms measured using the multiwire electrode were. He and Horst Baumgärtl therefore produced what I think must be the finest tipped Clark type needle electrode ever made in order to carry out measurements within tissue. He presented their results at his last ISOTT meeting in Nijmegen in 2000 [8]. He was able to show that with increasing distances between pO_2 histogram measurement points, the histogram remained unchanged thus showing that measurements of histograms with the MDO, which encompass several capillary supply units, do indeed represent the distribution within a single unit.

The citation for the award of the Diesel Gold Metal stated: "Of particular significance, then, is the fact that not only did he invent things that were highly innovative – at the same time he endeavoured to put his ideas into clinical practice" [2]. An excellent example of this is the development of the transcutaneous pO_2 electrode which he continued after his move to Dortmund. He, along with the Huchs, discovered that the blood supply of the skin of newborn babies is so high that it was possible to effectively measure arterial pO_2 across the skin. It was important, however, that the hyperaemia, which was induced by heating the skin, always remained sufficient for the pO_2 to remain independent of changes in blood flow. How this was achieved was the subject of yet another of Dietrich's patents and the transcutaneous pO_2 electrode [9] was adopted throughout the world in neonatal intensive care units for many years as an indispensable monitoring device until it was eventually superseded by the pulse oximeter.

Dietrich became interested in fluorescence lifetime measurements of pO_2 in the early 1970s whilst trying to study angiogenesis and oxygen supply in the dorsal skin fold chamber in the rat. Measuring pO_2 was a problem because they had to open the chamber to do so. He decided that this new optical technique would be suitable and with Norbert Opitz developed so-called optodes (optical electrodes) [10]. They found, however, that the fluorescence-pO_2 calibration was unstable when the indicator was placed directly in the tissue. His trick was to sandwich the optical sensor between an oxygen-permeable and an oxygen-impermeable membrane. In a further development, the indicator itself was bound into a membrane. pH and pCO_2 optodes followed and, in collaboration with a number of commercial companies, the technology has been incorporated into blood gas analysers, single use flow-through devices and catheter devices for continuous monitoring.

In typical fashion, Dietrich used his inventiveness to apply his fluorescence sensor technology to another physiological question. Some years earlier, his group had discovered, using pO_2 microelectrodes, that the pO_2 in the upper layers of skin is a function of depth. He wanted to know to what extent atmospheric air was the source of oxygen supply to the skin, and what clinical implications this might have. He therefore adapted the fluorescence sensor to measure the oxygen flux across the skin. In collaboration with Markus Stücker at the Dermatological Clinic in Bochum they carried out a number of investigations and were able to demonstrate that cutaneous blood flow contributes little to the oxygen supply of the upper layers of skin [11]. Recognising the clinical importance of the discovery – particularly for the treatment of diabetic, venous and ischaemic skin diseases – in collaboration with Dr Paul Hartmann of AVL (later to become part of Roche Diagnostics) an oxygen flux imaging system was developed [12].

However, never reliant on the results provided by a single methodology, it was at this time (2001) that I was recruited to the Dortmund team – albeit for just 2 weeks – to apply the transcutaneous hydrogen clearance technique that I had developed based on his transcutaneous pO_2 electrode. The idea was to test the reverse hypothesis, i.e. that if the blood flow contributed significantly to the oxygen supply at the surface of the skin at normal skin temperature, the freely diffusible, biologically inert hydrogen carried by the blood would be detected at the surface of the skin. Although we were unable to complete a full series of experiments at the time, for Dietrich our preliminary results provided further confirmation of the important role of atmospheric oxygen supply to the skin [13].

I have just used the example of skin, but through Dietrich's long scientific career he was a prolific publisher of some 450 papers reporting studies involving the oxygen supply of all of these organs, cells and organelles: heart, brain, liver, carotid body, kidney, eye, tumours, inner ear, lymphatics, olfactory lobe, placenta, capillaries, mitochondria, erythrocytes and many more.

On a personal note, Horst Baumgärtl [1] told me that it was always difficult for Dietrich to get away from the Institute to go on holiday with his family. With almost predictable regularity some sort of calamity occurred just before he was due to go on leave. On one occasion the garage door fell down on his head; on another he damaged his leg; on another he injured his wrist. In the end the members of the Institute assumed that these events had nothing to do with chance or accidents, but an unconscious reluctance to leave his scientific endeavours.

2.3 Honours and Awards

Dietrich Lübbers' achievements were recognised with a host of honours and awards: Member of the New York Academy of Science; Honorary Professor, Ruhr University, Bochum; Corresponding member of the Mainz Academy of Science and Literature, 1975; President of ISOTT, 1981–2; President of the German Physiological Society, 1984; Honorary member of the German Physiological Society 1986; Honorary member of the German Microcirculation Society, 1985 Honorary member of the Association for Occupational Physiology and Occupational Safety, 1977. He was awarded the Ratschow Medal of the German Society of Angiology in 1985 and, as mentioned above, the Diesel Gold Medal of the Diesel Trust at the German Institute for Inventions in 1997. The first Dietrich W Lübbers Award was awarded by ISOTT in 1994.

2.4 ISOTT

Dietrich was a member of the first International Committee of ISOTT [14] and attended almost every meeting until 2000. Indeed it was the workshop organised at the Max Planck Institute in Dortmund in July 1971 that was probably the inspiration for Duane Bruley to organise, with Melvin Knisely, what turned out to be the first ISOTT meeting in April 1973 in Charleston, South Carolina [14]. My first ISOTT meeting was in Dortmund in 1982 and as a young scientist I was in awe of Dietrich who always had a challenging question for the discussion. However, as I got to know him well, I learnt that it was simply a passionate searching for the scientific truth that inspired his questions.

Dietrich was usually accompanied by his wife Angela to the meetings and together they were very much part of the ISOTT "Family". In 1996 I had the honour of welcoming him to the ISOTT meeting I organised in Dundee. Manfred Kessler used to refer to Dietrich as his "scientific father" and my "scientific grandfather". It was a term I think that Dietrich himself didn't really approve of, mainly, I believe, because it made him feel old. His health started to make it difficult for him to travel long distances, and the Nijmegen meeting in 2000 was the last ISOTT he attended.

Many of us remember Dietrich as a dedicated scientist with a sharp mind, always ready to discuss new ideas and concepts. As someone who was always seeking after the scientific truth, it may be a surprise to learn that he was a shameless story-teller. He used to tell all sort of tall stories to his children about strange goings on in a castle they were passing, or the wildest tales about the river they were walking alongside [15]. He was also a religious man. Although professing to be unmusical, he and his family often sang together a particular song in German by Matthias Claudius "The moon has risen" (Tr. Catherine Winkworth, 1855) that begins:

Look up; the moon tonight
Shows us but half her light,
And yet we know her round and fair.
At other things how oft
We in our blindness scoffed
Because we saw not what was there.

The words of the whole song reveal that Dietrich was fascinated by the beauty and mysteries of God's creation – and this is clearly reflected in his scientific endeavours to understand it.

Dietrich felt greatly honoured by the Society's decision to present an annual award to young scientists bearing his name and it is indeed very fitting that through the Dietrich Lübbers Award, members of ISOTT will continue to recognise and celebrate his enormous contribution to research in the field of oxygen transport to tissue.

References

1. H. Baumgärtl, Personal communication, (2006).
2. P. Vaupel, Obituary for Dietrich Lübbers. *Yearbook of the Academy of Science and Literature, Mainz, 2005/2006* (Steiner Verlag, Stuttgart, 2006), pp. 127–130.
3. G. Thews, The scientific works of Dietrich Lübbers. *Commemorative Volume on the Occasion of the Retiral of Prof. Dr med. D. W. Lübbers, 31st May 1985*. edited by R. Kinne, H. Acker and E Leniger-Follert (Max-Plank-Institute for Systems Physiology, Dortmund, 1985).
4. D. W. Lübbers, Personal papers, ca 1996.
5. W. Niesel, D. W. Lübbers, D. Schneewolf, J. Richter and W. Botticher, Double beam spectrometer with 10-msec recording time. *Rev. Sci. Inst*. 35, 578–581 (1964).
6. U. Meschbrock, J. Hoffmann, L. Caspary, J. Huber, U. Schmicholy and D. W. Lübbers, Fast wavelength scanning spectrophotometer for non-invasive determination of hemoglobin oxygenation in human skin, *Int. J. Microcirc*. 14, 274–281 (1994).
7. L. C. Clark, Monitoring and control of blood and tissue oxygen, *Trans. Am. Soc. Artif. Organs* 2, 41–8 (1956).
8. H. Baumgärtl, W. Zimelka and D. Lübbers, Evaluation of profiles to describe the oxygen pressure field within the tissue. *Comp. Biochem. & Physiol. A* 132, 75–85 (2002).

9. R. Huch, D. W. Lübbers and A Huch, Quantitative continuous measurement of oxygen pressure on the skin of adults and newborn babies. *Pflüg. Arch. Ges. Physiol.* 337, 185–198 (1973).
10. D. W. Lübbers and N. Opitz, The pCO_2-/pO_2-optode: a new probe for measurement of pCO_2 or pO_2 in fluids and gases. *Z. Naturforsch. C Biosci.* 30, 532–533 (1975).
11. M. Stücker, A. Struk, P. Altmeyer, M. Herde, H. Baumgärtl and D. W. Lübbers, The cutaneous uptake of atmospheric oxygen contributes significantly to the oxygen supply of human dermis and epidermis, *J. Physiol.* 538, 985–994 (2002).
12. P. Hartmann, W. Ziegler, G. Holst and D. W. Lübbers, Oxygen flux fluorescence lifetime imaging, *Sens. Actuators B* 38–39, 110–115 (1997).
13. D. Harrison, D. W. Lübbers, H. Baumgärtl, C. Stoerb, S. Rapp, P. Altmeyer and M. Stücker, Capillary blood flow and cutaneous uptake of oxygen from the atmosphere. In: *Progress in Biomedical Optics and Imaging: Functional Monitoring and Drug-Tissue Interaction*, editors G. J. Müller and M. Kessler. *SPIE Proc Series* 4623, 195–205 (2002).
14. D. F. Bruley DF. The genesis of ISOTT, *Adv Exp Med Biol.* 454, 1–6 (1998).
15. A. Lübbers, Personal communication, (2006).

Part I
Oxygen Transport in Tissue

Chapter 3
Investigation of Frontal Cortex, Motor Cortex and Systemic Haemodynamic Changes During Anagram Solving

Ilias Tachtsidis[1], Terence S. Leung[1], Martin M. Tisdall[2], Presheena Devendra[1], Martin Smith[2], David T. Delpy[1], and Clare E. Elwell[1]

Abstract We have previously reported changes in the concentrations of oxy- ($\Delta[HbO_2]$) deoxy- ($\Delta[HHb]$) and total haemoglobin ($\Delta[HbT] = \Delta[HbO_2] + \Delta[HHb]$) measured using near infrared spectroscopy (NIRS) over the frontal cortex (FC) during an anagram solving task. These changes were associated with a significant increase in both mean blood pressure (MBP) and heart rate (HR). The aim of this study was to investigate whether the changes in MBP previously recorded during an anagram solving task produces associated changes in scalp blood flow (flux) measured by laser Doppler and whether any changes are seen in NIRS haemodynamic measurements over a control region of the brain (motor cortex: MC). During the 4-Letter anagram task significant changes were observed in the $\Delta[HbO_2]$, $\Delta[HHb]$ and $\Delta[HbT]$ in both the frontal and motor cortex ($n = 11$, FC $p<0.01$, MC $p<0.01$). These changes were accompanied by significant changes in both MBP ($n = 11$, $p<0.01$) and scalp flux ($n = 9$, $p = 0.01$). During the 7-Letter anagram task significant changes were observed in the $\Delta[HbO_2]$ and $\Delta[HbT]$ ($n = 11$, FC $p<0.01$, MC $p<0.01$), which were accompanied by significant changes in both MBP ($n = 11$, $p = 0.05$) and flux ($n = 9$, $p = 0.05$). The task-related changes seen in MBP and flux in this study appear to contribute to the changes in the NIRS signals over both the activated and control regions of the cortex.

3.1 Introduction

A major aim of functional mapping studies of the human brain is to monitor the magnitude and spatial distribution of activity associated with brain function. To that extent cranial functional near-infrared spectroscopy (NIRS) has been widely used to investigate the haemodynamic changes which occur in response

[1]Department of Medical Physics and Bioengineering, Malet Place Engineering Building, Gower Street, University College London, London, UK, WC1E 6BT.
[2]The National Hospital for Neurology and Neurosurgery, Queen Square, London, UK.

to functional activation of specific regions of the cerebral cortex. Based on the tight coupling of neuronal activity and oxygen delivery, changes in the concentration of oxygenated ($\Delta[HbO_2]$) and deoxygenated ($\Delta[HHb]$) haemoglobin as measured by NIRS are quantified and taken as indicators of cortical activation.

NIRS is increasingly being used to monitor the haemodynamic response over the frontal and prefrontal regions during cognitive tasks such as colour Stroop [1], working memory [2], Wisconsin card sorting test [3], calculations [4], mathematical problems [5], playing video games [6], and anagram solving tasks [7]. It is possible that some mental tasks used in these studies may elicit a systemic response which may affect the measured NIRS signals. We have previously reported that significant changes in mean blood pressure (MBP) and heart rate (HR) occur during anagram activation tasks and observed that NIRS haemodynamic changes were in some volunteers significantly correlated with these systemic changes [8].

The aim of this study is to investigate whether the changes in MBP during anagram solving tasks produce associated changes in scalp blood flow and whether any changes are seen in NIRS haemodynamic measurements over a control region of the brain.

3.2 Materials and Methods

3.2.1 Subjects

11 healthy volunteers (6 males and 5 females) all right handed with English as their first language (age 20 to 36 years; mean 25 years) took part in this study.

3.2.2 Instrumentation

A continuous wave near-infrared spectrometer with a sampling rate of 6Hz (NIRO 300, Hamamatsu Photonics KK) was used to measure changes in tissue $[HbO_2]$ and [HHb] using the modified Beer-Lambert law. The optodes from the dual channel system were placed on the head based on the 10/20 EEG electrode placement system. Channel 1 was placed on the left motor cortex (MC) responsible for finger and hand movement identified as the C3 position. Channel 2 was placed on the left frontal cortex (FC) identified as the Fp1 position. Both channels were shielded from ambient light by using an elastic bandage and a black cloth. An optode spacing of 4 or 5cm was used in order to optimise the detected light intensity. For the conversion of the optical attenuation changes to concentration changes a differential pathlength factor (DPF) of 6.26 was applied [9]. A Portapres® system (TNO Institute of Applied Physics) was used to continuously and non-invasively measure MBP and HR from the finger. Finally a laser Doppler probe (FloLab, Moore Instruments) was placed over the forehead to monitor the changes in scalp blood flow (flux) in nine of the eleven subjects.

3.2.3 Procedure

All the volunteers were positioned in a comfortable sitting position. Data were recorded during two minutes of the subject at rest (baseline), followed with one minute period of the subject solving 4-Letter anagrams (15 anagrams, 4 seconds per anagram) and then with one minute period of the subject solving 7-Letter anagrams (6 anagrams, 10 seconds per anagram). Each anagram-solving period was repeated a total of three times, with the study ending after a 2-minute rest period (total study time 10 minutes). In this study solving an anagram was defined as producing one coherent word using only the letters from another word (e.g. icon–coin; reserve–reverse).

The subjects were encouraged to solve as many anagrams as possible and were instructed to say possible solutions out loud (without moving); however, the subjects were not scored on their performance.

3.2.4 Analysis

The NIRS haemoglobin signals were first detrended to remove any slow drift, then all the signals including MBP, HR and flux, were low pass filtered at 0.08Hz to minimise the effects of other signal components. The filtering was carried out by a 5th order low pass Butterworth digital filter in forward backward directions to avoid introducing a phase delay (MatLab Mathworks Inc). The filtered signals from each volunteer were ensemble averaged over the repetition cycles (per volunteer two rest periods, three 4-Letter periods and three 7-Letter periods). Changes in total haemoglobin concentration (Δ[HbT]) were calculated from the sum of $\Delta[HbO_2]$ and Δ[HHb].

The response to stimulation was calculated as the difference between the average of 10 seconds worth of baseline data at the end of the rest period, and the average of 10 seconds of data commencing 20 seconds after the onset of the 4-Letter anagram solving period and the 7-Letter anagram solving period respectively. A 'Student's t-test' was used to assess the significance of these responses (the threshold of significance was set at $p \leq 0.05$ from baseline). Correlations between variables were analysed with the Pearson correlation model.

3.3 Results

3.3.1 Activation Results

Figure 3.1 shows the grand average of the NIRS, MBP and scalp flux data from all volunteers during the entire ten minute test. Table 3.1 shows the mean response of each signal during 4- and 7-Letter anagram solving.

During the 4-Letter anagram task significant changes were observed in the $\Delta[HbO_2]$ (n = 11, FC $p<0.01$, MC $p<0.01$), $\Delta[HHb]$ (n = 11, FC $p=0.05$, MC $p<0.01$) and $\Delta[HbT]$ (n = 11, FC $p<0.01$, MC $p<0.01$) in both the frontal and motor cortex. These changes were accompanied by significant changes in both MBP (n = 11, $p<0.01$) and flux (n = 9, $p=0.01$). During the 7-Letter anagram task significant changes were observed in the $\Delta[HbO_2]$ (n = 11, FC $p<0.01$, MC $p<0.01$) and $\Delta[HbT]$ (n = 11, FC $p<0.01$, MC $p<0.01$),

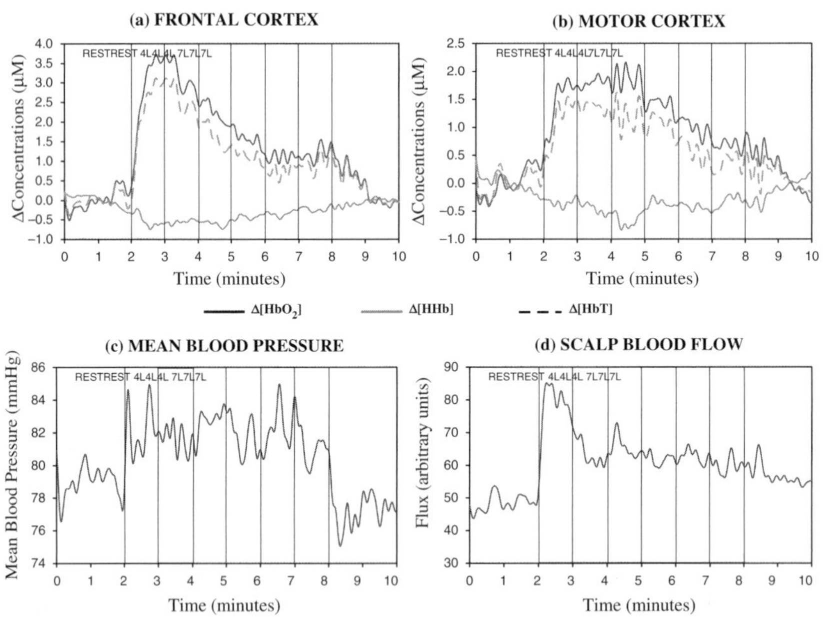

Fig. 3.1 Grand averaged responses for $\Delta[HbO_2]$, $\Delta[HHb]$ and $\Delta[HbT]$ for all 11 subjects measured over the (a) frontal cortex and (b) motor cortex; (c) average ($n = 11$) mean blood pressure; (d) average ($n = 9$) scalp blood flow. (4L: 4-Letter Anagrams, 7L: 7-Letter Anagrams.)

Table 3.1 Response of NIRS signals over the motor and frontal brain regions (MC: motor cortex; FC: frontal cortex) and MBP and Flux during 4- and 7-Letter anagram solving. Data from all volunteers are presented as means±SD

	No Subjects	4-Letters minus Rest		7-Letters minus Rest	
		MC	FC	MC	FC
$\Delta[HbO_2]$ (μM)	11	1.55±1.14*	2.04±1.37*	1.34±1.23*	1.83±1.26*
Δ[HHb] (μM)	11	−0.48±0.51*	−0.38±0.62‡	−0.28±0.66	−0.26±0.68
Δ[HbT] (μM)	11	1.08±1.23*	1.65±1.28*	1.07±1.20*	1.57±1.07*
MBP (mmHg)	11	4.7±4.4*		3.3±5.2‡	
ΔFlux (%)	9	50.2±56.5†		18.0±25.2‡	

(t-test *p<0.01; †p<0.03; ‡p≤0.05)

which were accompanied by similar significant changes in both MBP (n = 11, p = 0.05) and flux (n = 9, p = 0.05). The changes in Δ[HHb] during the 7-letter anagram task were not significant. No significant differences were found between the 4-Letter and 7-Letter anagram activation periods for the NIRS and MBP signals.

3.3.2 *Inter-subject Correlation*

The $\Delta[HbO_2]$ and Δ[HHb] signals measured over the frontal and motor cortex regions were found to have a varying association with the MBP and flux signals across different volunteers. In order to investigate this we calculated the correlation coefficient between the filtered $\Delta[HbO_2]$ and MBP, Δ[HHb] and MBP; $\Delta[HbO_2]$ and flux, and Δ[HHb] and flux for both frontal and motor cortex in all subjects. These results are shown in Figs. 3.2 and 3.3

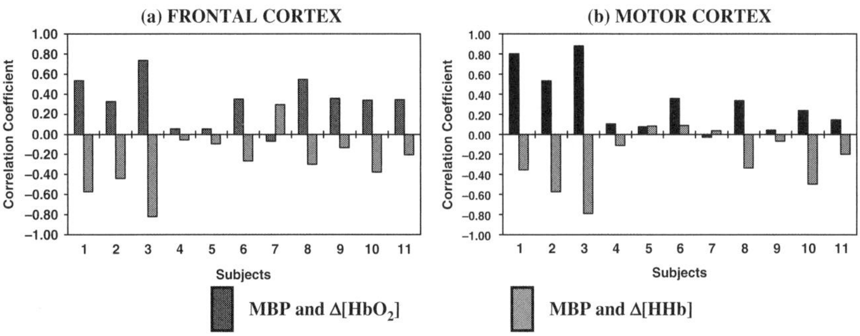

Fig. 3.2 Individual correlation coefficients between MBP and $\Delta[HbO_2]$ and MBP and Δ[HHb] for each subject for (a) the frontal cortex and (b) the motor cortex.

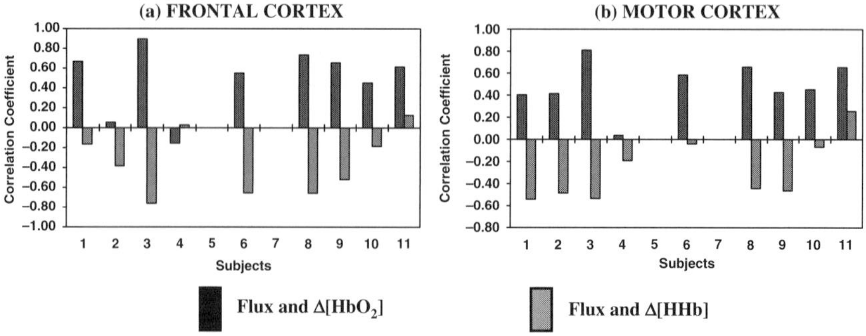

Fig. 3.3 Individual correlation coefficients between flux and $\Delta[HbO_2]$ and flux and $\Delta[HHb]$ for each subject. For subjects 5 and 7 the scalp flux signal was not collected.

3.4 Discussion

In this study we observed significant changes in the $[HbO_2]$, [HHb] and [HbT] measured over both the left frontal and motor cortex regions during a 4-letter anagram solving task. We also observed significant changes in the $[HbO_2]$ and [HbT] measurements during a 7-letter anagram solving task. Furthermore, in the group data, we observed a significant increase from rest in both MBP and scalp flux when the subjects were solving the 4- and 7-letter anagrams. We found that the haemoglobin changes measured by NIRS over the frontal and motor cortex during anagram activation were in some volunteers significantly correlated with the changes in MBP and scalp flux.

During the anagram task there is no reason to expect haemodynamic changes over the motor cortex. The task-related changes seen in MBP and flux in this study appear to contribute to the changes in the NIRS haemodynamic signals over the activated and control regions of the cortex. It is possible that the anagram task elicits an emotional response, which produces changes in blood pressure that are likely to cause passive changes in the scalp blood flow as observed in the laser Doppler flux signal. These changes in the scalp blood flow can produce small changes in the $[HbO_2]$ and [HHb] signals as measured by cranial NIRS.

When analysing cerebral haemodynamic activation data using functional neuroimaging the task-specific activation observed is due to the existence of a close coupling between regional changes in brain metabolism and regional cerebral blood flow. In order for this response to be monitored unambiguously it is important that the haemodynamic task-related activity is occurring on top of an unchanged global systemic and brain resting state. The blood pressure and scalp flux changes observed in this study suggest that systemic task related responses may also be present and that they may lead to haemodynamic changes characteristic of functional activation changes in a control region of the brain.

The relatively high correlation coefficient found in some subjects in this study between the NIRS haemodynamic measurements with the MBP and scalp flux signals suggest a global task-related haemodynamic response. In the absence of high resolution maps of haemodynamic response it is difficult to determine whether changes in the NIRS signals are due to the global changes in systemic variables or haemodynamic changes originating from specific regions of the cerebral cortex.

There are numerous recent publications using functional NIRS where differences in response have been reported, for example in frontal and prefrontal activation between healthy volunteers and schizophrenic patients [10], between healthy volunteers and adults with pervasive developmental disorders [11], between adults and preschool children [12], between men and women [13], and between different age groups from 20 to 90 years old [14]. In none of the studies mentioned above were systemic changes monitored. We suggest that caution should be exercised when analysing quantitatively the cerebrovascular response during frontal and prefrontal activation due to the unknown haemodynamic contribution from systemic alterations occurring during the stimulation.

Acknowledgment The authors would like to thank the UCL/UCLH trustees, the EPSRC/MRC, grant No GR/N14248/01 and Hamamatsu Photonics KK. This paper is dedicated to the memory of Grigoris Xatzieustratiou (22 July 1978–01 October 2006).

References

1. M. L. Schroeter, S. Zysset, F. Kruggel, and D. Y. von Cramon, Age dependency of the hemodynamic response as measured by functional near-infrared spectroscopy, *NeuroImage* **19**(3), 555–564 (2003).
2. Y. Hoshi, B. H. Tsou, V. A. Billock, M. Tanosaki, Y. Iguchi, M. Shimada, T. Shinba, Y. Yamada, and I. Oda, Spatiotemporal characteristics of hemodynamic changes in the human lateral prefrontal cortex during working memory tasks, *NeuroImage* **20**(3), 1493–1504 (2003).
3. S. Sumitani, T. Tanaka, S. Tayoshi, K. Ota, N. Kameoka, S. Ueno, and T. Ohmori, Activation of the prefrontal cortex during the wisconsin card sorting test as measured by multichannel near-infrared spectroscopy, *Neuropsychobiology* **53**(2), 70–76 (2006).
4. A. Villringer, J. Planck, C. Hock, L. Schleinkofer, and U. Dirnagl, Near Infrared Spectroscopy (NIRS): a new tool to study hemodynamic changes during activation of brain function in human adults, *Neurosci. Lett.* **154**(1–2), 101–104 (14-5-1993).
5. Y. Hoshi and M. Tamura, Near-infrared optical detection of sequential brain activation in the prefrontal cortex during mental tasks, *NeuroImage* **5**(4 Pt 1), 292–297 (1997).
6. S. Nagamitsu, M. Nagano, Y. Yamashita, S. Takashima, and T. Matsuishi, Prefrontal cerebral blood volume patterns while playing video games–a near-infrared spectroscopy study, *Brain Dev.* **28**(5), 315–321 (2006).
7. B. Chance, S. Nioka, S. Sadi, and C. Li, Oxygenation and blood concentration changes in human subject prefrontal activation by anagram solutions, *Adv. Exp. Med. Biol.* **510**, 397–401 (2003).

8. I. Tachtsidis, T. S. Leung, L. Devoto, D. T. Delpy, and C. E. Elwell, Measurement of frontal lobe functional activation and related systemic effects: a near-infrared spectroscopy investigation., *Adv. Exp. Med. Biol.* (2007).
9. A. Duncan, J. H. Meek, M. Clemence, C. E. Elwell, L. Tyszczuk, M. Cope, and D. T. Delpy, Optical pathlength measurements on adult head, calf and forearm and the head of the newborn infant using phase resolved optical spectroscopy, *Phys. Med. Biol.* **40**(2), 295–304 (1995).
10. Y. Kubota, M. Toichi, M. Shimizu, R. A. Mason, C. M. Coconcea, R. L. Findling, K. Yamamoto, and J. R. Calabrese, Prefrontal activation during verbal fluency tests in schizophrenia–a near-infrared spectroscopy (NIRS) study, *Schizophr. Res.* 77(1), 65–73 (2005).
11. H. Kuwabara, K. Kasai, R. Takizawa, Y. Kawakubo, H. Yamasue, M. A. Rogers, M. Ishijima, K. Watanabe, and N. Kato, Decreased prefrontal activation during letter fluency task in adults with pervasive developmental disorders: A near-infrared spectroscopy study, *Behav. Brain Res.* **172**(2), 272–277 (2006).
12. S. Tsujimoto, T. Yamamoto, H. Kawaguchi, H. Koizumi, and T. Sawaguchi, Prefrontal cortical activation associated with working memory in adults and preschool children: an event-related optical topography study, *Cereb. Cortex* **14**(7), 703–712 (2004).
13. J. Leon-Carrion, J. Damas, K. Izzetoglu, K. Pourrezai, J. F. Martin-Rodriguez, J. M. Martin, and M. R. Dominguez-Morales, Differential time course and intensity of PFC activation for men and women in response to emotional stimuli: a functional near-infrared spectroscopy (fNIRS) study, *Neurosci. Lett.* **403**(1–2), 90–95 (2006).
14. I. L. Kwee and T. Nakada, Dorsolateral prefrontal lobe activation declines significantly with age–functional NIRS study, *J. Neurol.* **250**(5), 525–529 (2003).

Chapter 4
Do Red Blood Cell-β-Amyloid Interactions Alter Oxygen Delivery in Alzheimer's Disease?

Joy G. Mohanty[1], D. Mark Eckley[2], J. D. Williamson[3], L. J. Launer[4], and Joseph M. Rifkind[1]

Abstract Oxygen delivery requires that Red Blood Cells (RBCs) must be deformable to pass through the microcirculation. Alzheimer's disease (AD) is a progressive neurodegenerative disorder characterized by abnormal extracellular deposition of β-amyloid peptide (Aβ) and neuronal loss. We have analyzed RBC morphology in blood from subjects with AD and found that >15% of the RBCs are elongated as compared to 5.9% in normal controls ($p<0.0001$). To determine whether these morphology changes can be associated with the greater exposure of RBCs to Aβ in AD subjects, we investigated the in vitro effect of Aβ fibrils on blood. Morphological analysis of RBCs treated with $A\beta_{1\text{-}40}$ or $A\beta_{1\text{-}42}$ fibrils show 8.6% or 11.1% elongated cells, respectively. In contrast, only 2.9% or 1.3% of RBCs are elongated when blood is treated with buffer or mock fibrils generated from $A\beta_{42\text{-}1}$. Elongated RBCs are expected to be less deformable. This prediction is consistent with our earlier studies showing impaired deformability of RBCs treated with Aβ fibrils. An additional factor previously reported by us, expected to impair the flow of RBCs through the microcirculation is their adherence to endothelial cells (ECs) when $A\beta_{1\text{-}40}$ fibrils are bound to either RBCs or ECs. This factor would be more pronounced in AD subjects with elevated levels of Aβ on the vasculature. These results suggest that Aβ interactions with RBCs in AD subjects can result in impaired oxygen transport and delivery, which will have important implications for AD.

4.1 Introduction

Alzheimer's disease (AD) is characterized by neuronal degeneration and synaptic loss [1, 2]. These changes have been associated with the extracellular deposi-

Corresponding author: Joy G. Mohanty, e-mail: MOHANTYJ@MAIL.NIH.GOV.
[1]Molecular Dynamics Section, [2]Image Informatics & Cell Biology Unit, Laboratory of Genetics, National Institute on Aging, Baltimore, MD, [3]Roena Kulynych Center for Memory and Cognition Research, Department of Medicine, Wake Forest University School of Medicine, Winston-Salem, NC, [4]Laboratory of Epidemiology, Demography and Biometry, National Institute on Aging, National Institutes of Health, Bethesda, MD.

tion of plaques containing β amyloid (Aβ) fibrils and neurofibrillary tangles. In addition to this, Aβ is also found in the vasculature causing amyloidosis in AD subjects [3–5].

Neuronal function uses ~20% of the total oxygen consumed by an individual. The potential contribution of impaired oxygen delivery to the brain causing neuronal dysfunction associated with AD has thus been considered an important factor in AD. It has in fact been proposed that AD may originate as a vascular disorder with the resultant impairment of oxygen delivery and oxidative changes initiating the cascade of neuronal changes found in AD [6].

The delivery of oxygen to the brain requires that Red Blood Cells (RBCs) carrying oxygen need to deform in order to pass through the narrow pores of the ~600km of capillaries that supply oxygen to the human brain. Vascular changes associated with amyloidosis are expected to directly impair blood flow. Furthermore, in the capillaries the RBCs come in intimate contact with the vasculature where it encounters Aβ involved in amyloidosis. Thus, the possible contribution of RBC interactions with Aβ to impaired oxygen delivery needs to be considered.

Aβ fibrils are formed by both $A\beta_{1\text{-}40}$ and $A\beta_{1\text{-}42}$ peptides. While $A\beta_{1\text{-}42}$ is the major constituent of senile plaques, $A\beta_{1\text{-}40}$ is the main component of vascular amyloidosis [4, 5]. Our earlier studies [7, 8] have shown that both $A\beta_{1\text{-}40}$ and $A\beta_{1\text{-}42}$ fibrils bind to RBCs increasing RBC volume and decreasing RBC deformability. Thus, we hypothesized that perhaps RBCs in Alzheimer's subjects experience increased binding to the available beta amyloid peptides or fibrils in blood resulting in a decrease in the RBC deformability thus hindering the passage of RBCs through the microvasculature. In this study, morphological characteristics of RBC samples from Alzheimer's subjects and their age-matched controls have been investigated. The potential influence of Aβ binding to RBCs leading to their morphological changes in Alzheimer's subjects has also been investigated following incubation of $A\beta_{1\text{-}40}$ and $A\beta_{1\text{-}42}$ fibrils with blood from normal healthy subjects.

4.2 Materials and Methods

Human blood samples from six Alzheimer's subjects and ten age-matched controls were collected in heparinized (green cap) tubes at Roena Kulynych Center, Wake Forest University School of Medicine, Winston-Salem, NC, following an approved protocol and sent overnight on icepack to Molecular Dynamics Section (MDS), National Institute on Aging at Baltimore, MD for processing. Subjects ($N = 16$) in this approved ancillary study were members of the Ginkgo Evaluation of Memory Study (GEMS, $N = 3{,}072$) recruited between September 2000 and May 2004 [9] and at study entry they were free of dementia based on full neuropsychological assessment. Subjects with significant neurological or neurodegenerative diseases that by themselves would affect cognitive function or carried a higher risk of dementia (e.g., Parkinson's disease) were excluded.

Similarly, anyone on cognitive enhancers or treatments for AD (cholinesterase inhibitors) was excluded at baseline. Data on cognitive function were collected every 6 months. According to the GEMS protocol, if the subject failed the full NPB on scores below cutoff in memory and one other cognitive domain, or failed five subtests across the battery (including one test in the memory domain) they were referred for neurological evaluation and MR scan and dementia adjudication (Alzheimer's dementia, vascular dementia, other dementia, no dementia) to an expert panel composed of neurologist and neuropsychologist from the University of Pittsburgh Alzheimer's Disease Research Center. This ancillary study was also approved by the Human Subjects Review Board of Wake Forest University. Participants referred for adjudication of possible dementia and their proxy provided informed consent for one additional blood draw and a control from within the GEMS study remaining free of dementia was also asked to be a part of this ancillary study, and after informed consent, blood was drawn for analysis.

Unless otherwise indicated all chemicals including Citrate-Phosphate-Dextrose solution with Adenine (CPDA) were obtained from Sigma Aldrich, St. Lois, MO.

4.2.1 Processing of Blood Samples and Microscopy of RBCs

Blood samples upon arrival at MDS were immediately centrifuged at $\sim$1125 $\times$g for 10min at 4 °C. The buffy coat and plasma were carefully collected from the top of the tube and discarded. The RBC pellet was washed 3 times by resuspending the cells in CPDA buffer and centrifuging them as above. Cells were resuspended in CPDA buffer to 5% hematocrit and cell morphology was recorded in an Olympus IX-70 with 40X objective using Deltavision software. Microscopic images were saved as TIFF files, opened by Adobe Photoshop and elongated cells were counted manually in each frame. Then all the calculation and statistical analysis was performed using MS Excel as well as Origin software.

4.2.2 Preparation of Amyloid Fibrils

Amyloid fibrils from $A\beta_{1\text{-}40}$, $A\beta_{1\text{-}42}$ peptides (BioSource International, Inc., Camarillo, CA), were prepared as described earlier [7,8] by reconstituting the lyophilized powders obtained from the commercial source in phosphate buffered saline at 1mg/ml and incubating them for 72 hours in a water bath at 37 °C. Since Aβ having a reversed amino acid sequence ($A\beta_{42\text{-}1}$) do not form fibrils, a sample of $A\beta_{42\text{-}1}$ peptides were also prepared in a similar manner as a negative control (mock fibrils) for the amyloid fibrils. Following incubation, amyloid samples were stored at –80 °C until use.

4.2.3 *Reaction of Amyloid Fibrils with RBCs and Their Morphological Analysis*

Blood samples collected in heparinized tubes from normal healthy human subjects were used for binding to amyloid fibril samples prepared as above. Blood (0.9ml) was incubated with either A$\beta_{1\text{-}40}$/A$\beta_{1\text{-}42}$ fibrils, mock fibrils from A$\beta_{42\text{-}1}$ preparation or phosphate buffered saline (0.1ml) for 30min at 37 °C with slow stirring. Following incubation, samples were pelleted at ~1125×g for 10min at 4 °C, supernatant plasma and buffy coat were carefully removed and discarded. Cells were washed three times with CPDA buffer and their morphological analysis was performed as described above.

4.3 Results

A microscopic morphological analysis of a sample of RBCs from an Alzheimer's subject and a healthy individual is shown in Fig. 4.1. It can be clearly observed that RBCs from the healthy individual contain more of smooth biconcave cells, while there is an increased number of RBCs with altered morphology or elongated RBCs in the sample from the Alzheimer's subject.

Analysis of elongated RBCs in all the blood samples is shown in Fig. 4.2. As can be seen, the number of elongated RBCs in blood samples from Alzheimer's subjects is 15.2% while those in control subjects are only 5.9%. The difference in these two percentages is statistically significant ($P < 0.0001$).

To test the possibility that the above observation of a higher percentage of elongated RBCs seen in blood samples from Alzheimer's subjects could be due to interactions of Aβ peptide aggregates present in their blood, an in vitro

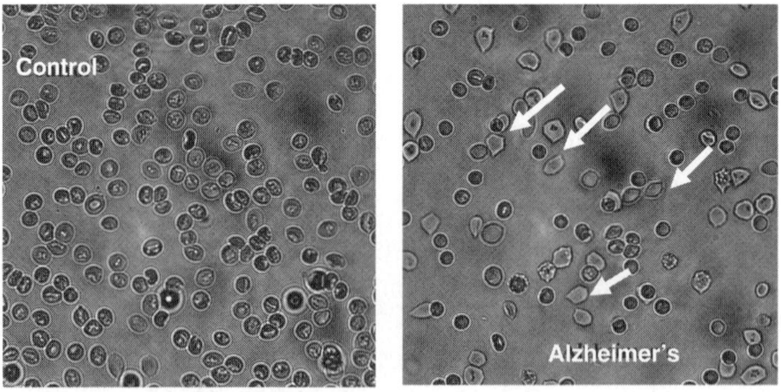

Fig. 4.1 RBCs from an Alzheimer subject (*right panel*) show altered morphology with many more elongated shapes than those from a control (*left panel*) subject.

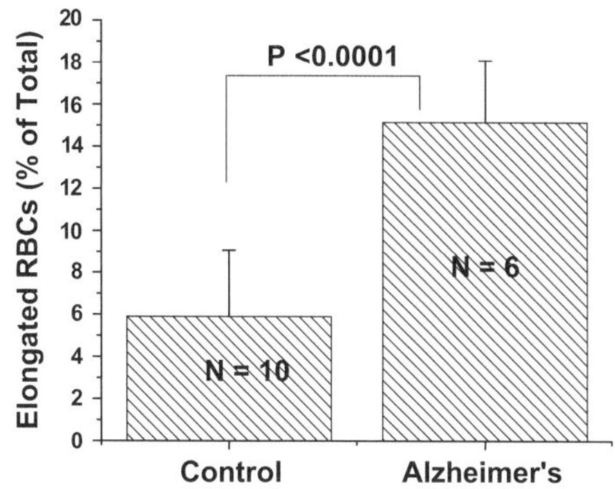

Fig. 4.2 The percent of elongated RBCs in samples from Alzheimer's subjects (15.2%) is higher than those in samples from control subjects (5.9%). The difference is statistically significant ($P < 0.0001$).

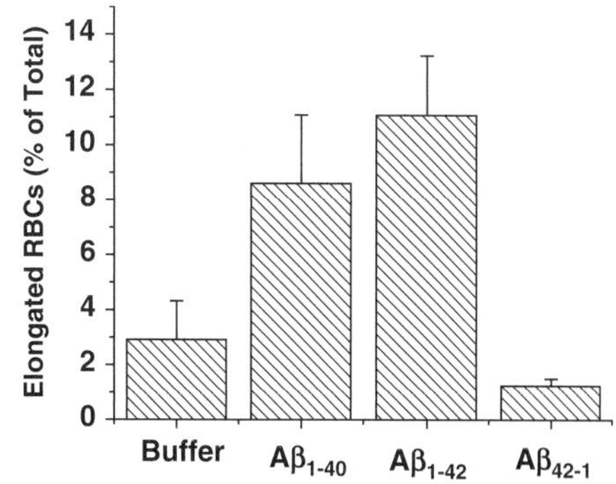

Fig. 4.3 Percentage of elongated RBCs observed in blood samples treated with either $A\beta_{1-40}$ or $A\beta_{1-42}$ fibrils are higher than those treated with control buffer or mock fibrils prepared from $A\beta_{42-1}$ peptides.

experiment was performed. Samples of blood from healthy individuals were incubated with $A\beta_{1-40}$, or $A\beta_{1-42}$ fibrils and then the morphological analysis was performed. For comparison, blood samples from healthy individuals were also incubated with either CPDA buffer as a control or mock fibrils prepared from $A\beta_{42-1}$ peptides. The results are shown in Fig. 4.3. Interestingly, the percentage of elongated RBCs in samples treated with either $A\beta_{1-40}$ or $A\beta_{1-42}$ fibrils (8.6% and 11.1 %, respectively) is much higher than the samples treated with buffer only (2.9%) or with mock fibrils (1.3%). This suggests that the morphological changes observed in RBCs from Alzheimer's subjects can be attributed to the binding of $A\beta_{1-40}$ or $A\beta_{1-42}$ fibrils to the RBCs.

4.4 Discussion

Our earlier studies [7, 8] have shown that both $A\beta_{1-40}$ and $A\beta_{1-42}$ fibrils bind to RBCs. β-Amyloids have been reported to be associated with cerebral blood vessels, particularly the capillaries (microvasculature), and have been reported

Table 4.1 Measurement of RBC deformability

β-Amyloid (μM)	Mean Cell Vol. (fl)	Mean Cell Transit Time (msec)*	Cells/sec passing thru pores*	% of Slow cells (<6.4 msec)*
0	81.56	3.73	11.67	2.64
5	84.19	3.58	8.62	1.64
15	85.14	4.08	5.79	4.29
20	85.98	4.13	4.86	5.53

* Transit of RBCs through 5μ pores using a Cell Transit Analyzer. (Data from: J.M. Rifkind et al., Adv. Cell Aging Gerontol., 2002, Vol. 11, 283–307)

to be transferred from the brain into blood across the Blood Brain Barrier. In addition, β-Amyloids are also produced in platelets and thus are in circulation in the peripheral blood. Therefore, β-Amyloids are readily available to RBCs in the circulation, particularly in AD subjects.

The RBC morphological changes observed in blood samples for Alzheimer's subjects in Figs. 4.1 and 4.2 suggest that the exposure of RBCs to Aβ in vivo may be the cause of these changes. While other changes in AD subjects can affect the RBCs, the demonstration that similar changes are produced when Aβ fibrils interact in vitro with RBCs from normal subjects support the hypothesis that these morphological changes originate from Aβ interactions with RBCs.

The passage of RBCs through capillaries with a pore size smaller than the mean diameter of the RBC requires RBC deformability. RBC deformability depends on both the properties of the red cell membrane and the excess surface area of the RBC. The elongated RBCs found in Alzheimer subjects indicate a decrease in the excess surface area as well as possible membrane changes suggesting decreased RBC deformability. In support of this suggestion, our earlier studies investigating the effects of amyloids on RBCs have also found an increase in mean cell volume and a decrease in deformability (Table 4.1) [10].

An additional factor that can contribute to impaired blood flow and oxygen delivery in AD subjects involves increased adherence of RBCs to the brain microvasculature. We have recently reported that $A\beta_{1\text{-}40}$ bound to either the RBC or endothelial cells increase the adherence of the RBCs to the endothelial cells. This implies that RBCs bound to amyloid will adhere to the endothelium slowing down blood flow. This effect would be particularly pronounced in the brain microvasculature with vascular amyloidosis.

Acknowledgment This research was supported (in part) by the Intramural Research Program of the NIH, National Institute on Aging, grant # NIH-HC-99-260 from National Center for Complimentary and Alternative Medicine (NCCAM), and the Tab Williams Fund for Dementia Research and the Roena Kulynych Center for Memory and Cognition Research, Wake Forest University School of Medicine, Winston-Salem, NC.

References

1. J. Carter and C. F. Lippa, Beta-amyloid, neuronal death and Alzheimer's disease, *Curr. Mol. Med.,* **1**(6), 733–737 (2001).
2. J. Hardy and D. J. Selkoe, The amyloid hypothesis of Alzheimer's disease: progress and problems on the road to therapeutics, *Science,* **297**(5580), 353–356 (2002).
3. J. C. de la Torre, Vascular basis of Alzheimer's pathogenesis, *Ann. N. Y. Acad. Sci.,* **977**, 196–215 (2002).
4. K. Ozawa, T. Tomiyama, M. L. Maat-Schieman, R. A. Roos and H. Mori, Enhanced Abeta40 deposition was associated with increased Abeta42-43 in cerebral vasculature with Dutch-type hereditary cerebral hemorrhage with amyloidosis (HCHWA-D), *Ann. N. Y. Acad. Sci.,* **977**, 149–154 (2002).
5. J. B. Mackic, M. H. Weiss, W. Miao, E. Kirkman, J. Ghiso, M. Calero, J. Bading, B. Frangione and B. V. Zlokovic, Cerebrovascular accumulation and increased blood-brain barrier permeability to circulating Alzheimer's amyloid beta peptide in aged squirrel monkey with cerebral amyloid angiopathy, *J. Neurochem.,* **70**(1), 210–215 (1998).
6. S. Varadarajan, S. Yatin, M. Aksenova and D. A. Butterfield, Review: Alzheimer's amyloid beta-peptide-associated free radical oxidative stress and neurotoxicity, *J. Struct. Biol.,* **130**(2–3), 184–208 (2000).
7. R. Jayakumar, J. W. Kusiak, F. J. Chrest, A. A. Demehin, J. Murali, R. P. Wersto, E. Nagababu, L. Ravi and J. M. Rifkind, Red cell perturbations by amyloid beta-protein, *Biochim. Biophys. Acta,* **1622**(1), 20–28 (2003).
8. L. B. Ravi, J. G. Mohanty, F. J. Chrest, R. Jayakumar, E. Nagababu, P. V. Usatyuk, V. Natarajan and J. M. Rifkind, Influence of beta-amyloid fibrils on the interactions between red blood cells and endothelial cells, *Neurol. Res.,* **26**(5), 579–585 (2004).
9. S. T. Dekosky, A. Fitzpatrick, D. G. Ives, J. Saxton, J. Williamson, O. L. Lopez, G. Burke, L. Fried, L. H. Kuller, J. Robbins, R. Tracy, N. Woolard, L. Dunn, R. Kronmal, R. Nahin and C. Furberg, The Ginkgo Evaluation of Memory (GEM) study: design and baseline data of a randomized trial of Ginkgo biloba extract in prevention of dementia, *Contemp. Clin. Trials,* **27**(3), 238–253 (2006).
10. J. M. Rifkind, O. O. Abugo, E. Nagababu, S. Ramasamy, A. Demehin and R. Jayakumar, In: *Advances in Cell Aging and Gerontology*, edited by T. Hagen, (Elsevier, New York, 2002), pp. 283–307.

Chapter 5
Uncoupling Protein-2 in Diabetic Kidneys

Increased Protein Expression Correlates to Increased Non-transport Related Oxygen Consumption

Malou Friederich[1], Johan Olerud[1], Angelica Fasching[1], Per Liss[2], Peter Hansell[1], and Fredrik Palm[1]

Abstract Diabetic patients have an elevated risk to develop renal dysfunction and it has been postulated that altered energy metabolism is involved. We have previously shown that diabetic rats have markedly decreased oxygen availability in the kidney, resulting from increased oxygen consumption. A substantial part of the increased oxygen consumption is unrelated to tubular transport, suggesting decreased mitochondrial efficiency. In this study, we investigated the protein expression of mitochondrial uncoupling protein (UCP)-2 in kidney tissue from control and streptozotocin (STZ)-induced diabetic rats.

Protein levels of UCP-2 were measured in adult male control and STZ-diabetic Wistar Furth as well as Sprague Dawley rats in both the kidney cortex and medulla by Western blot technique.

Two weeks of hyperglycemia resulted in increased protein levels of UCP-2 in kidneys from both Wistar Furth and Sprague Dawley rats. Both cortical and medullary UCP-2 levels were elevated 2–3 fold above control levels.

We conclude that sustained STZ-induced hyperglycemia increases the kidney levels of mitochondrial UCP-2, which could explain the previously reported increase in non-transport related oxygen consumption in diabetic kidneys. The elevated UCP-2 levels may represent an effort to reduce the increased production of superoxide radicals which is evident during diabetes.

5.1 Introduction

Mitochondrial oxidative phosphorylation is the cellular source of energy (adenosine triphosphate; ATP) and thus a prerequisite for normal cell function. In summary, protons (H^+) are pumped through the mitochondrial inner

[1]Department of Medical Cell Biology, Uppsala University, BMC, PO 571, 751 23 Uppsala, Sweden.
[2]Department of Oncology, Radiology and Clinical Immunology, University Hospital, 751 23 Uppsala, Sweden.

membrane by complex I–IV of the electron transport chain. This results in an electrochemical (potential) gradient across the mitochondrial membrane which is used by the ATP-synthase (complex V) to create ATP from adenosine diphosphate (ADP) and inorganic phosphate (P_i). This gradient is also used to transport ions and metabolites [1]. However, not all H^+ are used by the ATP-synthase since the membrane is partially permeable to H^+. This results in a basal H^+ leak, which regulation is dependent on the electrochemical gradient and the energy situation in that specific cell. If ADP and P_i is present, i.e. energy demand is high, the H^+ permeability is usually low and if ATP is present, i.e. energy demand is low, the H^+ permeability increases [1, 2]. A high H^+ permeability is energy wasting and has been shown to account for up to 25% of the standard metabolic rate [3].

Mitochondria are a significant source of reactive oxygen species (ROS), which can cause oxidative stress [4]. Mitochondria primarily produce superoxide (O_2^-), which is rapidly converted to hydrogen peroxide (H_2O_2) by intracellular and mitochondrial superoxide dismutase (SOD). Studies have shown that complex I of the electron transport chain is the major source of ROS, however the mechanisms are unclear [5] . It is believed that a high and stable mitochondrial potential gradient is the cause of the increased O_2^- production. The different complexes of the electron transport chain are regulated by their own redox-status and therefore differently regulated by changes in the membrane potential and pH. Both O_2^- and H_2O_2 are highly reactive and quickly interacts with DNA, proteins and lipids, resulting in oxidative damage. Oxidative damage is believed to play a major role in many pathological conditions such as arteriosclerosis, ischemia-reperfusion damage and diabetic nephropathy [6, 7]

A suggested cytoprotective mechanism against increased mitochondrial radical production is increased levels of uncoupling proteins (UCP). The protonophoric properties of UCP can increase the H^+ leak back across the mitochondrial inner membrane. This is called inducible H^+ leak and should be considered as an additional mechanism for the cells to regulate the H^+ permeability across the mitochondrial membrane [1]. Studies have shown that UCP decreases ROS production by lowering the mitochondrial H^+ gradient, supporting the theory that UCP has antioxidant properties [8]. A negative feedback-loop is believed to exist between ROS and UCP-2, and a signaling role for 4-hydroxy-2-nonenal (a by-product in lipid peroxidation) in the regulation of UCP has recently been reported [9].

Uncoupling proteins are a subfamily of proteins in the family of mitochondrial anion carriers [7]. Five different UCPs have been identified so far. UCP-1 is expressed exclusively in brown adipose tissue and it appears that its only function is to generate heat. The very low level of ATP-synthase in these cells is believed to activate UCP-1 resulting in excessive heat generation [10]. UCP-2 is expressed ubiquitously throughout the body. UCP-3 is expressed mostly in muscles. Neither UCP-2 nor 3 have been shown to have any significant thermogenic effect. This is believed to be a result of the presence of ATP-synthase in these cells. UCP-4 and UCP-5 is expressed exclusively in the brain [10, 11].

We have previously shown that sustained hyperglycemia results in increased oxygen consumption by both isolated cortical and medullary tubular cells [6, 12]. The main part of the diabetes-induced increase in oxygen consumption was unrelated to tubular electrolyte transport; the latter measured as ouabain-sensitive oxygen consumption. In the present study, we therefore investigated if streptozotocin (STZ)-induced diabetic rats have elevated protein expression of mitochondrial UCP-2 in the kidneys since this might provide a feasible explanation for the previously reported increase in non-transport related oxygen consumption. Furthermore, in order to rule out possible strain differences, we investigated the UCP-2 levels in kidneys from both Wistar Furth and Sprague Dawley rats.

5.2 Materials and Methods

Adult male Wistar-Furth and Sprague-Dawley rats (250–300 g; B&K Universal, Sollentuna, Sweden) were randomly divided into normoglycemic controls ($n = 5$ per strain) and hyperglycemics ($n = 5$ per strain). The local animal ethics committee at the University of Uppsala approved all experiments.

5.2.1 Diabetes Induction and Surgical Procedures

Diabetes mellitus was induced by an injection of streptozotocin (STZ; 45 mg/kg BW dissolved in saline for WF and 55 mg/kg for SD; Sigma-Aldrich, St. Louis, MO) in the tail vein. Animals were considered diabetic if blood glucose concentrations increased to $\geq$ 15 mmol/l within 24 hours after STZ-injection and remained elevated. Blood glucose concentrations were determined with test reagent strips (MediSense, Bedford, MA) from blood samples obtained from the cut tip of the tail.

All rats were anaesthetized by an intraperitoneal injection of thiobutabarbital sodium, (Inactin, 120 mg/kg BW for non-diabetic rats and 80 mg/kg BW for diabetic rats; Sigma-Aldrich) and placed on a servo-regulated heating pad. Tracheotomy was performed to ensure sufficient breathing. A polyethylene catheter was placed in a carotid artery and perfused with 30 ml of phosphate buffer saline (PBS, Medicago AB, Uppsala, Sweden), and the right renal vein was cut open in order to facilitate complete perfusion of the kidneys. Rats were sacrificed by administration of saturated KCl.

5.2.2 Protein Extraction and Western Blot

All procedures were performed on ice. Both kidneys were dissected out and prepared for protein extraction. Renal cortex and medulla were separated under microscope and placed in separate homogenizers and 700 μl of RIPA buffer (1.0% NP40, 0.5% sodium deoxycholate, 0.1% SDS, 10 mM NaF, 80mM Tris pH 7.5) with protein inhibitors (Protease Arrest (10 μl/ml; GBiosciences, St. Louis, MO, USA), Phosphatase inhibitor cocktail-2 (10 μl/ml; Sigma-Aldrich) and Complete Mini (1 tablet/1.5 ml; Roche Diagnostics, Mannheim, Germany)) was added and the tissue homogenized. After centrifugation (15 min, 5000 G at +4 °C), the supernatant was stored at –70 °C.

Protein samples were mixed 1:1 with Laemmli + 2-merkaptoethanol buffer (BioRad Laboratories). Samples were boiled and 200 μg protein was loaded onto precast 10% Tris-HCL gels set in a Criterion Cell with Tris/glycine/SDS buffer (BioRad Laboratories). The proteins were transfer to a nitrocellulose membrane, which was blocked with 5% non-fat dry milk in Tris-HCL Tris-base, NaCl, pH 7.4 for 1 h. The membrane was incubated with goat anti-rat UCP-2 antibody (1:1000; Santa-Cruz Biotechnology, Santa Cruz, USA) overnight at 4°C. Horseradish peroxidase (HRP)-conjugated secondary antibody (1:5000; sheep anti-goat, Santa-Cruz Biotechnology) was used to detect the specific band. Detection was performed according to the manufacturers' instructions (Chemiglow West; Alpha Innotech, San Leandro, CT, USA) and analyzed using an ECL- camera (Kodak image station 2000; New Haven, CT, USA) detecting the HRP-emitted light. Kodak 1DIM 3.6.3 software was used for analyzing Western blots.

5.2.3 Statistical Analysis

All values are expressed as means±SEM. Student's t-test was used to compare means of two groups. Statistical analysis was performed using Graph Pad Prism software (Graph Pad Inc., San Diego, CA, USA), and $P<0.05$ was considered statistically significant.

5.3 Results

5.3.1 Wistar Furth Rats

STZ-injection resulted in a sustained hyperglycemia in all injected Wistar Furth rats (24.5±2.5 vs. 6.1±0.4 mM for control animals; n = 5 each). Diabetic Wistar Furth rats had 3.7-fold higher UCP-2 levels in the renal cortex and 3.0-fold higher in the medulla compared to normoglycemic control animals (Fig. 5.1).

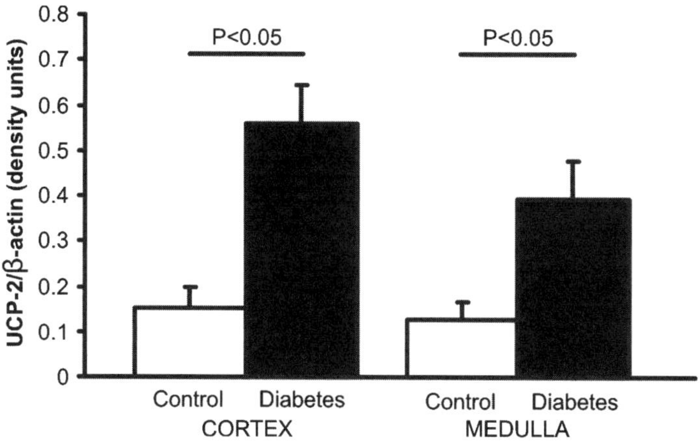

Fig. 5.1 Protein expression of mitochondrial uncoupling protein-2 (UCP-2) in renal cortex and medulla of normoglycemic and hyperglycemic Wistar Furth rats ($n=5$ per group).

5.3.2 Sprague Dawley Rats

STZ-injection resulted in a sustained hyperglycemia in all injected Sprague Dawley rats (27.6±0.1 vs. 6.9±0.4 mM for control animals; $n=5$ each). Diabetic Sprague Dawley rats had 2.5-fold higher UCP-2 levels in the renal cortex and 2.1-fold higher in the medulla compared to normoglycemic control animals (Fig. 5.2).

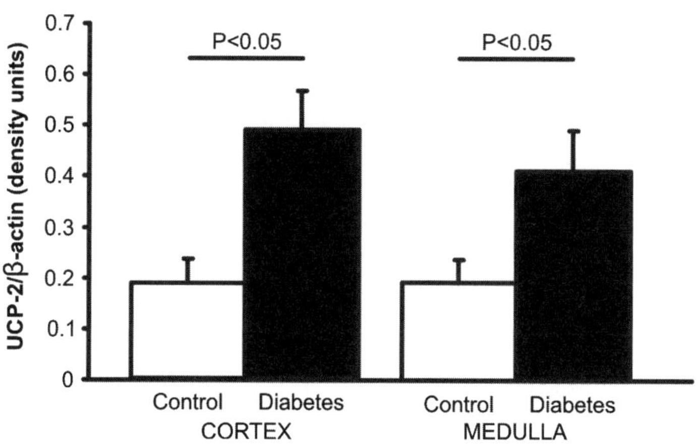

Fig. 5.2 Protein expression of mitochondrial uncoupling protein-2 (UCP-2) in renal cortex and medulla of normoglycemic and hyperglycemic Sprague Dawley rats ($n=5$ per group).

5.4 Discussion

The main finding of the present study is that UCP-2 levels in both renal cortex and medulla are increased in STZ-induced diabetic rats compared to normoglycemic controls. This supports our theory that UCP-2 is a protective mechanism counteracting diabetes-induced ROS production. However, an unwanted side effect of the increased UCP-2 levels should in theory be increased oxygen consumption. Oxygen consumption related to mitochondrial uncoupling is unrelated to tubular transport processes, i.e. ouabain insensitive. Indeed, oxygen consumption by freshly isolated ouabain-treated tubular cells from diabetic rats is markedly increased compared to control cells with corresponding treatment [6, 12]. Renal oxygen tension is decreased in STZ-induced diabetes and the mechanism has been identified as increased renal mitochondrial oxygen consumption resulting from increased oxidative stress [6]. We now further clarify the mechanism accounting for the diabetes-induced increase in renal oxygen consumption to involve increased UCP-2 protein abundance, which has been shown to increase mitochondrial uncoupling.

Previous studies have shown that the degree of hyperglycemia is a good predictor of the progression of diabetes complications [13], including the development of renal dysfunction [14]. It is also well known that sustained hyperglycemia increases the ROS production. ROS decreases NO bioavailability, resulting in both increased vascular tone and increased oxygen consumption [14]. Taken together, the decreased renal oxygen tension and increased ROS production could be an important pathway leading to the development of diabetic nephropathy [14].

It is believed that an altered UCP regulation can contribute to several pathological conditions such as arteriosclerosis, insulin resistance and diabetic nephropathy [7]. Pioneering work by Mattiasson and co-workers showed that UCP-2 protects against long-term damage resulting from stroke and brain trauma [15]. In vitro, UCP-2 over-expressing cells showed significantly lower mortality when subjected to 90 minutes of oxygen-glucose deprivation compared to controls [15]. The suggested mechanism for the UCP-2 protective properties in this study was that UCP-2 prevents activation of apoptotic proteins by lowering the electrochemical gradient and thereby reducing ROS production.

5.5 Summary

Sustained STZ-induced hyperglycemia increases the kidney levels of mitochondrial UCP-2 in both Wistar Furth and Sprague-Dawley rats, which might be the reason for the previously reported increase in non-transport related renal oxygen consumption. The elevated UCP-2 levels may represent an effort to reduce the increased production of reactive oxygen species (ROS) which is evident during diabetes.

Acknowledgment This work was funded by The Swedish Research Council, The Marcus and Amalia Wallenberg Foundation, The Linné Foundation for Medical Research, The Swedish Diabetes Association, and The Swedish Society for Medical Research.

References

1. P. S. Brookes, Mitochondrial H(+) leak and ROS generation: an odd couple, *Free Radic Biol Med* **38**(1), 12–23 (2005).
2. D. G. Nicholls, The non-Ohmic proton leak–25 years on, *Biosci Rep* **17**(3), 251–57 (1997).
3. D. F. Rolfe, and G. C. Brown, Cellular energy utilization and molecular origin of standard metabolic rate in mammals, *Physiol Rev* 77(3), 731–58 (1997).
4. J. F. Turrens, Superoxide production by the mitochondrial respiratory chain, *Biosci Rep* **17**(1), 3–8 (1997).
5. J. St-Pierre, J. A. Buckingham, S. J. Roebuck, and M. D. Brand, Topology of superoxide production from different sites in the mitochondrial electron transport chain, *J Biol Chem* **277**(47), 44784–90 (2002).
6. F. Palm, J. Cederberg, P. Hansell, P. Liss, and P. O. Carlsson, Reactive oxygen species cause diabetes-induced decrease in renal oxygen tension, *Diabetologia* **46**(8), 1153–1160 (2003).
7. P. Jezek, Possible physiological roles of mitochondrial uncoupling proteins–UCPn, *Int J Biochem Cell Biol* **34**(10), 1190–206 (2002).
8. S. Papa, and V. P. Skulachev, Reactive oxygen species, mitochondria, apoptosis and aging, *Mol Cell Biochem* **174**(1–2), 305–19 (1997).
9. K. S. Echtay, T. C. Esteves, J. L. Pakay, M. B. Jekabsons, A. J. Lambert, M. Portero-Otin, R. Pamplona, A. J. Vidal-Puig, S. Wang, S. J. Roebuck, and M. D. Brand, A signalling role for 4-hydroxy-2-nonenal in regulation of mitochondrial uncoupling, *Embo J* **22**(16), 4103–10 (2003).
10. S. Krauss, C. Y. Zhang, and B. B. Lowell, The mitochondrial uncoupling-protein homologues, *Nat Rev Mol Cell Biol* **6**(3), 248–261 (2005).
11. P. Jezek, and E. Urbankova, Specific sequence of motifs of mitochondrial uncoupling proteins, *IUBMB Life* **49**(1), 63–70 (2000).
12. F. Palm, P. Hansell, G. Ronquist, A. Waldenstrom, P. Liss, and P. O. Carlsson, Polyol-pathway-dependent disturbances in renal medullary metabolism in experimental insulin-deficient diabetes mellitus in rats, *Diabetologia* **47**(7), 1223–1231 (2004).
13. The effect of intensive treatment of diabetes on the development and progression of long-term complications in insulin-dependent diabetes mellitus. The Diabetes Control and Complications Trial Research Group, *N Engl J Med* **329**(14), 977–86 (1993).
14. F. Palm, Intrarenal oxygen in diabetes and a possible link to diabetic nephropathy, *Clin Exp Pharmacol Physiol* **33**(997–1001) (2006).
15. G. Mattiasson, M. Shamloo, G. Gido, K. Mathi, G. Tomasevic, S. Yi, C. H. Warden, R. F. Castilho, T. Melcher, M. Gonzalez-Zulueta, K. Nikolich, and T. Wieloch, Uncoupling protein-2 prevents neuronal death and diminishes brain dysfunction after stroke and brain trauma, *Nat Med* **9**(8), 1062–68 (2003).

Chapter 6
Measurement of Oxygenation at the Site of Stem Cell Therapy in a Murine Model of Myocardial Infarction

Mahmood Khan[1], Vijay Kumar Kutala[1], Sheik Wisel[2], Simi M. Chacko[1], M. Lakshmi Kuppusamy[1], Pawel Kwiatkowski[2], and Periannan Kuppusamy[1]

Abstract We have developed a noninvasive EPR (electron paramagnetic resonance) oximetry, based on a new class of oxygen-sensing nano-particulate probe (LiNc-BuO), for simultaneous monitoring of stem-cell therapy and in situ oxygenation (partial pressure of oxygen, pO_2) in a mouse model of acute myocardial infarction (AMI). AMI was induced by a permanent occlusion of left-anterior-descending (LAD) coronary artery. Skeletal myoblast (SM) cells were used for therapy. The oximetry probe was implanted in the mid-ventricular region using a needle. Tissue histological studies after 3 weeks of implantation of the probe revealed significant fibrosis, which was solely due to the needle track and not due to the probe particles. The feasibility of long-term monitoring of pO_2 was established in control (non-infarct) group of hearts (> 3 months; $pO_2 = 15.0 \pm 1.2$ mmHg,). A mixture of the probe with/without SM cells (1×10^5) was implanted as a single injection in the infarcted region and the myocardial tissue pO_2 at the site of cell therapy was measured for 4 weeks. The pO_2 was significantly higher in infarcted hearts treated with SM cells ($pO_2 = 3.5 \pm 0.9$ mmHg) compared to untreated hearts ($pO_2 = 1.6 \pm 0.7$ mmHg). We have demonstrated, for the first time, the feasibility of monitoring pO_2 in mouse hearts after stem cell therapy.

6.1 Introduction

Ischemic heart disease is a leading cause of cardiac failure and a clinically disabling condition worldwide. A number of studies have confirmed the beneficial effects of stem cell transplantation on cardiac function after acute

[1]Center for Biomedical EPR Spectroscopy and Imaging, Davis Heart and Lung Research Institute, Division of Cardiovascular Medicine, Department of Internal Medicine;
[2]Division of Cardiothoracic Surgery, Department of Surgery, The Ohio State University, 420 W. 12th Ave, Columbus, OH 43210.
Corresponding author: Periannan Kuppusamy, e-mail: kuppusamy.1@osu.edu;
Tel: 1-614-292-8998; Fax: 614-292-8454

myocardial infarction [1–3]. Skeletal muscle satellite cells are myogenic precursor cells located between the basal lamina and sarcolemma of adult skeletal muscle. Upon activation, as in the case of an injury, these cells have a tendency to divide mitotically into myoblasts [4]. Transplantation of skeletal myoblast (SM) cells in animal models of acute myocardial infarction has been reported to result in improved cardiac performance and graft survival [5–10].

It is, however, uncertain whether or not the oxygen concentration in the infarcted myocardium could play a role in the process of engraftment and survival. Hence, the monitoring of *in situ* tissue oxygenation after SM transplantation is vital for the understanding of the effects of stem cell therapy. We have recently developed a probe, made of lithium octa-*n*-butoxy-naphthalocyanine (LiNc-BuO) radicals, in the form of submicron-sized (270 ± 120 nm) crystals (hereafter referred to as OxySpin), which can be directly detected by electron paramagnetic resonance (EPR) spectroscopy with high sensitivity and specificity [11, 12]. This prompted us to further develop this unique technology for the evaluation of stem cell therapy, whereby long-term monitoring of local tissue pO_2 at the transplanted site could be performed. Accordingly, the goal of this study was to monitor myocardial tissue oxygenation in infarcted murine hearts immediately following transplantation of SM cells, and to subsequently monitor *in situ* pO_2 for several weeks after SM cell transplantation using EPR spectroscopy.

6.2 Materials and Methods

6.2.1 Reagents

The cell-culture medium contained myoblast basal growth medium (SkBM, Clonetics, San Diego, CA) with 20% fetal bovine serum (HyClone, CO), recombinant human epidermal growth factor, dexamethasone, and antibiotics. Trypsin/EDTA and collagenase were obtained from Invitrogen (CA) and Worthington Biochemicals (NJ), respectively. The OxySpin probes were synthesized as reported [12].

6.2.2 Isolation and Characterization of Murine Skeletal Myoblasts

Approximately, 6 g of hind limb skeletal muscle from mice was obtained and the connective tissue and tendons were removed. The biopsies were minced into a slurry and subjected to several cycles of enzymatic digestion at 37 °C with trypsin/EDTA (0.5 mg/ml) and collagenase (0.5 mg/ml) to release the myoblasts. The skeletal myoblasts were cultured in myoblast basal growth medium containing 20% fetal bovine serum, recombinant human epidermal growth factor (10 ng/ml), dexamethasone (3 μg/ml) and antibiotics. The cells were limited to 60–70% confluency to prevent myotube formation. Periodically, the purity of cells was checked using a specific monoclonal CD56 antibody by flow cytometry.

6.2.3 Transplantation of Skeletal Myoblasts with OxySpin

Myoblasts, at 70% confluency (1×10^4 cells/35-mm dish), in 3 ml of medium containing 10% FBS were trypsinized and the suspension of cells was centrifuged at 2500g. The pellet was collected and resuspended in 1 ml of sterile PBS containing glucose (0.1%). OxySpin (100 μg/ml) was added to the cell suspension (2×10^5 cells/ml) and mixed well with a pipette prior to transplantation into the myocardium.

6.2.4 Preparation of Mice

C57BL/6 male mice, weighing 25–30 g, were anesthetized with a mixture of ketamine (55 mg/kg) and xylazine (15 mg/kg) that was injected intraperitoneally. The intubation tube consisted of a 20-gauge intravenous catheter attached to a connector. The ventilator was set at 120 breaths/min with a tidal volume of 250 μL (Harvard Apparatus, Hollister, MA). The body temperature was maintained at 37±1°C using an isothermal heating pad (Braintree Scientific, Braintree, MA). All of the procedures were performed with approval of the Institutional Animal Care and Use Committee of the Ohio State University and conformed to the Guide for the Care and Use of Laboratory Animals (NIH Publication No. 86–23).

6.2.5 Induction of Myocardial Infarction

Acute myocardial infarction (AMI) was created by permanently occluding the left anterior descending (LAD) coronary artery. An oblique 8-mm incision was made 2-mm away from the left sternal border toward the left arm pit. The chest cavity was opened with scissors by a small incision (5 mm long) at the level of the third or fourth intercostal space 2 to 3 mm from the left sternal border. The LAD coronary artery was visualized as a pulsating bright red spike, running through the midst of the heart wall from underneath the left atrium toward the apex. The LAD artery was ligated 1 to 2 mm below the tip of the left auricle with a tapered needle using 8-0 polypropylene ligature. The needle was passed underneath the LAD coronary artery and a double knot was made to occlude the LAD. Occlusion was confirmed by a sudden change in color (pale) of the anterior wall of the left ventricle (LV). The chest cavity was closed by bringing together the third and fourth ribs with one 6-0 polypropylene silk suture. The layers of muscle and skin were closed with 5-0 polypropylene suture.

6.2.6 *Implantation of Skeletal Myoblasts in the Infarcted Heart*

A single intramyocardial injection of sonicated OxySpin (100 μg/ml) with (experimental group) or without (control group) skeletal myoblasts (1×10^5 cells/15 μl) was injected into the mid-ventricular region of the LAD-ligated mice. The chest was closed after implantation of the cells, and EPR measurements were performed immediately, and then every week for several weeks.

6.2.7 *pO_2 Measurements in the Heart*

The sensitivity of the EPR linewidth of OxySpin to oxygen was calibrated as described previously [12]. After a thoracotomy, 15 μl of a suspension of OxySpin in saline was injected into the mid-ventricular region of mouse hearts using a 29 ½-gauge needle. The mouse (with/without LAD ligation) was placed in the L-band EPR spectrometer (Magnettech, Germany) with its heart close to the loop of the surface coil resonator. The instrument settings were: microwave power, 4 mW; modulation amplitude, 180 mG, modulation frequency 100 kHz; receiver time constant, 0.2 msec; and acquisition time, 30 sec. EPR spectra were acquired as single scans. The peak-to-peak linewidth was used to calculate pO_2 using the standard calibration curve.

6.2.8 *Histological Analysis of Needle-track Injury*

OxySpin (20 μg) was implanted in the mid-myocardium of the mouse heart using a 25-gauge needle. Identification of needle-track injury in control (needle only) and OxySpin-implanted hearts was done after 1, 7 and 21 days after needle-track injury. The hearts were excised and washed several times with PBS and fixed in 10% formalin. Histological analysis was done on 4-μm sections stained with Hematoxylin and Eosin (H&E) or Masson Trichrome (MT). The latter stain identifies collagen deposition, which is an indication of fibrosis.

6.3 Results and Discussion

6.3.1 Histology of Mouse Heart with Needle-track Injury

To determine whether the histological changes caused to the heart is due the OxySpin implantation or by the needle itself, we studied the heart histology in twodifferent groups: one with needle track only and the other with OxySpin implantation. The histological studies performed in tissues obtained after three weeks of implantation revealed significant fibrosis in both the groups (Fig. 6.1). The results also showed that the injury was solely due to the needle track and not due to the particles.

6.3.2 Measurement of pO_2 in Mouse Heart by EPR Spectroscopy

The OxySpin mixed with SM cells and transplanted in the infarcted heart was monitored noninvasively, using in vivo EPR spectroscopy. The EPR spectrum of the OxySpin probes is characterized by a single, narrow peak. The OxySpin

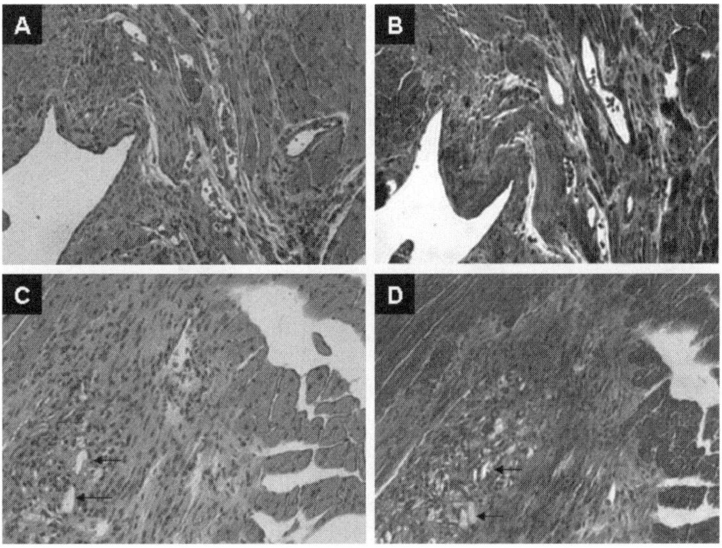

Fig. 6.1 Histology showing needle-track injury in mouse hearts at 3 weeks after implantation (200x). (**A**) Hematoxylin and Eosin (H & E) staining, (**B**) Masson Trichrome (MT) staining showing collagen deposition and intense fibrosis (blue), (**C**) H & E staining after OxySpin implantation, and (**D**) MT staining after OxySpin implantation showing fibrosis. The arrows in (**C**) and (**D**) show the OxySpin probes in the tissue. (See also color insert.)

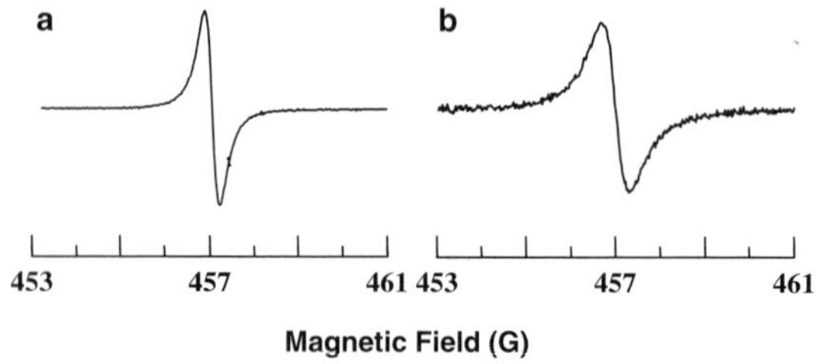

Fig. 6.2 EPR spectrum of OxySpin. (**a**) The probe suspended in PBS showed a linewidth of 0.37 G under anoxic conditions. (**b**) The spectrum obtained from a beating heart at 4 weeks following the transplantation of SM cells mixed with OxySpin in the infarcted region.

suspended in PBS showed a linewidth 0.37 G under anoxic conditions (Fig. 6.2a). Figure 6.2b shows the EPR spectrum obtained from the beating heart at 4 weeks following the transplantation of SM cells mixed with OxySpin in the infarcted region. The results suggest that the probe was retained in the heart for four weeks or possibly longer, enabling precise measurement of in situ oxygenation in the heart.

6.3.3 Long-term Monitoring of in vivo Myocardial pO_2 After SM Transplantation

Noninvasive measurement of myocardial tissue pO_2 was performed using an L-band EPR spectrometer. OxySpin was implanted in the area of risk, prior to the induction of infarction (by ligation of LAD coronary artery). Figure 6.3a

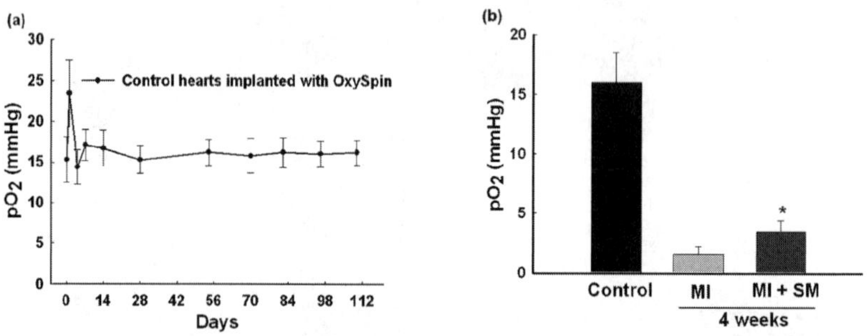

Fig. 6.3 Long-term monitoring of in situ pO_2 at the site of transplanted skeletal myoblasts (SM) in infarcted mouse hearts. (**a**) Myocardial tissue pO_2 from mice (N = 5) implanted with OxySpin (only) in the mid-ventricular region without LAD coronary artery ligation. Data show the feasibility of pO_2 measurements for more than 3 months after implantation. (**b**) Myocardial tissue pO_2 in mice transplanted with skeletal myoblasts in the mid-ventricular (ischemic) region after LAD artery ligation. Legends: **Control**, un-infarcted; **MI**, infarcted; **MI+SM**, infarcted hearts treated with skeletal myoblasts. Values are expressed as mean ± SD (N = 5). The tissue pO_2 is higher in the infarcted hearts treated with the SM cells (*$p<0.05$) compared to hearts not treated with SM cells.

shows the myocardial tissue pO_2 measured noninvasively for 16 weeks (112 days) from non-infarcted (control) hearts. Animals without LAD artery ligation served as controls to obtain baseline pO_2 values. The mean pO_2 in the non-ligated hearts was 15.0±1.2 mmHg. This result demonstrates the stability of OxySpin in the tissue and feasibility to perform noninvasive measurement of oxygen concentration in beating hearts over a period of more than 3 months. Figure 6.3b shows the myocardial pO_2 measured from infarcted hearts. The data revealed a marked decrease in myocardial pO_2 (1.6±0.7 mmHg; $p<0.001$ *versus* control) in the infarct region of untreated hearts at 4 weeks after infarction. The hearts transplanted with SM cells showed a significant increase in myocardial pO_2 (3.5±0.9 mmHg, $p<0.05$) compared to the MI only control group at 4 weeks post transplantation of skeletal myoblasts.

6.4 Conclusions

The results of this study demonstrated our ability to noninvasively monitor changes in situ tissue oxygenation during the process of stem cell therapy for several weeks, in vivo, using EPR oximetry. The EPR technique is advantageous in offering high sensitivity and repeated measurements of myocardial tissue oxygenation in the beating hearts.

Acknowledgment We thank Brian Rivera for critical reading of the manuscript. This work was supported by NIH grant R01 EB004031 and a State of Ohio Third Frontier grant BRTT/CBE.

References

1. D. Orlic, J. Kajstura, S, Chimenti, D. M. Bodine, A. Leri, and P. Anversa, Transplanted adult bone marrow cells repair myocardial infarcts in mice, *Ann. N. Y. Acad. Sci.* **938**, 221–229 (2001).
2. J. Y. Min, M. F. Sullivan, Y. Yang, J. P. Zhang, K. L. Converso, J. P. Morgan, and Y. F. Xiao, Significant improvement of heart function by cotransplantation of human mesenchymal stem cells and fetal cardiomyocytes in postinfarcted pigs, *Ann. Thorac. Surg.* **74**(5), 1568–1575 (2002).
3. A. A. Kocher, M. D. Schuster, M. J. Szabolcs, S. Takuma, D. Burkhoff, J. Wang, S, Homma, N. M. Edwards, and S. Itescu, Neovascularization of ischemic myocardium by human bone-marrow-derived angioblasts prevents cardiomyocyte apoptosis, reduces remodeling and improves cardiac function, *Nat. Med.* **7**(4), 430–436 (2001).
4. C. E. Murry, R. W. Wiseman, S. M. Schwartz, and S. D. Hauschka, Skeletal myoblast transplantation for repair of myocardial necrosis, *J. Clin. Inves.* **98**(11), 2512–2523 (1996).
5. K. A. Hutcheson, B. Z. Atkins, M. T. Hueman, M. B. Hopkins, D. D. Glower, and D. A. Taylor, Comparison of benefits on myocardial performance of cellular cardiomyoplasty with skeletal myoblasts and fibroblasts, *Cell Transplant.* **9**(3), 359–368 (2000).
6. M. Jain, H. Dersimonian, D. A. Brenner, S. Ngoy, P. Teller, A. S. Edge, A. Zawadzka, K. Wetzel, D. B. Sawyer, W. S. Colucci, C. S. Apstein, and R. Liao, Cell therapy attenuates

deleterious ventricular remodeling and improves cardiac performance after myocardial infarction, *Circulation* **103**(14), 1920–1927 (2001).
7. B. Pouzet, A. A. Hagege, J. T. Vilquin, M. Desnos, D. Duboc, J. P. Marolleau, and P. Menasche, Transplantation of autologous skeletal myoblasts in ischemic cardiac insufficiency, *J. Soc. Biol.* **195**(1), 47–49 (2001).
8. B. Pouzet, J. T. Vilquin, A. A. Hagege, M. Scorsin, E. Messas, M. Fiszman, K. Schwartz, and P. Menasche, Intramyocardial transplantation of autologous myoblasts: can tissue processing be optimized? *Circulation* **102**(19 Suppl 3), III 210–215 (2000).
9. D. A. Taylor, B. Z. Atkins, P. Hungspreugs, T. R. Jones, M. C. Reedy, K. A. Hutcheson, D. D. Glower, and W. E. Kraus, Regenerating functional myocardium: improved performance after skeletal myoblast transplantation, *Nat. Med.* **4**(8), 929–933 (1998).
10. M. Scorsin, A. Hagege, J. T. Vilquin, M. Fiszman, F. Marotte, J. L. Samuel, L. Rappaport, K. Schwartz, P. Menasche, Comparison of the effects of fetal cardiomyocyte and skeletal myoblast transplantation on postinfarction left ventricular function, *J. Thorac. Cardiovasc. Surg.* **119**(6), 1169–1175 (2000).
11. V. K. Kutala, N. L. Parinandi, R. P. Pandian, and P. Kuppusamy, Simultaneous measurement of oxygenation in intracellular and extracellular compartments of lung microvascular endothelial cells, *Antioxid. Redox Signal.* **6**(3), 597–603 (2004).
12. R. P. Pandian, N. L. Parinandi, G. Ilangovan, J. L. Zweier, and P. Kuppusamy, Novel particulate spin probe for targeted determination of oxygen in cells and tissues, *Free Radic. Biol. Med.* **35**(9), 1138–1148 (2003).

Chapter 7
Oxygen Pressures in the Interstitial Space of Skeletal Muscle and Tumors in vivo

David F. Wilson, William M.F. Lee, Sosina Makonnen, Sophia Apreleva, and Sergei A. Vinogradov[1]

Abstract A new Oxyphor (Oxyphor G3) has been used to selectively determine the oxygen pressure in interstitial (pericellular) spaces. Oxyphor G3 is a Pd-tetrabenzoporphyrin, encapsulated inside generation 2 poly-arylglycine (AG) dendrimer, and therefore is a true near infrared oxygen sensor, having a strong absorption band at 636nm and emission near 800nm. The periphery of the dendrimer is modified with oligoethylene glycol residues (Av. MW 350) to make the probe water soluble and biologically inert. Oxyphor G3 was injected along "tracks" in the tissue using a small needle (30gage or less) and remained in the pericellular space, allowing oxygen measurements for several hours with a single injection. The oxygen pressure distributions (histograms) were compared with those for Oxyphor G2 in the intravascular (blood plasma) space. In normal muscle, in the lower oxygen pressure region of the histograms (capillary bed) the oxygen pressure difference was small. At higher oxygen pressures in the histograms there were differences consistent with the presence of high flow vessels with oxygen pressures substantially above those of the surrounding interstitial space. In tumors, the oxygen pressures in the two spaces were similar but with large differences among tumors.

In mice, anesthesia with ketamine plus xylazine markedly decreased oxygen pressures in the interstitial and intravascular spaces compared to awake or isoflurane anesthetized mice.

7.1 Introduction

Oxygen transported to tissue, after reaching the tissue microcirculation, diffuses from the blood plasma through the walls of the micro-vessels into the interstitial (pericellular) space and then from interstitial space into the cells and

[1]David F. Wilson, William M.F. Lee, Sosina Makonnen, Sophia Apreleva, and Sergei A. Vinogradov, Department of Biochemistry and Biophysics (DFW, SA, SAV) and Department of Medicine (WMFL, SM), Medical School, University of Pennsylvania, Philadelphia, PA 19104.

finally to the mitochondria. As it diffuses, from the source (blood plasma) to a sink (mitochondria), an oxygen pressure gradient is formed in which the pressure is lower at the sink than at the source. The difference in oxygen pressure between the blood plasma and the mitochondria increases with increase in the rate of oxygen consumption by the mitochondria and the distance from the vessel to the mitochondria. The distance over which oxygen can be supplied to the mitochondria is, therefore, determined by a) the rate of oxygen consumption by the mitochondria, b) the distance from the blood plasma (the oxygen source) to the mitochondria and c) the oxygen pressure in the blood plasma.

Oxygen dependent quenching of phosphorescence is a minimally non-invasive optical method that can quantitate oxygen pressures in biological and other samples [1–4]. Although it has been widely used [1–17] for measurements in vivo, focus has been on the intravascular space. The available oxygen sensitive phosphors, such as Oxyphors R0, R2 and G2 (Oxygen Enterprises, Ltd, Philadelphia, PA), contained Pd-porphyrin cores that are at least partially exposed to the medium. As a result, the oxygen sensitivity is dependent on the microenvironment of the porphyrin and therefore on the macromolecule to which it is bound, and on the fraction of the Oxyphor bound to that macromolecule. In blood plasma, Oxyphors R0, R2 and G2 are essentially quantitatively bound to albumin. Albumin plays an important role, helping both to limit access of oxygen to the porphyrin core, facilitating oxygen measurements in the physiological range (0–120 Torr), and to provide a relatively homogeneous microenvironment for the phosphor.

A new family of Oxyphors has been synthesized that can be used in a much wider range of media, particularly in highly heterogeneous environments such as the interstitial space. The porphyrin core is first coated with dendrons and then the external surface of the dendrimer modified with oligoethylene glycol fragments [18–20]. Oxyphor G3 is a member of this oxygen sensor family. Not only are its oxygen quenching properties unaffected by biological macromolecules such as albumin, but also its oxygen quenching constant and phosphorescent lifetimes are well suited for measuring oxygen in vivo and in vitro.

7.2 Materials and Methods

7.2.1 Measurement of Oxygen Pressure Histograms

Phosphorescence lifetime measurements were performed using a PMOD-5000 phosphorometer (Oxygen Enterprises, Ltd., Philadelphia, PA, USA) [4], a frequency domain instrument with a range of 100–100,000 Hz. Phosphorescence lifetimes are independent of local phosphor concentration and insensitive to endogenous tissue fluorophores and chromophores. The PMOD-5000 was used in multifrequency mode [4] in order to determine distributions of phosphorescence lifetimes. The lifetime distributions were used to calculate distributions of oxygen pressures, i.e. oxygen histograms [21, 22]. The excitation light (635 nm) was modulated by a waveform consisting of 37 sinusoids with equal amplitudes and frequencies ranging from 100 Hz to 38 kHz. The tips of the light guides were brought into contact with the skin but care was taken not to apply pressure that might restrict flow in the surface blood vessels. The obtained signal was used to calculate the dependence of the phosphorescence amplitude and phase on the modulation frequency. The resulting phase/amplitude dependence was analyzed using the Maximal Entropy Method [21, 22] to yield the distribution of phosphorescence lifetimes. This distribution was converted into the distribution of oxygen pressure in the sample as described previously [21, 22]. The basis for the conversion is the Stern-Volmer relationship:

$$I^o/I = T^o/T = 1 + k_Q{}^*T^o\,{}^*pO_2, \qquad (7.1)$$

where, I^o, T^o and I, T are the phosphorescence intensities and lifetimes in the absence of oxygen and at oxygen pressure pO_2, respectively. The quenching constant, k_Q, is a second order rate constant, describing the quenching of the excited state of the phosphor by oxygen. The values of T^o and k_Q have been determined for each phosphor for the experimental conditions [4] (temperature etc. as appropriate).

According to (1), intensities (amplitudes) of phosphorescent signals decrease with increasing oxygen pressures. Thus, for equal volumes of tissue, containing equal amounts of the phosphorescent probe and excited by equal numbers of photons, the accuracy in determination of lifetimes and/or amplitudes will be higher for volumes with lower oxygen pressures. The decrease in accuracy (decrease in signal level) causes asymmetric broadening of oxygen histograms. This broadening increase with increasing oxygen pressure (decreasing signal) and this is responsible for the "tail" effect on the high oxygen end of the histogram. This broadening is intrinsic to the MEM analysis, reflecting the fact that uncertainty in determination of phosphorescence lifetimes increases as the signal-to-noise ratio (S/N) decreases. At lower oxygen pressures there is little broadening, less than 3 Torr for pressures below 20 Torr, but for oxygen pressures above about 80 Torr the histograms are substantially broadened are only qualitative. The presented histograms were arbitrarily truncated at 140 Torr.

7.2.2 Phosphorescent Probes Oxyphor G2 and Oxyphor G3

Both Oxyphors G2 [23] and G3 are based on Pd-tetrabenzoporphyrin cores [20]. The structure of G3 is published in Wilson et al [24]. and synthesis of similar dendritic porphyrins has been reported [25]. Pd tetrabenzoporphyrin (PdTBP) dendrimers G2 and G3 differ by the dendrimer composition (G2 – polyglutamate; G3 – polyarylglycine) and surface coatings (G2 – none;

G3 - PEG, Av. MW 350). G2 (MW 2,642) is designed to be used in combination with albumin, which provides a uniform microenvironment for the phosphor. In contrast, G3 (MW 16,100) is not affected by albumin and other biomolecules due to the surface layer of polyethyleneglycols (PEG's). The absorption and the phosphorescence spectra of G2 and G3 are nearly identical. Both phosphors have quantum yields of about 10% and lifetimes of about 270 μs in deoxygenated aqueous solutions. Oxygen quenching constants (k_Q) of G2 and G3 in aqueous buffered solutions at pH 7.2 at 38°C are 2,800 $Torr^{-1}s^{-1}$ and 180 $Torr^{-1}s^{-1}$ respectively. Unbound Oxyphor G2 cannot be used to measure oxygen in physiological range. In the blood, however, it binds tightly to albumin, and the oxygen quenching constant (k_Q) of the G2-albumin complex at 38°C is 280 $Torr^{-1}sec^{-1}$. Phosphorescence lifetime and oxygen quenching constant of Oxyphor G3 are insensitive to the presence of albumin (at 1–5% by weight) as well as changes in pH and ionic strength throughout the physiological range.

7.2.3 Measurements of Oxygen in the Blood Plasma and Interstitial Space of Muscle

Mouse preparation. The fur on the right and left rear quarters was removed by first using electrical clippers and then depilated. Care was taken not to cause any abrasions to the skin. The oxygen measurements were made non-invasively through the undisturbed skin. The fur was removed because in dark colored mice the fur absorbs both the excitation light and the emitted phosphorescence, greatly attenuating the phosphorescence signal.

Measuring oxygen histograms in the blood plasma (Oxyphor G2). Anesthesia was induced with 1.5% isoflurane in air and 0.1 ml of a solution of Oxyphor G2 (3.2 mg/ml) in physiological saline was injected into the tail vein. As soon as anesthesia was induced, isoflurane was decreased to 1.2% and the oxygen histograms were measured about 10 min after injection of the Oxyphor. It has been previously noted [17, 24] that induction of anesthesia with isoflurane causes a transient decrease in tissue oxygen pressures that recovers within 10 min of continuing anesthesia. After measuring the oxygen histograms (anesthetized), the nose cone supplying the isoflurane was removed and the mice replaced in their cage. After about 40 min without inhaled anesthetic, the oxygen histograms were again measured (awake).

Throughout the periods of anesthesia, body temperature was maintained by laying the mice on a 38o isothermal pad covered with a terry cloth towel to be sure they did not overheat.

Measuring oxygen in the interstitial space (Oxyphor G3). The mice were shaved and depilated as described above. They were anesthetized with isoflurane (nose cone, 1.5% mixed with air) and given injections of Oxyphor G3 solution (80 micromolar in physiological saline) along 3 different 1 cm tracks (20 μL containing 1.6 nmoles of Oxyphor per track) in the thigh muscle using a 30 gage needle. The nose cone was removed and the mice returned to their cage. They were allowed to wake up and run about in the cage for 70–90 min to help distribute the phosphor within the interstitial space of the muscle and then the oxygen histograms measured in the awake mouse. Each mouse was then anesthetized with either isoflurane or ketamine xylazine and the oxygen histograms measured described above. The amount of Oxyphor G3 injected into the muscle was about 4% of that required to give the concentration of Oxyphor G2 injected into the blood. Thus, the measured phosphorescence comes from the interstitial space.

The experiments were carried out by investigators trained to handle mice. All of the experimental procedures were reviewed and approved by the local IACUC committee. At the end of the experiment the mice were euthanized according to guidelines established by the AVMA Panel on Euthanasia.

7.3 Results

Preliminary measurements have been made in subcutaneous tumors grown on the hind quarter of mice. These tumors grow under the skin and the measurements can be made that are selective for tumor tissue since the tumor tissue is readily separated from the underlying muscle tissue. Illustrative measurements of the oxygen histograms for the interstitial space of muscle and tumors are shown in Fig. 7.1A and B. For Fig. 7.1A measurements were made for Oxyphor G3 in the interstitial space of a RENCA tumor and muscle measured on the same mouse. In this case the mouse was awake, illustrating that the

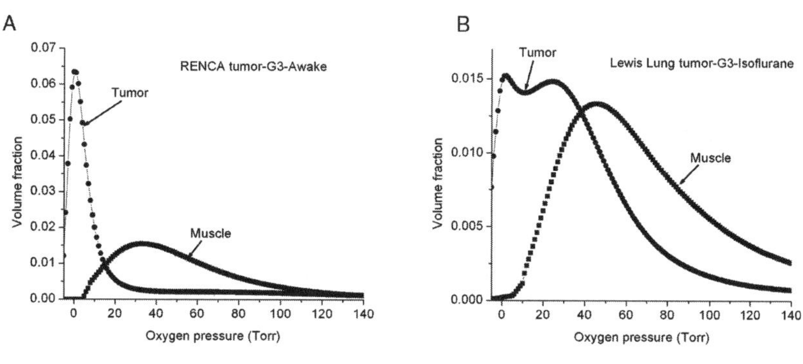

Fig. 7.1 (**A**) Oxygen pressure histograms for the interstitial space in RENCA tumors and normal muscle tissue. The Oxyphor G3 was microinjected and the oxygen histograms measured as described in Methods. The measurements were made while the mouse was awake and held in the hand. (**B**) Oxygen pressure histograms for the interstitial space in a Lewis Lung carcinoma and in normal muscle. The mouse was anesthetized with 1.2% isoflurane while the measurements were made. The histograms have been normalized to the same area under the curve for both tumor and muscle in order to eliminate differences in the total Oxyphor and illumination intensities.

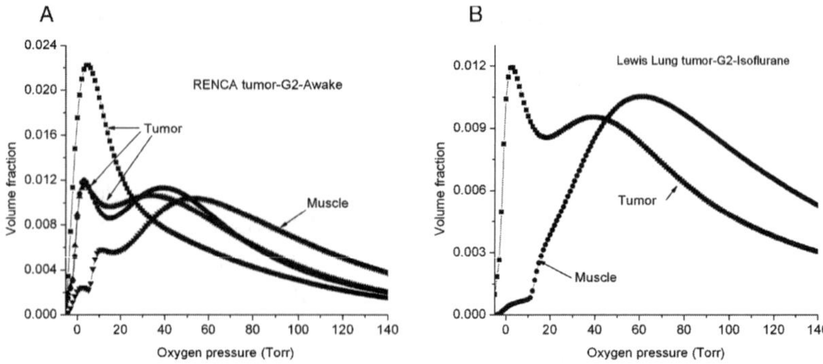

Fig. 7.2 **(A)** Oxygen pressure in the intravascular space in RENCA tumors and normal muscle. Oxyphor G2 was injected i.v. in the tail vein and then the oxygen pressure histograms measured in an awake mouse. Three histograms are presented, each measured for a different region of the tumor to emphasize the heterogeneity of this tumor [17]. (**B**) Oxygen histograms from a Lewis Lung carcinoma and normal muscle in an isoflurane anesthetized mouse.

measurements can be made in awake animals. It is important, however, that the animals be preconditioned to not become anxious when being handled. Although their becoming agitated does not affect the tumor measurements very much, if the leg muscles are being used to try to escape, or if the mice are stressed, this alters vascular regulation, blood pressure, and local blood flow. As a result, the tissue oxygen pressures are altered.

In normal muscle essentially all of the interstitial space of normal muscle has oxygen pressures greater than 10 Torr and there is a very small fraction with oxygen pressures less than 15 Torr. This is consistent with the results published earlier as part of a comparison of the oxygen pressures in the interstitial space and the vascular space in resting muscle [24]. In both the RENCA and Lewis Lung tumors the interstitial space oxygenation (Fig. 7.2A, B) is heterogeneous and generally lower than those in normal tissue. Particularly evident, for these two tumor types, is that a substantial part of both the interstitial and the vascular spaces have oxygen pressures less than 15 Torr. The tumor oxygen pressure distributions are, however, sufficiently different among tumors of the same type that much more detailed studies will be required to determine if further generalizations can be made. In addition, our preliminary measurements indicate that the anesthetic may also affect tissue oxygen pressures in the tumors more than in the muscle, and this needs to be studied in more detail.

7.4 Discussion

Oxygen pressures in the interstitial space can not be measured by other methods, making it impossible to compare the measured values with values from the literature. Micro-oxygen electrodes and solid EPR probes [26] measure a mixture of the interstitial space and capillary oxygenation, whereas nitroimidazole binding measures intracellular oxygenation. Most micro oxygen electrode measurements for normal tissue have been made in softer tissue, such as the kidney, liver and brain. Baumgärtl and coworkers [27] published histograms of the oxygen distribution in dog kidney with mean PO_2 values of 36.8 ± 6.0 (± SD) Torr, but did not indicate the anesthetic that was used. Oxygen measurements have been made in rodent muscles using oxygen electrodes and phosphorescence quenching. The electrode measurements were, however, typically made in urethane and/or barbiturate anesthetized animals and the muscle tissues were surgically exposed. Whalen and coworkers [28, 29] used electrodes with very small tips to measure oxygen pressures within the cells in living tissue in animals anesthetized with urethane and barbiturate. They reported 75% of the values were between 0 and 5 Torr in guinea pig gracilus and cat heart muscles whereas those in the cat soleus muscle were higher, having a mean value of 18.9 ± 1.8 Torr. The influence of the anesthetic on oxygen pressure in the tissue was not appreciated, and, partly for this reason early oxygen electrode measurements gave rise to the erroneous, view that the oxygen pressures in normal tissue are very low and there were significant volumes with effectively zero oxygen pressures. Later measurements have given higher values, and mean values reported for muscle tissue include 19 [30] and 26.8 [31] for the rat cremaster muscle, and 31.4 [14] Torr for the rat spinotrapezius microvasculature. These are still much lower than the 46.2 Torr (awake) or 36.9 Torr (isoflurane anesthesia) values obtained with phosphorescence quenching for the interstitial space [24], but are more consistent with those for ketamine plus xylazine anesthesia.

Tissue oxygen measurements using EPR active particles injected into the tissue [26], are reported to give oxygen pressures in the rat brain of 39.3 ± 4.1 Torr in isoflurane anesthetized rats [32, 33].

Nitroimidazole binding has been used to measure intracellular oxygenation (for review see [34]). Binding is small in normoxic tissue but increases strongly with decreasing oxygen pressures. Normal muscle and other tissues show little binding of the nitroimidazole, EF5 [2-(2-nitro-1H-imidazol-1-yl)-N-(2,2,3,3,3-pentafluoropropyl) acetamide] in awake and isoflurane anesthetized animals, indicating there are few cells with intracellular oxygen pressures less than about 15 Torr.

We conclude that the currently available data are consistent with mean oxygen pressures in normal skeletal muscle interstitium of 35–45 Torr and with there being negligible volumes with oxygen pressures less than 15 Torr. Further, direct measurements of oxygen pressures in the intravascular and interstitial spaces (see Wilson et al [24].) shows that the difference in oxygen pressure across the capillary walls under resting conditions is very small, less than 1.5 Torr. Thus, the capillary walls consume insignificant oxygen and provide very little resistance for oxygen movement from the blood plasma to the pericellular space. This contrasts with the suggestion by Tsai et al [35]. that the walls of small arterioles consume a substantial fraction of the available oxygen, resulting in a difference in oxygen pressure across the wall of tens of Torr.

Tumors, in contrast to normal tissue, are now well recognized as having substantial heterogeneity within individual tumors and among different tumor types. Preliminary measurements have shown that the oxygen pressures measured in the intravascular space and the interstitial spaces are very similar, although this is expected to depend on the tumor being measured. Particularly important will be the extent of tumor necrosis, as necrotic volumes will contribute to the interstitial space, but not the vascular space, oxygen measurements. In tumors, there seems no alternative to making the oxygen measurements in the tumor at the time of treatment if this important parameter is to be useful for developing therapeutic protocols. It is clear that conclusions concerning the efficacy of therapeutic protocols based on experiments in which the tumor tissue oxygenation was not measured must be interpreted with great caution.

Acknowledgment Supported in part by U54 CA105008-01 (WMFL and DFW), NS-31465 (DFW), HL081273 (DFW & SAV).

References

1. Vanderkooi JM, Maniara G, Green TJ, and Wilson DF. An optical method for measurement of dioxygen concentration based on quenching of phosphorescence, *J. Biol. Chem.* 262: 5476–5482, 1987.
2. Wilson DF, Rumsey WL, Green TJ, and Vanderkooi JM. The oxygen dependence of mitochondrial oxidative phosphorylation measured by a new optical method for measuring oxygen. *J. Biol. Chem.* 263: 2712–2718, 1988.
3. Dunphy I, Vinogradov SA, and Wilson DF. Oxyphor R2 and G2: Phosphors for measuring oxygen by oxygen dependent quenching of phosphorescence. *Analy. Biochem.* 310: 191–198, 2002.
4. Vinogradov SA, Fernandez-Seara MA, Dugan BW, and Wilson DF Frequency domain instrument for measuring phosphorescence lifetime distributions in heterogeneous samples, *Rev. Sci. Instruments* 72: 3396–3406, 2001.
5. Rumsey WL, Vanderkooi JM, and Wilson DF. Imaging of phosphorescence: A novel method for measuring the distribution of oxygen in perfused tissue. *Science 241*: 1649–1651, 1988.
6. Rumsey WL, Pawlowski M, Lejavardi N, and Wilson DF. Oxygen pressure distribution in the heart in vivo and evaluation of the ischemic "border zone". *Am. J. Physiol.* 266(4 Pt 2): H1676–80, 1994.
7. Shonat RD and Johnson PC. Oxygen tension gradients and heterogeneity in venous microcirculation: a phosphorescence quenching study. *Am. J. Physiol. Heart Circ. Physiol.* 272: H2233–H2240, 1997.
8. Buerk DG, Tsai AG, Intaglietta M, and Johnson PC. Comparing tissue PO_2 measurements by recessed microelectrode and phosphorescence quenching. *Adv. Exp. Biol. Med.* 454: 367–374, 1998.
9. Shonat, RD, Wilson DF, Riva CE, and Pawlowski M. Oxygen distribution in the retinal and choroidal vessels of the cat as measured by a new phosphorescence imaging method. *Applied Optics* 31: 3711–3718, 1992.
10. Vinogradov SA, Lo L-W, Jenkins WT, Evans SM, Koch C, and Wilson DF. Noninvasive imaging of the distribution of oxygen in tissue in vivo using near infra-red phosphors, *Biophys. J.* 70: 1609–1617, 1996.

11. Sinaasappel M, Donkersloot C, van Bommel J, and Ince C. PO_2 measurements in the rat intestinal microcirculation. *Amer. J. Physiol.* 276: G1515–20, 1999.
12. Richmond KN, Shonat RD, Lynch RM, and Johnson PC. Critical PO_2 of skeletal muscle in vivo. *Am. J. Physiol. Heart Circ. Physiol.* 277: H1831–H1840, 1999.
13. Dewhirst MW, Ong ET, Braun RD, Smith B, Klitzman B, Evans SM, and Wilson DF. Quantification of longitudinal tissue pO2 gradients in window chamber tumours: impact on tumour hypoxia, *Br. J. Cancer* 79: 1717–1722, 1999.
14. Behnke BJ, Kindig CA, Musch TI, Koga S, and Poole DC. Dynamics of microvascular oxygen pressure across the rest-exercise transition in rat skeletal muscle. *Resp. Physiol.* 126(1): 53–63, 2001.
15. Poole DC, Behnke BJ, McDonough P, McAllister RM, and Wilson DF. Measurement of muscle microvascular oxygen pressures: compartmentalization of phosphorescent probe. *Microcirculation.* 11(4): 317–26, 2004.
16. Wilson DF, Vinogradov SA, Grosul P, Vaccarezza MN, Kuroki A, and Bennett J. Oxygen distribution and vascular injury in the mouse eye measured by phosphorescence lifetime imaging. *Appl. Optics* 44: 1–10, 2005.
17. Ziemer L, Lee WMF, Vinogradov SA, Sehgal C, and Wilson DF. Oxygen distribution in murine tumors: characterization using oxygen-dependent quenching of phosphorescence. *J. Appl. Physiol.* 98: 1503–1510, 2005.
18. Rozhkov V, Wilson DF, and Vinogradov SA. Tuning oxygen quenching constants using dendritic encapsulation of phosphorescent Pd-porphyrins. *Polymeric Materials: Sci. & Eng.* 85: 601–603, 2001.
19. Rozhkov V, Wilson DF, and Vinogradov SA. Phosphorescent Pd porphyrin-dendrimers: Tuning core accessibility by varying the hydrophobicity of the dendritic matrix. *Macromolecules* 35: 1991–1993, 2002.
20. Rietveld IB, Kim E, and Vinogradov, SA. Dendrimers with tetrabenzoporphyrin cores: near infrared phosphors for in vivo oxygen imaging. *Tetrahedron* 59: 3821–3831, 2003.
21. Vinogradov SA and Wilson DF. Phosphorescence lifetime analysis with a quadratic programming algorithm for determining quencher distributions in heterogeneous systems. *Biophys. J.* 67: 2048–2059, 1994.
22. Vinogradov SA and Wilson DF Recursive maximum entropy algorithm and its application to the luminescence lifetime distribution recovery. *Applied Spectroscopy* 54: 849–855, 2000.
23. Vinogradov SA and Wilson DF. Metallotetrabenzoporphyrins. New phosphorescent probes for oxygen measurements. *J. Chem. Soc., Perkin Trans. II*, 103–111, 1994.
24. Wilson DF, Lee WMF, Makonnen S, Finikova O, Apreleva S, and Vinogradov SA Oxygen pressures in the interstitial space and their relationship to those in the blood plasma in resting skeletal muscle. *J. Appl. Physiol.* 101: 1648–1656, 2006.
25. Vinogradov SA Arylamide dendrimers with flexible linkers via haloacyl halide method. *Organic Letters* 7: 1761–1764, 2005.
26. Swartz HM. Using EPR to measure a critical but often unmeasured component of oxidative damage: oxygen. [Review] *Antioxidants & Redox Signaling* 6(3): 677–686, 2004.
27. Baumgärtl H, Zimelka W, and Lübbers D. Evaluation of PO_2 profiles to describe the oxygen pressure field within the tissue. *Comp. Biochem. & Physiol. Part A* 132: 75–85, 2002.
28. Whalen WJ. Intracellular PO2 in heart and skeletal muscle. *Physiologist* 14(2): 69–82, 1971
29. Whalen WJ, Nair P, and Ganfield RA. Measurements of oxygen tension in tissues with a micro oxygen electrode. *Microvascular Research.* 5(3): 254–262, 1973.
30. Prewitt RL, and Johnson PC. The effect of oxygen on arteriolar red cell velocity and capillary density in the rat cremaster muscle. *Microvasc. Res.* 12: 59–70, 1976.

31. Johnson PC, Vandegriff K, Tsai AG, and Intaglietta M. Effect of acute hypoxia on microcirculatory and tissue oxygen levels in rat cremaster muscle. *J. Appl. Physiol.* 98: 1177–1184, 2005.
32. Swartz HM, Taie S, Miyake M, Grinberg OY, Hou H, el-Kadi H, and Dunn JF. The effects of anesthesia on cerebral tissue oxygen tension: use of EPR oximetry to make repeated measurements. *Adv. Exptl. Med. & Biol.* 530: 569–575, 2003.
33. O'Hara JA, Hou H, Demidenko E, Springett RJ, Khan N, and Swartz HM. Simultaneous measurement of rat brain cortex PtO_2 using EPR oximetry and a fluorescence fiber-optic sensor during normoxia and hyperoxia. *Physiol. Measur.* 26(3): 203–213, 2005.
34. Koch CJ. Measurement of absolute oxygen levels in cells and tissue using oxygen sensors and EF5. *Meth. in Enz.* 352: 3–31, 2002.
35. Tsai AG, Friesenecker B, Mazzoni MC, Kerger H, Buerk DG, Johnson PC, and Intaglietta M. Microvascular and tissue oxygen gradients in the rat mesentery. *Proc. Natl. Acad. Sci. USA* 95(12): 6590–6595, 1998.

Part II
Other Metabolite Transport in Tissue

Chapter 8
Adjuvant Induced Glucose Uptake by Activated T Cells is not Correlated with Increased Survival

Sadhak Sengupta*, Rebecca J. Vitale*, Paula M. Chilton[1], and Thomas C. Mitchell[1]

Abstract Authors contributed equally to this manuscript Natural adjuvants, such as bacterial lipopolysaccharide (LPS), activate antigen presenting cells via Toll-like receptors and, indirectly, increase the survival of antigen-activated T cells. The molecular mechanisms leading to increased survival remain poorly defined. Because T cell clonal expansion leads to high energy demands, we hypothesized that increased glucose uptake and/or utilization in adjuvant-activated T cells could be important molecular event(s) that would lead to adjuvant-associated T cell survival advantage. Using a fluorescent analog of 2-deoxyglucose, 2-NBDG, we measured glucose accumulation and rate of uptake in T cells from mice treated with antigen in the absence or presence of LPS. Although adjuvant activated T cells increased the accumulation of 2-NBDG, the rate of uptake was unchanged compared to cells activated with only antigen. Furthermore, glucose transport inhibitors, cytochalasin B or phloretin, decreased the accumulation of glucose in adjuvant-treated T cells, but this decrease did not impair adjuvant-associated survival advantages. Together, these data indicate that increased glucose uptake through glucose transporters is not required for increased survival of activated T cells.

8.1 Introduction

T cell proliferation initiated by activation with antigen is followed by rapid death due to a loss of antigen exposure and acute shortage of pro-survival cytokine signals. This deletion, although important for reducing the risk of autoimmune responses, hampers immunity because complete elimination of the

[1]Institute for Cellular Therapeutics, University of Louisville School of Medicine, 570, S. Preston Street, Louisville, KY 40202, e-mail: tom.mitchell@louisville.edu
*Authors contributed equally to this manuscript

responding T cells would reduce the ability of a host to resist subsequent infection. Vella et al. first showed that in vivo activated T cells were protected from growth-factor withdrawal induced death by the natural adjuvant LPS [1]. The mechanism(s) by which natural adjuvants keep activated T cells alive in an immune response is still not fully understood. Co-stimulatory factors and known pro-survival molecules like Bcl2 and Bcl-X_L were previously shown to be insufficient for adjuvant mediated survival [2, 3]. Activated T cell survival is often linked to the activation of PI3-kinase during clonal expansion because it enhances proliferation and short-term survival through increased expression of cytokines [4, 5]. We recently reported that although transiently activated by adjuvant exposure, PI3-kinase stimulation was insufficient to account for adjuvant-induced survival and suggested that induction of post PI3-kinase signaling effects must be involved [6].

PI3-kinase activity is reported to induce increased localization of the glucose transporter GLUT1 to the cell surface [7]. Recent reports also show that PI3-kinase stimulates increases in GLUT1 expression in B cells upon activation with antigen, which results in increased glucose uptake [8]. Increased uptake of glucose has also been reported to be required by activated T cells for maintenance of aerobic glycolyis, a metabolic system that is proposed to provide the carbon source needed by cells as they prepare for several rounds of division [9]. Moreover, activated T cells harvested near the peak of clonal expansion in immunized mice showed an increase in glucose accumulation [6]. We therefore decided to study whether this accumulation was due to increased rates of glucose uptake and whether or not it was correlated with adjuvant-mediated survival advantages.

Using 2-NBDG, a fluorescinated derivative of 2-deoxyglucose and known glucose-transport inhibitors, cytochalasin B and phloretin [10, 11], we measured glucose uptake and survival of T cells activated either in presence or absence of adjuvant effects induced by LPS. The results indicate that adjuvant-mediated survival effects do not require increased rates of uptake or intracellular accumulation of glucose.

8.2 Materials and Methods

8.2.1 T Cell Activation and Primary Cell Culture

Activated T cells were harvested from antigen treated mice as described elsewhere [6]. Briefly, Vβ3$^+$ TCR bearing T cells were activated by injecting B10.BR mice via the tail vein with 0.1μg of the T cell superantigen Staphylococcal enterotoxin A (SEA; Toxin Technologies, Sarasota, FL) and 16 h later with 10 μg of bacterial lipopolyaccharide (LPS; from *Salmonella typhosa*; Sigma Aldrich, St. Louis, MO). Spleens were harvested 40 h after activation, red blood cells were lysed with ACK buffer (160 mM NH_4Cl, 10 mM $KHCO_3$,

0.1 mM EDTA) and splenocytes were resuspended to 5×10^6 cells/ml in RPMI-1640 tissue culture medium (Invitrogen, Carlsbad, CA) supplemented and L-glutamine (Invitrogen). Fetal bovine serum was not used in any of these short-term cell culture experiments to avoid survival effects due to entrinsic growth factors.

8.2.2 Glucose Uptake by Activated T Cells

Activated splenocytes harvested from mice after 40 h of SEA (± LPS) activation were cultured in each well of a 96-well tissue culture plate (BD Falcon, Bedford, MA) in serum-free and glucose-free RPMI (Invitrogen) up to 1 h. The cells were pulsed with 100 μM of 2-NBDG in glucose-free medium for 10 mins either before or after the 1 h incubation period. After washing the cells to remove excess fluorescent dye they were surface stained with anti-CD4 and anti-Vβ3 monoclonal antibodies (BD Pharmingen, San Diego, CA). The levels of 2-NBDG taken up by $V\beta3^+$ $CD4^+$ T cells were analyzed using a FACScalibur flow cytometer (BD Immunocytometry System, San Jose, CA).

8.2.3 Glucose Transport Inhibitors and Measurement of Glucose Uptake and Survival of Activated T Cells

To test for the inhibition of glucose transport activity, 5×10^5 splenocytes harvested from SEA (± LPS) treated mice were plated in a 96-well tissue culture plate and incubated in RPMI-1640 medium supplemented with glucose transporter blockers cytochalasin B (0–20 μM; Sigma Aldrich, St. Louis, MO) or Phloretin (0–20 μM; Sigma Aldrich, St. Louis, MO). Following 20 h culture with the inhibitors, cells were washed and incubated for 1 hour with 2-NBDG (30 μM) in glucose-free medium and mean fluorescence intensity (MFI) of 2-NBDG in $V\beta3^+$ $CD4^+$ T cells was assessed by flow cytometry. Portions of each 20 h culture were reserved to measure T cell survival. Briefly, cells were washed after 20 h culture and stained for CD4 and Vβ3; survival was determined using a flow cytometer by measuring the proportion of $V\beta3^+$ $CD4^+$ T cells whose light scatter properties showed they were alive or dead [6].

8.3 Results and Discussion

We often use SEA, which is a T cell superantigen (SAg), as a tool to study antigen and adjuvant-specific effects on primary T cell activation. SAg treatment allows for the activation of limited but detectable amounts of the available repertoire of normal, primary T cells and does not need to be processed by APC before being 'presented' to T cells. This allows adjuvant-induced survival effects to be restricted to post-presentation events in order to better isolate and study the underlying mechanism(s). Acute exposure of responsive T cells to purified SAg results in activation, expansion and then deletion of these activated cells by apoptosis unless adjuvants are added to the system [1, 12]. In order to measure glucose uptake by activated T cells 2-NBDG was used. 2-NBDG (Molecular probes, Eugene, OR) is a fluorescent analog of 2-deoxyglucose which is taken by the glucose transporters and emits fluorescence at a peak E_m of 550 nm. 2-deoxyglucose is phosphorylated by hexokinase into 2-deoxyglucose-6-phosphate which cannot be metabolized further or transported out and therefore accumulates in the cells, which makes 2-NBDG a useful tool for measuring glucose uptake via flow cytometry and confocal microscopy [6, 13, 14].

Rates of glucose uptake were measured in activated T cells. B10.BR mice were injected with SEA in the presence or absence of LPS and activated splenocytes were harvested 40 h after SEA injection. The cells were subjected to a 10-minute pulse with 100 μm 2-NBDG before or after 1 h of glucose starvation. Initial levels of 2-NBDG taken up by freshly harvested activated T cells were low. After 60 mins of glucose deprivation, the amount of 2-NBDG taken in during a 10-min pulse increased significantly. However, cells from both different treatment groups took up the same amount of 2-NBDG indicating that previous exposure to adjuvant had no effect on T cell's ability to upregulate glucose transport activity (Fig. 8.1).

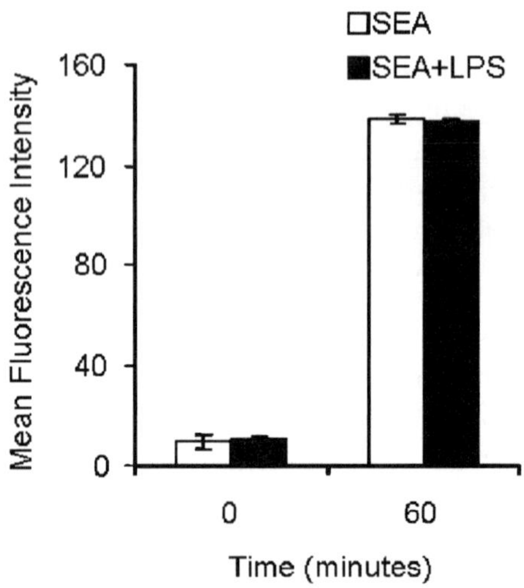

Fig. 8.1 Rate of Glucose uptake in activated T cells is not changed upon adjuvant stimulation. Splenocytes were harvested from B10.BR mice injected with either Staphylococcal enterotoxin A (SEA) alone or along with LPS (SEA+LPS) and subjected to a 10-minute pulse with 100 μM 2-NBDG before or after 1 hr culture in glucose-free condition. Rate of 2-NBDG uptake in activated CD4 T cells were analyzed by flow cytometry. No difference in rate of 2-NBDG uptake was observed in adjuvant stimulated T cells both before or after glucose-starvation indicating that adjuvant treatment had no effect on the glucose transporters.

We next assessed the effect of glucose transport inhibitors on glucose accumulation in activated T cells. Splenocytes harvested from SEA ± LPS treated mice were cultured with cytochalasin B and phloretin for 20 h. Cytochalasin B is a cell-permeable fungal toxin which inhibits cytoplasmic division by blocking the formation of contractile microfilaments [15]. It inhibits cell movement by shortening actin filaments by blocking monomer addition at the fast-growing end of polymers [16]. Cytochalasin B also inhibits glucose uptake by competitive inhibition with D-glucose by binding specifically to GLUT1; other members of the GLUT family are less affected [10, 17]. Phloretin also competitively inhibits glucose uptake and affects all GLUT isoforms [11, 18].

Following incubation with the inhibitors, the cells were washed and incubated with 30 μM of 2-NBDG for 60 mins in glucose-free medium. In the absence of inhibitors, accumulation of intracellular 2-NBDG was much greater in T cells from SEA+LPS treated mice when compared to T cells from mice given SEA alone (Fig. 8.2). T cells that had been treated with increasing amounts of phloretin and

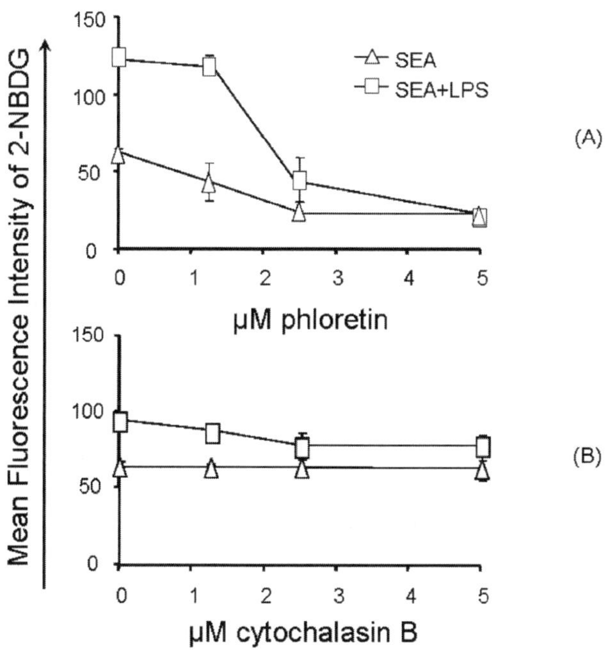

Fig. 8.2 Glucose accumulation is inhibited upon treatment of activated T cells with phloretin and cytochalasin B. Splenocytes were harvested from B10.BR mice injected with either Staphylococcal enterotoxin A (SEA) alone or along with LPS (SEA+LPS) and incubated overnight with increasing concentrations of phloretin **(A)** or cytochalasin B **(B)**. The cells were then incubated for 1 hr with 30 μM 2-NBDG in glucose-free condition. Levels of 2-NBDG in activated CD4 T cells were analyzed by flow cytometry. Uptake of 2-NBDG was inhibited with both blockers. The effect was visible with 2.5 μM cytochalasin B which however did not affect the uptake in SEA only T cells. Phloretin inhibited 2-NBDG uptake in both SEA and SEA+LPS treated T cells.

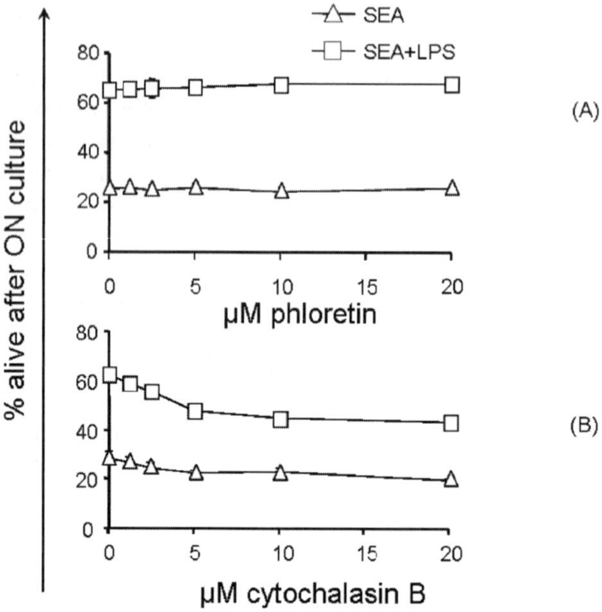

Fig. 8.3 Adjuvant-mediated survival advantage is not reduced by treatment of activated T cells with cytochalasin B or phloretin. Splenocytes were harvested from B10.BR mice injected with either Staphylococcal enterotoxin A (SEA) alone or along with LPS (SEA+LPS) and incubated overnight with increasing concentrations of phloretin **(A)** and cytochalsin B **(B)**. Survival analysis by flow cytometry showed an overall decrease in viability of activated T cells but it did not affect the adjuvant-mediated survival advantage of the LPS treated cells even with 20 μM cytochalasin B . Incubation with phloretin did not affect the general viability of the activated T cells and the adjuvant-mediated survival advantage was maintained with excess of phloretin.

cytochalasin B showed decreasing levels of 2-NBDG, such that the adjuvant-associated advantage narrowed in a dose-dependent manner. 5 μM of either inhibitor was sufficient to reduce glucose accumulation to the same levels in both populations. Dose-adjusted vehicle controls (methanol or DMSO) had no effect on 2-NBDG uptake or retention (data not shown). Therefore the glucose transport inhibitors phloretin and cytochalasin B prevented preferential accumulation of 2-NBDG induced by adjuvant.

To test whether or not glucose uptake was necessary for adjuvant-induced survival effects, splenocytes from SEA or SEA+LPS mice were cultured in RPMI for 20 h with phloretin and cytochalasin B up to 20 μM; which is in excess of the concentration needed to reduce glucose uptake to minimal levels (Fig. 8.2). As shown in Fig. 8.3, phloretin showed no ability to diminish the adjuvant-induced survival advantage of T cells isolated from SEA+LPS treated mice at any dose tested. T cell survival was moderately decreased by low concentrations of cytochalasin B, but further decreases in survival were not detectable in higher concentrations. Therefore, both phloretin and cytochalasin B failed to prevent adjuvant-induced survival effects even at concentrations 2–4 folds those needed to inhibit the preferential glucose accumulation by the adjuvant treated T cells.

These observations indicate that the adjuvant-mediated survival advantage in activated T cells is not dependent upon efficient glucose uptake. Furthermore, the phenomenon of increased accumulation of 2-NBDG in adjuvant-stimulated T cells is not due to increased glucose uptake but is probably caused by increased hexokinase-mediated phosphorylation in T cells of the glucose that enters the cells. Hexokinase activity in T cells has been reported to be induced by PI3-kinase/pAkt signaling [9]. Therefore we propose that PI3-kinase/pAkt-mediated stimulation of hexokinase is likely to be one of many mechanisms by which adjuvant alter the metabolism of activated T cells.

References

1. A.T. Vella, J.E. McCormack, P.S. Linsley, J.W. Kappler, and P. Marrack. "Lipopolysaccharide interferes with the induction of peripheral T cell death." Immunity 2, no. 3(March 1995):261–70.
2. T. Mitchell, J. Kappler, and P. Marrack. "Bystander virus infection prolongs activated T cell survival." J. Immunol. 162, no. 8 (April 1999):4527–35.
3. T.C. Mitchell, T.K. Teague, D.A. Hildeman, J. Bender, W.A. Rees, R.M. Kedl, B. Swanson, J.W. Kappler, and P. Marrack. "Stronger correlation of bcl-3 than bcl-2, bcl-xL, costimulation, or antioxidants with adjuvant-induced T cell survival." Ann. N. Y. Acad. Sci. 975, (December 2002):114–31.
4. L.P. Kane and A. Weiss. "The PI-3 kinase/Akt pathway and T cell activation: pleiotropic pathways downstream of PIP3." Immunol. Rev. 192, (April 2003):7–20.
5. F.V. Lali, J. Crawley, D.A. McCulloch, and B.M. Foxwell. "A late, prolonged activation of the phosphatidylinositol 3-kinase pathway is required for T cell proliferation." J. Immunol. 172, no. 6 (March 2004):3527–34.
6. S. Sengupta, P.M. Chilton, and T.C. Mitchell. "Adjuvant-induced survival signaling in clonally expanded T cells is associated with transient increases in pAkt levels and sustained uptake of glucose." Immunobiology 210, no. 9 (2005):647–59.
7. J.C. Rathmell, C.J. Fox, D.R. Plas, P.S. Hammerman, R.M. Cinalli, and C.B. Thompson. "Akt-directed glucose metabolism can prevent Bax conformation change and promote growth factor-independent survival." Mol. Cell Biol. 23, no. 20 (October 2003):7315–28.
8. C.A. Doughty, B.F. Bleiman, D.J. Wagner, F.J. Dufort, J.M. Mataraza, M.F. Roberts, and T.C. Chiles. "Antigen receptor-mediated changes in glucose metabolism in B lymphocytes: role of phosphatidylinositol 3-kinase signaling in the glycolytic control of growth." Blood 107, no. 11 (June 2006):4458–65.
9. K.A. Frauwirth and C.B. Thompson. "Regulation of T lymphocyte metabolism." J. Immunol. 172, no. 8 (April 2004):4661–65.
10. B. Hellwig and H.G. Joost. "Differentiation of erythrocyte-(GLUT1), liver-(GLUT2), and adipocyte-type (GLUT4) glucose transporters by binding of the inhibitory ligands cytochalasin B, forskolin, dipyridamole, and isobutylmethylxanthine." Mol. Pharmacol. 40, no. 3 (September 1991):383–89.

11. M. Kobori, H. Shinmoto, T. Tsushida, and K. Shinohara. "Phloretin-induced apoptosis in B16 melanoma 4A5 cells by inhibition of glucose transmembrane transport." Cancer Lett 119, no. 2 (November 1997):207–12.
12. J.E. McCormack, J.E. Callahan, J. Kappler, and P.C. Marrack. "Profound deletion of mature T cells in vivo by chronic exposure to exogenous superantigen." J. Immunol. 150, no. 9 (May 1993):3785–92.
13. A. Virkamaki, E. Rissanen, S. Hamalainen, T. Utriainen, and H. Yki-Jarvinen. "Incorporation of [3-3H]glucose and 2-[1-14C]deoxyglucose into glycogen in heart and skeletal muscle in vivo: implications for the quantitation of tissue glucose uptake." Diabetes 46, no. 7 (July 1997):1106–10.
14. K. Yoshioka, H. Takahashi, T. Homma, M. Saito, K.B. Oh, Y. Nemoto, and H. Matsuoka. "A novel fluorescent derivative of glucose applicable to the assessment of glucose uptake activity of Escherichia coli." Biochim. Biophys Acta 1289, no. 1 (February 1996):5–9.
15. A. Ghosh, J. Maniloff, and D.A. Gerling. "Inhibition of mycoplasma cell division by cytochalasin B." Cell 13, no. 1 (January 1978):57–64.
16. P.A. Theodoropoulos, A. Gravanis, A. Tsapara, A.N. Margioris, E. Papadogiorgaki, V. Galanopoulos, and C. Stournaras. "Cytochalasin B may shorten actin filaments by a mechanism independent of barbed end capping." Biochem Pharmacol. 47, no. 10 (May 1994):1875–81.
17. J.F. Griffin, A.L. Rampal, and C.Y. Jung. "Inhibition of glucose transport in human erythrocytes by cytochalasins: A model based on diffraction studies." Proc. Natl. Acad. Sci. U. S. A. 79, no. 12 (June 1982):3759–63.
18. R.M. Krupka. "Asymmetrical binding of phloretin to the glucose transport system of human erythrocytes." J. Membr. Biol. 83, no. 1–2 (1985):71–80.

Chapter 9
Lactate, with Oxygen, Incites Angiogenesis

Thomas K. Hunt, Rummana Aslam, Zamir Hussain, and Stefan Beckert[1]

Abstract Lactate has been reconsidered! As we now know, most is produced aerobically We report that lactate accumulation commonly occurs in the presence of oxygen and is sufficient to instigate signals for angiogenesis and connective tissue deposition. These include vascular endothelial growth factor (VEGF), transforming growth factor beta (TGF beta), interleukin-1 (IL-1), and hypoxia-inducible factor (hif-1alpha). This paper, a mini-review, is occasioned by new data showing increased presence of VEGF and angiogenesis in an oxygenated site by adding a slow-release source of lactate into Matrigel® and implanting the Matrigel subcutaneously in mice.

9.1 Introduction

Intuition is sufficient to convince that metabolic need is a stimulant to angiogenesis. The well-known hypoxic induction of angiogenic signals is the most dramatic example. By that reasoning, lactate, at times an expression of metabolic need, should be a subsidiary signal; and, indeed, several investigators have demonstrated that accumulated lactate, enhances collagen deposition, proteoglycans deposition, and endothelial migration (via stimulation of VEGF) [1–4].

Lactate in the presence of oxygen? Hypoxia is not even the major source of lactate nor does added oxygen necessarily enhance its metabolism! Authoritative investigators have now agreed that lactate, long thought to be merely the end product of hypoxia, has far greater significance! In fact, lactate has many aerobic sources including aerobic glycolysis, and activated leukocytes, and performs many important and previously unanticipated functions. These new views have been well discussed in a recent review [5].

[1]Thomas K. Hunt, Rummana Aslam, Zamir Hussain, and Stefan Beckert, Department of Surgery at the University of California Medical Center, San Francisco, CA 94143

K.A. Kang et al. (eds.), *Oxygen Transport to Tissue XXIX*

Since lactate, even if aerobically derived, has these properties which heretofore were ascribed to hypoxia, it follows that lactate-derived angiogenesis might have many sources other than hypoxia. For instance, lactate accumulates in wounds regardless of PO_2. Albina [6] has found that hypoxia-inducible factor hif-1alpha detected by pimonidazol adduct formation is maximal in wounds at 6 hours, the time-frame in which lactate accumulates to its usual 5–10 mM level, and well before PO_2 falls. Wounds are inflammatory lesions, and leukocytes are major sources of aerobic lactate [7]. Might lactate be the dominant source of angiogenesis that accompanies inflammation of all sorts? Wounds, even well oxygenated wounds, are a major source and reservoir of lactate. Wound angiogenesis is arguably the most robust and most rapid found in nature.

We have previously demonstrated that wound concentrations of lactate monomer, induce collagen production in fibroblasts [2, 8]. With oxygen, it induces post-transcriptional hydroxylation of collagen [8]. VEGF release follows the lactate-induced reduction of the NAD+/NADH via reduction of ADPRibosylations [1]. With or without oxygen, lactate induces VEGF from macrophages [2]. Adding exogenous lactate to wounds increases the presence of TGF beta, VEGF, interleukin-1 (Il-1) and collagen deposition [8]. Liu [9] has shown that all the major products of glycolysis, i.e. acetoacetate, pyruvate, and lactate increase the presence of hif-1alpha, a precursor of VEGF, even when oxygen is present.[1]

To complicate the situation further, hyperoxia significantly accelerates angiogenesis and collagen deposition. Therefore, a mechanism for angiogenesis must exist that tolerates high concentrations of oxygen. We propose that lactate accumulation is a potent stimulator of angiogenesis, and that oxygen is a required partner in the tissue response.

9.2 Materials and Methods

We chose the Matrigel® (BD biosciences) model for this study because as other investigators have found, it produces only slight inflammation and no angiogenesis on implantation whether or not it is the reduced growth factor variety. Yet, it does support angiogenesis when, for instance, a sufficient amount of VEGF is incorporated. The PO_2 in such implants is 50 to 60 mm Hg in animals breathing air at ambient pressure [10].

Preliminarily, we tested various molecular weight polymers of lactate-co-glycolide to determine if any released sufficient monomer (by non-enzymatic hydrolysis) in Matrigel to raise and sustain the level of lactate monomer from

[1] Since all of these products are in a dynamic equilibrium with each other, it is difficult to know which of them is/are responsible. However, this distinction is not important for present purposes.

about 3 mM in controls to about 6 mM (roughly the characteristic wound level). Polylactate co-glycolide 50:50 (MW 40,000–75,000) (Sigma) met that expectation. Higher molecular weight polymers were not hydrolysable, did not induce angiogenesis, and were used as negative controls. To obtain the most uniform lactate concentration as possible, the polymer crystals were finely powdered (in a coffee grinder).[2]

9.2.1 Conditions

The lactate polymer powder was mixed into cold Matrigel (30 mg/ml), and 0.2 ml of the mixture was injected subcutaneously into each flank in each of 50 animals. Fifty controls were implanted with Matrigel alone. Animals were sacrificed at 11 days. The implants were removed, and sections taken from mid-implant were stained with hematoxylin and eosin, Mallory's trichrome, and were evaluated for angiogenesis and collagen presence by two independent observers. The criteria for quantification were: No cells or blood vessel = 0. A few scattered endothelial or other cells but no alignments of cells in linear or circular tubes = 1. A few circular alignments of cells but no RBC containing areas = 2. The presence of tubular structures containing RBC = 3. Presence of small and medium sized blood vessels with RBC = 4.

9.2.1.1 Experiment 1

Hypothesis: Implantation of lactated Matrigel in animals that subsequently breathed ambient air at 1 ATA will demonstrate angiogenesis.

Matrigel pellets were removed at 11 days and those containing lactate (six animals) were compared with pellets not containing lactate that had been implanted in 6 other animals, and 6 in which insoluble polymer had been incorporated.

9.2.1.2 Experiment 2

Hypothesis: A quantitative relationship develops between the concentrations of lactate monomer and VEGF in lactated implants.

The rationale was that our prior findings in vitro demonstrated that lactate induces VEGF release from macrophages. Scattered inflammatory cells, some

---

[2] The polymer releases lactate, not lactic acid. Lactate is a weak base and thus alkalinizes the solution slightly. Hydrolysis does not produce local acidosis! The result of hydrolysis is a racemic mixture. Glycolide is quickly metabolized, and some is converted to lactate [3, 8]. L-lactate and d-lactate equally influence endothelial migration, and both chelate with iron ion.

of them macrophages, normally are present at the edge of Matrigel implants. They are increased in lactated implants.

Two lactated implants were injected into each of 10 animals. Ten controls were injected with Matrigel containing no additive. After their removal at 11 days the pellets were removed and quickly frozen. When all samples had been collected, they were weighed, thawed, quickly centrifuged free of residual lactate polymer, and the liquid phase was analyzed for lactate and VEGF (ELISA, R&D Systems Inc.).

9.2.1.3 Experiment 3

Hypothesis: Oxamate, an inhibitor of LDH and ADPRibosylation, will abolish the ability of lactate to induce angiogenesis.

The rationale was that prior data had demonstrated that decreased ADPRibosylation due to lactate induced collagen deposition and VEGF production. The mechanism was traced to diminished NAD+ due to increased conversion of NAD+ to NADH. This happens because ADPRibose (Adenosine Diphospho-Ribose) and subsequent ADPRibosylations are derived only from NAD+. As a consequence, the ADPR that inhibits collagen prolyl hydroxylase is withdrawn, thus increasing prolyl hydroxylation and collagen deposition. Inhibition of lactate dehydrogenase prevents the reduction of NAD+, withdrawal of ADPRibose from prolyl hydroxylase, and greatly inhibits the effect of lactate [1]. A similar effect regulates VEGF. Oxamate was, therefore expected to abrogate lactate-induced angiogenesis.

Six animals were injected with lactated Matrigel. Six others were injected with lactated implants that also contained 3 mM of oxamate (Sigma).

9.3 Results

Experiment 1: The addition of soluble lactate polymer induced angiogenesis in every implant. The average score was 3.5. Neither non-lactated controls nor implants with insoluble lactate polymer induced any significant angiogenesis in any implant. The average score was 0.5. There was no overlap ($p < 0.01$). Vessels in the lactated implants ranged from small, one-cell layer thick capillaries to about 50 μ, medium sized vessels with what appear to be surrounding pericytes, Collagen deposition was evident in and around the perivascular structures.

Experiment 2: A three phase statistical relationship to VEGF was found. VEGF rose in parallel with the lactate concentration until high levels indicated lactate toxicity, and VEGF fell (significance, $p<0.05$). The effect of lactate was abrogated by Anti-VEGF polyclonal antibody [2].

Experiment 3: Oxamate greatly reduced angiogenesis when given together with lactate. (Average score 1.0 vs. 3.5). Again, there was no overlap in the results.

9.4 Discussion

In previous studies we established that addition of lactate polymer to implanted wire mesh cylinder wounds raised lactate monomer concentration from 6 to 9 mM ($p<0.01$) collagen deposition by 50% ($p<0.02$) and also raised VEGF, TGF-B, and Il-1 (transiently) while lowering IGF-1 to almost zero. PO_2, pH, and PCO2 were unchanged [8]. In unpublished studies, we have demonstrated enhanced superoxide flux that was present focally in cultured human endothelial cells after the cells were exposed to nitroblue tetrazolium (Sigma) and 15 mM lactate monomer at standard cell culture oxygen conditions of PO2, i.e. about 35 mm Hg.

Also in prior studies, we demonstrated that angiogenesis occurs in proportion to oxygen tension. Hyperoxia more than doubled the number of "mature" vessels [10]. Hypoxia reduced mature vessel count to zero even when VEGF was implanted.

These results, taken together, strongly imply that accumulation of lactate to the 5–10 mM level, in the presence of oxygen, is by itself, but only in the presence of oxygen, sufficient to initiate a healing reaction, i.e., increased cell migration, angiogenesis, proteoglycans production, and collagen synthesis and deposition. Elevation of oxygen tension, as we have shown increases both angiogenesis and VEGF [10, 11].

The aggregate results are made congruent by five recent discoveries all of which point to a redox control mechanism of the lactate effect: 1) $Fe^{++,+++}$ form a chelate with lactate that in the presence of H_2O_2 enhances OH- [12]. 2) Iron-containing structures that participate in redox reactions involving H_2O_2 occur in or on the endoplasmic reticulum [9]. In Liu's experiments, hif-1 gene was internalized into the nucleus when H_2O_2 concentration was increased. The peroxide concentration was proportional to the PO_2. 3) Pyruvate, lactate and acetoacetate enhance the presence of hif-1 when added to benign or malignant cells in the presence of oxygen [13], The investigators propose that this manifests as a "feed forward," an amplification, loop. 4) As reported by Sen et al. peroxide binds to the SP-1 site on the VEGF reporter leading to hif-1-independent VEGF production [14], and 5) lactate diminishes NAD^+ thus decreasing ADPRibosylation and activating collagen gene transcription and collagen prolyl hydroxylase activity [15]. All of these manifestations are likely to be increased in moderate hyperoxia. When excessively high, oxygen is also likely to become toxic due to redox stress as has been found by Hopf [10].

These findings indicate that lactate participates in a complex redox control system that has at least three pathways. We postulate that lactate chelates iron ion in highly specific sites whereupon H_2O_2, itself derived from oxygen, is diverted from other functions to produce OH-, or perhaps better said, diverted to a focal area in which redox flux is intensified. It seems likely that these sites coincide with those found by Liu [9]. In this circumstance, lactate essentially reverses the expected reduction of hif-1alpha and allows the cell to behave as if it is hypoxic while other roles of oxygen, collagen prolyl hydroxylation, for instance, remain uninfluenced and even enhanced.

Additionally, The ADPRibose system, diminished by rising lactate, activates angiogenesis and collagen deposition. Collagen deposition supports "maturation" of vessels in the sense that endothelial cells, anchored by integrins to collagen, can enlarge and withstand the pressure of circulating blood. Without collagen, new vessels weaken and rupture as seen in scurvy (ascorbic acid depletion).

Thus, instead of lactate being merely the end of the line for anaerobic metabolism, it becomes, by its many aerobic sources, a stimulant to increase anaerobic metabolism as well as to activate (some) "hypoxia response genes" and their down stream effects such as activation of angiogenesis, collagen synthesis and deposition, and matrix endothelial migration. Our composite view of the overall mechanism is as follows.

Although all the steps in Fig. 9.1 are supported by one or another investigator, questions remain. Among them: Does the oxidant flux stimulate VEGF and hif at one or separate sites? Does lactate inhibit hif-1alpha decay? Does lactate enhance the transcription of hif? Could the elevated tissue lactate found in hyperglycemia be prominent in the causation of vascular disease?

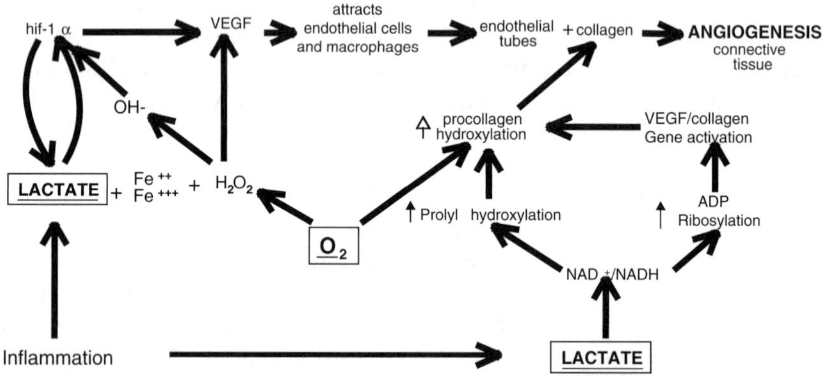

Fig. 9.1 A provisional schema for the composite actions of lactate in the deposition of angiogenesis and connective tissue has at least three limbs. The ADPRibosylation pathway is on the right. The lactate diversion occurs at the level of the Fe:lactate chelate with the diversion of hydroxyl radical toward the hif mechanism thus activating it [9]. Alternatively, more H_2O_2 produces these effect by activating the VEGF promoter [14]. Note also that addition of VEGF to Matrigel leads to vascularization and collagen formation *only* if oxygen is present [10]. The pathway of hyalouranan is not known.

9.5 Conclusions

The unique feature of lactate is that it has both aerobic and anaerobic sources; and, therefore, widens the scope of possibilities by which angiogenesis is stimulated in aerobic conditions. Its ability to increase collagen deposition (in the presence of oxygen) also widens the scope of vessel maturation. Lactate, occasioned by mechanical or chemical injury alone, may be the initial source of angiogenic stimuli in wounds and may be a part of tumor angiogenesis in which lactate is constitutively present.

Acknowledgment Supported by NIH NIGMS GM27345 and NIH NRSA GM08258.

References

1. Q. P. Ghani, S. Wagner, H. D. Becker, T. K. Hunt, M. Z. Hussain. Regulatory role of lactate in wound repair. Methods Enzymol. 2004; 381:565–75.
2. J. S. Constant, J. J. Feng, D. D. Zabel, H. Yuan, D. Y. Suh, H. Scheuenstuhl, T. K. Hunt, M. Z. Hussain. Lactate elicits vascular endothelial growth factor from macrophages: a possible alternative to hypoxia. Wound Repair Regen. 2000 Sep–Oct; 8(5):353–60.
3. S. Beckert, F. Farrahi, R. S. Aslam, H. Scheuenstuhl, A. Konigsrainer, M. Z. Hussain, T. K. Hunt. Lactate stimulates endothelial cell migration. Wound Repair Regen. 2006 May–Jun; 14(3):321–4.
4. B. Formby, R. Stern. Lactate-sensitive response elements in genes involved in hyaluronan catabolism. Biochem Biophys Res Commun. 2003 May 23; 305(1):203–8.
5. L. B. Gladden. Lactate metabolism: a new paradigm for the third millennium. J Physiol. 2004 Jul 1; 558(Pt 1):5–30. Review.
6. J. E. Albina, B. Mastrofrancesco, J. A. Vessella, C. A. Louis, W. L. Henry, Jr., J. S. Reichner. HIF-1 expression in healing wounds: HIF-1alpha induction in primary inflammatory cells by TNF-alpha. Am J Cell Physiology. 2001 Dec; 281(6):C1971–7.
7. S. Biswas, M. Ray, S. Misra, D. P. Dutta, S. Ray. Is absence of pyruvate dehydrogenase complex in mitochondria a possible explanation of significant aerobic glycolysis by normal human leukocytes? FEBS Lett. 1998 Apr 3; 425(3):411–4.
8. O. Trabold, W. Wagner, C. Wicke, H. Scheuenstuhl, M. Z. Hussain, N. Rosen, A. Seremetiev, H. D. Becker, T. K. Hunt. Lactate and oxygen constitute a fundamental regulatory mechanism in wound healing. Wound Repair Regen. 2003 Nov–Dec;11(6):504–9.
9. Q. Liu, U. Berchner-Pfannschmidt, U. Moller, M. Brecht, C. Wotzlaw, H. Acker, K. Jungermann, T. A. Kietzmann. Fenton reaction at the endoplasmic reticulum is involved in the redox control of hypoxia-inducible gene expression. Proc Natl Acad Sci U S A. 2004 Mar 23;101(12):4302–7.
10. H. W. Hopf, J. J. Gibson, A. P. Angeles, J. S. Constant, J.J. Feng, M. D. Rollins, M. Z. Hussain, T. K. Hunt. Hyperoxia and angiogenesis. Wound Repair Regen. 2005 Nov–Dec;13(6):558–64.
11. A. Y. Sheikh, J. J. Gibson, M. D. Rollins, H. W. Hopf, Z. Hussain, T. K. Hunt. Effect of hyperoxia on vascular endothelial growth factor levels in a wound model. Arch Surg. 2000 Nov;135 (11):1293–7
12. M.A. Ali, F. Yasui, S. Matsugo, T. Konishi. The lactate-dependent enhancement of hydroxyl radical generation by the Fenton reaction. Free Radic Res. 2000 May;32(5):429–38.

13. H. Lu, C. L. Dalgard, A. Mohyeldin, T. McFate, A. S. Tait, A. Verma. Reversible inactivation of HIF-1 prolyl hydroxylases allows cell metabolism to control basal HIF-1. J Biol Chem. 2005 Dec 23;280(51):41928–39.
14. C. K. Sen, S. Khanna, B. M. Babior, T. K. Hunt, E. C. Ellison, S. Roy. Oxidant-induced vascular endothelial growth factor expression in human keratinocytes and cutaneous wound healing. J Biol Chem. 2002 Sep 6;277(36):33284–90.
15. D. D. Zabel, J. J. Feng, H. Scheuenstuhl, T. K. Hunt, M. Z. Hussain. Lactate stimulation of macrophage-derived angiogenic activity is associated with inhibition of Poly (ADP-ribose) synthesis. Lab Invest. 1996; 74:644–9.

Part III
Blood, Hemostasis and Hemodynamics

Chapter 10
Activated Protein C Modulates Chemokine Response and Tissue Injury in Experimental Sepsis

Ganesh R. Sharma[1], Bruce Gerlitz[1], David T. Berg[1], Martin S. Cramer[1], Joseph A. Jakubowski[1], Elizabeth J. Galbreath[2], Josef G. Heuer[1], and Brian W. Grinnell[1]

Abstract The protein C (PC) pathway plays an important role in vascular function, and acquired deficiency during sepsis is associated with increased mortality. We have explored the role of PC suppression in modulating early inflammatory events in a model of polymicrobial sepsis. We show that increased levels of organ damage and dysfunction are associated with decreased levels of endogenous PC. Notably, animals with low PC had correspondingly high levels of pulmonary iNOS expression, which correlated with chemokines KC/Gro and MIP2, previously shown to predict outcome in this model. Treatment with activated protein C (aPC) not only reduced the pathology score, leukocyte infiltration and markers of organ dysfunction, but also suppressed the induction of iNOS, and the chemokine response (including KC/Gro, MIP2, IP-10, RANTES, GCP-2 and lymphotactin), and increased apoA1. aPC treatment also suppressed the induction of VEGF, a marker recently suggested to play a pathophysiological role in sepsis. These data demonstrate a clear link between low protein C and degree of organ damage and dysfunction in sepsis, as well as the early reversal with aPC treatment. Moreover, our data show a direct role of aPC in broadly modulating monocyte and T-cell chemokines following systemic inflammatory response.

10.1 Introduction

Sepsis syndrome occurs from a complex host response to insult following infection, and carries a high mortality rate [1]. In severe sepsis, the innate immune response becomes dysregulated, resulting in a cascade of inflammatory activation, microvascular coagulation, endothelial cell dysfunction and ultimately organ failure and death (reviewed in [2]). While many soluble factors

[1]Biotechnology Discovery Research, Lilly Research Laboratories, Lilly Corporate Center, Indianapolis, IN 46285.
[2]Pathology Lilly Research Laboratories, Lilly Corporate Center, Indianapolis, IN 46285.

change during infection and sepsis, the suppression of endogenous protein C (PC) has been associated with increased mortality and is prognostic for sepsis and sepsis severity [3, 4]. Moreover, low protein C levels are predictive of early death in a rat model of polymicrobial sepsis [5, 6], and clinically have been associated with early death resulting from refractory shock and multiple organ failure in severe sepsis [4].

In settings of thrombotic stress, the zymogen Protein C is converted to its active form by thrombin in complex with endothelial surface thrombomodulin (TM). Activated protein C (aPC) functions as a feedback inhibitor of thrombin generation [7] and has receptor-mediated anti-inflammatory and cytoprotective effects; [8–10] recent studies have also shown that aPC can inhibit leukocyte rolling and adhesion [11, 12]. In this study we have examined early markers of inflammation and organ function following induction of polymicrobial sepsis, both as a function of endogenous PC suppression and treatment with aPC. Our studies show that acquired PC deficiency in polymicrobial sepsis is highly correlated with early activation of chemokine response, tissue infiltration and organ damage. Moreover, treatment with aPC results in suppression of the chemokine response and improved organ function.

10.2 Experimental Procedures

The protocol for the rat cecal ligation and puncture (CLP) model of sepsis has been previously described in detail [5]. Sprague Dawley rats were purchased from Harlan (Indianapolis, IN). Sham rats received identical surgery (except for CLP) and post-operative management. All experimental methods were approved by the Institutional Animal Care and Use Committee and were in accordance with the institutional guidelines for the care and use of laboratory animals. Mortality studies were as previously described [5]. Recombinant rat PC was produced in AV12-664 cells, then activated with recombinant rat thrombomodulin/bovine thrombin complex essentially as described previously for human aPC [13]. Infusion syringes were pre-coated overnight with sterile 7.5% BSA, then continuous infusion of either vehicle (5% dextrose, 0.9% saline) or recombinant rat aPC (200 ug/kg/hr in 5% dextrose, 0.9% saline) was begun at 10 hrs post-CLP and continued until sacrifice of animals at 22 hrs post-CLP for collection of tissue samples. This resulted in a blood level of $\sim$100 ng/ml aPC. Blood sampling of animals pre-CLP and 10 hrs post-CLP was via retro-orbital bleed; cardiac puncture was used for the blood draw at 22 hrs post-CLP (time of sacrifice). Clinical chemistry (alanine aminotranferease [ALT], aspartate aminotransferase [AST], blood urea nitrogen [BUN] and creatinine [CRE]) was performed on serum, while cytokine/chemokine analysis was performed on EDTA-plasma using Rodent Multi-Analyte Profile (Rules Based Medicine; Austin, TX), Rat ELISA kits (R&D Systems, Minneapolis, MN) or the Rat Cytokine/chemokine Immunoassay Panel (Linco, St. Charles, MO).

Endogenous PC levels were determined using a rat-PC ELISA as described previously [5]. Total RNA was purified from rat tissue using RNeasy (Qiagen); quantitative real-time PCR was performed with an ABI Prism 7900HT Sequence Detection System.

Tissue was also collected for histopathology as previously described [14]. A composite pathology score was derived from scoring of lung, liver and kidney as follow. Lung pathology was scored by the number of myeloperoxidase (MPO) labeled cells with pavementing: Grade 1 = Mild intravascular margination (IM) of WBC; Grade 2 = Moderate IM of WBC; Grade 3 = Marked IM of WBC; Grade 4 = Marker IM with pavementing of WBC; Grade 5 = Marked IM with pavementing & extravascular WBC. Liver was scored by increasing numbers of MPO labeled cells: Grade 1 = <10 / high powered field (hpf); Grade 2 = 10–20/hpf; Grade 3 = 20–30/hpf; Grade 4 = 30–40/hpf; Grade 5 = >40/hpf,. Kidney was scored by increasing corticomedullary acute tubular necrosis (ATN): Grade 1 = proteinuria, rouleaux, individual tubular cell apoptosis /necrosis; Grade 2 = proteinuria, rouleaux, mild focally extensive (<10%) ATN; Grade 3 = proteinuria, rouleaux, mild multifocal (10–20%) ATN; Grade 4 = proteinuria, rouleaux, moderate multifocal (>20%) ATN; Grade 5 = proteinuria, rouleaux, marked multifocal to regional ATN, casts.

ANOVA was used to determine statistical significance, and multivariate regression analysis by pairwise comparison was determined using JMP5.1 software (SAS Institute). A p value of < 0.05 was considered significant. mRNA changes elicited by CLP and analyzed by qPCR, were expressed in terms of percent change relative to the control group (defined as 100%). The SEM for ratios was derived using the delta method [15].

10.3 Results

We examined early changes in inflammatory markers and tissue pathology following induction of systemic inflammatory response in the CLP model. Following CLP animals were observed for 96 hrs, with early death defined as being before 30 hrs and late death thereafter. Shown in Fig. 10.1 is a clustering of plasma markers that differentiated outcome, from a panel of 46 cytokines and chemokines measured 22 hrs post-CLP. Notably, low levels of endogenous PC and APO-A1, as well as elevated levels of proinflammatory chemokines predicted early death relative to sham animals.

We further analyzed this suppression of PC relative to known markers of tissue injury and observed significant negative correlations between PC and the following markers: AST $r = -0.81$, $p < 0.0001$; ALT $r = -0.83$, $p < 0.0001$; creatinine $r = -0.76$, $p < 0.002$. We next examined the tissue pathology at 22 hrs post CLP and observed a graded response using scoring that differentiated lung infiltration and margination of leukocytes (Fig. 10.2A), tubular damage in the kidney and degree of myeloperoxidase-positive cells in the liver. We observed that the animals having low PC at 22 hrs had a higher mean pathology score

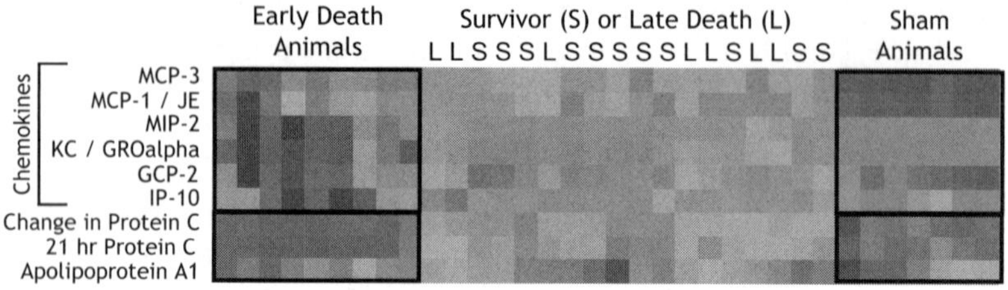

Fig. 10.1 Clustering of plasma markers as a function of outcome in the CLP model. The levels of various plasma markers from the Rodent Multi-Analyte Profile were clustered in JMP 5.1 using the Ward method. Dark blue to dark red change indicates low to high relative levels, respectively, for each marker. (See also color insert.)

than animals whose PC did not decrease below normal (defined as < 60% of baseline [6]) (Low PC = 4.5 +/–0.7 vs. Normal PC = 2.6 +/–0.3, p< 0.05). We observed significant correlations with degree of tissue pathology and the neutrophil chemotactic CXC chemokines MIP2 and KC/Gro (Fig. 10.2B). Moreover, the pathology score was significantly correlated with tissue injury markers (plasma AST p < 0.008 ; ALT p < 0.01). A key factor shown to correlate with inflammatory response, especially in acute lung injury, is inducible nitric oxide

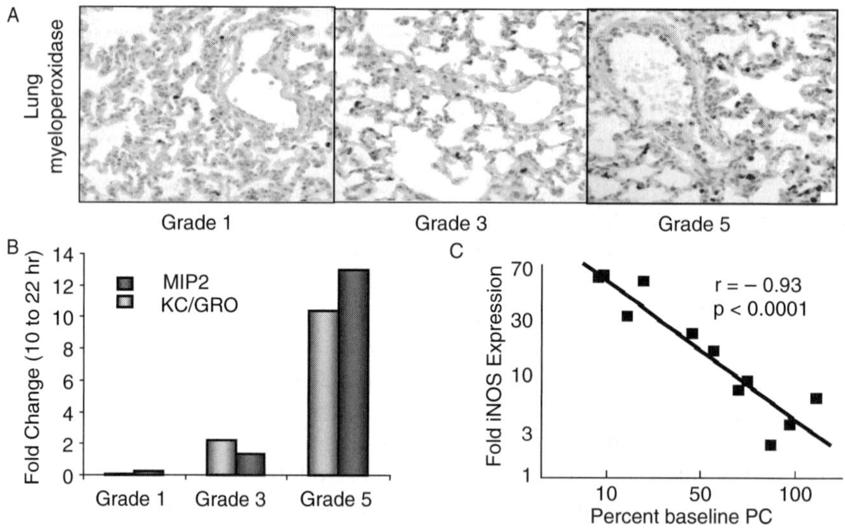

Fig. 10.2 Tissue injury, inflammatory markers and protein C suppression in the rat CLP. (A) Example of pathology in the lung by severity grade. (B) Analysis of change in protein C plasma level as a function of MIP2 and KC/Gro. Arteriole blood draws were performed at approximately 10 and 22h and plasma was analyzed for various analytes as described previously [5]. (C) Relationship of the change in PC as a function of lung iNOS expression, analyzed by quantitative real-time PCR (TaqMan®) n = 11 animals. (See also color insert.)

synthase (iNOS) expression [16–18]. As shown in Fig. 10.2C, increased lung iNOS expression was highly correlated with reduction in baseline PC level. Moreover, the increase in iNOS was highly correlated with the increase in both MIP2 ($r = 0.84$, $p < 0.001$) and KC/Gro ($r = 0.80$, $p < 0.005$).

The above data suggested a strong relationship between the level of endogenous PC and markers of tissue injury. To provide evidence of a causal relationship, we administered activated PC during disease progression, to determine any effect on tissue pathology and inflammatory mediators. As shown in Fig. 10.3A, infusion of aPC for 12 hrs, starting 10 hrs after CLP, significantly reduced the mean pathology score. Moreover, the mean increase in the level of ALT (Fig. 10.3B) and AST (not shown) following CLP was significantly reduced by aPC treatment. Treatment with aPC significantly restored plasma apoA1 and fibrinogen levels, both of which are markers of liver synthetic capacity (Fig. 10.3C). In view of apoA1's anti-inflammatory activity [19], blocking its suppression may play a role in aPC's anti-inflammatory activity, possibly complementing the suppression of the chemokine response.

An examination of changes in chemokine levels showed a significant suppression of neutrophil chemotactic CXC chemokines (Fig. 10.4A) as well as the T-cell/NK cell chemokines IP-10, RANTES and lymphotactin. We also observed significant inhibition of OSM, $MIP1_{\alpha}$, and $MIP1_{\beta}$ (data not shown). Of interest, aPC treatment had no significant effect on TNF_{α}, IL-1 or IL-6 in

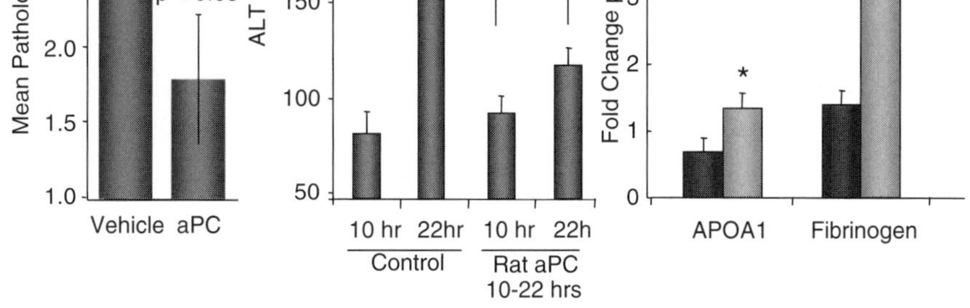

Fig. 10.3 Effect of aPC infusion on organ pathology and functional markers. Recombinant rat aPC or vehicle control (in 5% dextrose, 0.9% saline) was administered by continuous infusion begun at 10 hrs post-CLP and continued until sacrifice of animal at 22 hrs post-CLP for collection of tissue and plasma samples. (A) Effect of aPC on a composite mean pathology score ($n = 20$) (B) Effect of aPC treatment on ALT levels and (C) on liver markers apoA1 and fibrinogen. Data are mean +/– SE, n= 22 vehicle, 19 aPC treated.

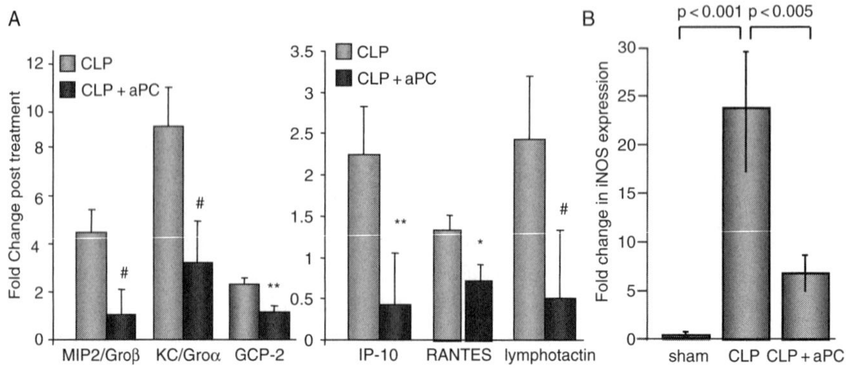

Fig. 10.4 Effect of aPC administration (described in legend to Fig. 10.3) on (A) chemokine response and (B) iNOS. Data are mean +/- SE from n = 22 vehicle, 19 aPC treated. * $p < 0.05$, ** $p < 0.02$, # $p < 0.01$.

this model (data not shown [5]). The up-regulation of iNOS, which correlated highly with PC suppression and chemokine activation, was also dramatically reduced by the aPC treatment.

Recent studies have suggested that increases in circulating levels of vascular endothelial growth factor (VEGF) may play a pathophysiologic role in mediating the sepsis phenotype [20]. In our rat model, we observed an increase in VEGF from 215 +/–7 pg/ml to 323 +/–35 pg/ml by 22 hrs post-CLP. We examined the

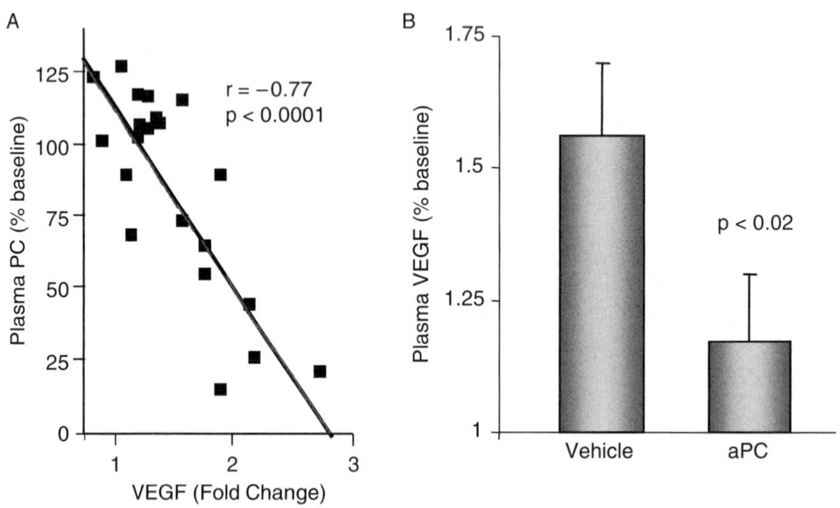

Fig. 10.5 (A) Relationship between endogenous PC and circulating plasma VEGF. (B) Effect of aPC administration (legend to Fig. 10.3) on VEGF levels. Data are mean +/– SE, n = 22 vehicle, 19 aPC treated.

relationship between endogenous PC and the increase in VEGF, and observed a significant negative correlation ($p < 0.0001$) (Fig. 10.5A). Moreover, we observed a significant reduction in the level of circulating VEGF with aPC treatment (Fig. 10.5B).

10.4 Discussion

Recent studies have begun to elucidate the role of the protein C pathway in controlling normal physiology of the vasculature and of the innate immune system. aPC has been shown to modulate endothelial function by inhibiting cell adhesion and apoptosis, and by promoting cell survival/angiogenesis [8, 9, 21–24]. These effects are likely due a combination of the ability of aPC to inhibit thrombin generation, and through specific receptor-mediated signaling at the PAR-1 receptor in conjunction with the endothelial protein C receptor. This receptor complex is present not only on the endothelium, but also appears to be present on monocytes, natural killer cells [11], neutrophils [25], and eosinophils [26]. Thus, the emerging data suggest that aPC plays a key role in modulating the endothelial/leukocyte interface in response to stress.

Our results demonstrate a clear role of the PC pathway in the tissue pathology and early chemokine response in the CLP model. The effect of aPC was predominately on suppression of chemokines critical for amplification of the response to infection, and on chemokines critical for both neutrophil and T-cell modulatory factors (reviewed in [27]). Of interest, previous studies demonstrated that KC/Gro and MIP2 were good predictors of early death in CLP, and we have now shown that these correlate highly with pathology in this model, and were significantly suppressed by aPC treatment. In addition, the high correlation of these chemokines with iNOS expression is of interest as Kim et al. [28] have suggested that iNOS may regulate certain chemokines, including RANTES and MIP1α. Possibly, suppression of iNOS by aPC may be driving the observed reduction of the chemokine response seen with treatment. Our results also suggest that aPC suppresses VEGF in the setting of sepsis. VEGF is known to sensitize the vasculature to the effect of cytokines, and thus likely plays a role in enhancing the activation of the endothelium during sepsis [20]. In light of the effect of aPC on modulating both VEGF and the chemokine response, it is notable that chemokines can regulate vascular/angiogenic function by modulating VEGF function (reviewed in [29]).

While the cause and effect relationship of low PC and clinical outcome has not been proven, the data presented here suggest that low endogenous PC levels during systemic inflammatory response may be pathophysiologically related to poor outcome. The low level of PC likely compromises the ability to naturally generate aPC, which results in a reduction in the natural protective mechanism of the vasculature to limit inflammatory and ischemic injury. Our data further suggest that aPC's ability to modulate not only coagulopathic dysfunction [4], but also the cascading inflammatory responses following infection, plays a key role in reversing vascular injury, poor tissue perfusion and resulting organ dysfunction.

Acknowledgment We gratefully acknowledge Eddie J. Stephens, Renee L. Grubbs, Kimberly C. Holmes, Kelly Fynboe, and Dominick Montani for assistance with animal care and CLP studies. We thank Joe Brunson, Sherri L. Hilligoss, and Don B. McClure for assistance with cell culture in producing rat aPC.

The authors disclose that they are employees of Eli Lilly and Co, who produce recombinant human protein C (drotrecogin alfa [activated]).

References

1. D. Angus, W.T. Linde-Zwirble, J. Lidicker, G. Clermont, J. Carcillo, M.R. Pinsky, Epidemiology of severe sepsis in the United States: analysis of incidence, outcome, and associated costs of care, Crit Care Med **291**;303–1310 (2001).
2. B. Grinnell and D.E. Joyce, Recombinant human activated protein C: A system modulator of vascular function for treatment of severe sepsis, Crit Care Med **29;**S53–S61 (2001).
3. C.J. Fisher and S.B. Yan, Protein C levels as a prognostic indicator of outcome in sepsis and related diseases, Crit Care Med **28**(9);S49–S56 (2000).
4. W.L. Macias and D.R. Nelson, Severes protein C deficiency predicts early death in severe sepsis, Crit Care Med **32**;S223–S228 (2004).
5. J.G. Heuer, G.R. Sharma, B. Gerlitz, T. Zhang, D.L. Bailey, C. Ding, D.T. Berg, D. Perkins, E.J. Stephens, K.C. Holmes, R.L. Grubbs, K.A. Fynboe, Y.F. Chen, B. Grinnell, and J.A. Jakubowski, Evaluation of protein C and other biomarkers as predictors of mortality in a rat cecal ligation and puncture model of sepsis. [see comment], Crit Care Med **32**(7);1570–1578 (2004).
6. D. Berg, B. Gerlitz, G. Sharma, M. Richardson, E. Stephens, R. Grubbs, K. Holmes, D. Montani, T. Zhang, M. Cramer, S. Engle, J. Jakubowski, H. JG, and B. Grinnell, FoxA2 Involvement in Suppression of Protein C, an Outcome Predictor in Experimental Sepsis, Clinical Vaccine Immunol. **13**;426–432 (2006).
7. C.T. Esmon, J.M. Gu, J. Xu, D. Qu, D.J. Stearns-Kurosawa, and S. Kurosawa, Regulation and functions of the protein C anticoagulant pathway, Haematologica **84**(4);363–8 (1999).
8. D.E. Joyce, L. Gelbert, A. Ciaccia, B. Dehoff, and B.W. Grinnell, Gene Expression Profile of Antithrombotic Protein C Defines New Mechanisms Modulating Inflammation and Apoptosis, J Biol Chem **276;**11199–11203 (2001).
9. L.O. Mosnier and J.H. Griffin, Inhibition of staurosporine-induced apoptosis of endothelial cells by activated protein C requires protease activated receptor-1 and endothelial cell protein C receptor, Biochem J 8(2003).
10. M. Riewald, R. Petrovan, A. Donner, B. Mueller, and W. Ruf, Activation of endothelial cell protease activated receptor 1 by the protein C pathway, Science **296**;1880–1882 (2002).
11. D.E. Joyce, D.R. Nelson, and B.W. Grinnell, Leukocyte and endothelial cell interactions in sepsis: relevance of the protein C pathway, Crit Care Med **32**(5 Suppl),(2004).
12. J.N. Hoffmann, B. Vollmar, M.W. Laschke, D. Inthorn, J. Fertmann, F.W. Schildberg, and M.D. Menger, Microhemodynamic and cellular mechanisms of activated protein C action during endotoxemia.[see comment], Crit Care Med **32**(4);1011–1017 (2004).
13. B. Gerlitz and B.W. Grinnell, Mutation of protease domain residues Lys^{37-39} in human Protein C inhibits activation by the thrombomodulin-thrombin complex without affecting activation by free thrombin, J Biol Chem **271**(37);22285–22288 (1996).

14. D.T. Berg, L.J. Myers, M.A. Richardson, G. Sandusky, and B.W. Grinnell, Smad6s regulates plasminogen activator inhibitor-1 through a protein kinase C-beta-dependent up-regulation of transforming growth factor-beta, J Biol Chem **280**(15);14943–7 (2005).
15. Y. Bishop, S. Feinberg, and H. PW, Discrete multivariate analysis: Theory and practice, Cambridge: The MIT Press (1975).
16. L. Dugo, S. Marzocco, E. Mazzon, R. Di Paola, T. Genovese, A.P. Caputi, and S. Cuzzocrea, Effects of GW274150, a novel and selective inhibitor of iNOS activity, in acute lung inflammation, Br J Pharmacol **141**(6);979–987 (2004).
17. H. Toga, T. Tobe, Y. Ueda, G.H. Yang, K. Osanai, M. Ishigaki, H. Okazaki, S. Katsuda, K. Takahashi, and N. Ohya, Inducible nitric oxide synthase expression and nuclear factor-kappaB activation in alveolar type II cells in lung injury, Experimental Lung Research **27**(6);485–504 (2001).
18. L.W. Chen, B. Hwang, W.J. Chang, J.S. Wang, J.S. Chen, and C.M. Hsu, Inducible nitric oxide synthase inhibitor reverses exacerbating effects of hypertonic saline on lung injury in burn, Shock **22**(5);472–477 (2004).
19. M. Navab, G. Anantharamaiah, and A. Fogelman, The role of high-density lipoprotein in inflammation, Trends Cardiovasc Med **15**;158–161 (2005).
20. K. Yano, P. Liaw, J. Mullington, S. Shih, H. Okada, N. Bodyak, P. Kang, L. Tolt, B. Belikoff, J. Buras, B. Simms, J. Mizgerd, P. Carmeliet, S. Karumanchi, and W. Aird, Vascular endothelial growth factor is an important determinant of sepsis morbidity and mortality, J Exp Med **203**(6);1447–1458 (2006).
21. D.E. Joyce and B.W. Grinnell, Recombinant human activated protein C attenuates the inflammatory response in endothelium and monocytes by modulating nuclear factor-kappaB, Crit Care Med **30**;S288–293 (2002).
22. T. Cheng, D. Liu, J. Griffin, J. Fernandez, F. Castellino, E. Rosen, K. Fukudome, and B. Zlokovic, Activated protein C blocks p53-mediated apoptosis in ischemic human brain endothelium and is neuroprotective, Nat Med **9;**338–342 (2003).
23. M. Uchiba, K. Okajima, Y. Oike, Y. Ito, K. Fukudome, H. Isobe, and T. Suda, Activated protein C induces endothelial cell proliferation by mitogen-activated protein kinase activation in vitro and angiogenesis in vivo, Circulation Research **95**(1);34–41 (2004).
24. K. Okajima, Prevention of endothelial cell injury by activated protein C: the molecular mechanism(s) and therapeutic implications, Current Vascular Pharmacology **2**(2);125–133 (2004).
25. D.H. Sturn, N.C. Kaneider, C. Feistritzer, A. Djanani, K. Fukudome, and C.J. Wiedermann, Expression and function of the endothelial protein C receptor in human neutrophils, Blood **102**(4);1499–1505 (2003).
26. C. Feistritzer, D.H. Sturn, N.C. Kaneider, A. Djanani, and C.J. Wiedermann, Endothelial protein C receptor-dependent inhibition of human eosinophil chemotaxis by protein C, J Allergy Clin Immunol **112**(2);375–381 (2003).
27. A. Mantovani, A. Sica, S. Sozzani, P. Allavena, A. Vecchi, and M. Locati, The chemokine system in diverse forms of macrophage activation and polarization, Trends in Immunology **25**(12);677–686 (2004).
28. J.Y. Kim, D. Kim, E.M. Lee, I. Choi, C.G. Park, K.S. Kim, J. Ha, S.J. Kim, J. Yang, Y.S. Kim, J.S. Han, S. Kim, J.S. Lee, and C. Ahn, Inducible nitric oxide synthase inhibitors prolonged the survival of skin xenografts through selective down-regulation of pro-inflammatory cytokine and CC-chemokine expressions, Transplant Immunology **12**(1);63–72 (2003).
29. M. Rosenkilde and T. Schwartz, The chemokine system – a major regulator of angiogenesis in health and disease, APMIS **112**(7–8);481–495 (2004).

Chapter 11
Manipulation of the Affinity Between Protein and Metal Ions by Imidazole and PH for Metal Affinity Purification of Protein c from Cohn Fraction IV-1

James J. Lee[1], Duane F. Bruley[2], and Kyung A. Kang[1]

Abstract Protein C (PC) is an important anticoagulant in blood plasma. Cohn Fraction IV-1 (CFIV-1) is an inexpensive PC source but contains a large amount of factor II (FII). Immobilized metal affinity chromatography (IMAC) utilizes metal ions to adsorb proteins primarily *via* their surface histidine. Two major operation parameters for IMAC are imidazole concentration and pH: imidazole is a histidine analog and pH controls the protein surface protonation level. The effects of these two parameters on the adsorption and elution of PC and FII were studied for each protein individually and also together as a mixture. For the individual proteins, low FII (16%) and high PC (98%) adsorption were achieved at 8 mM imidazole, pH 8.0. At 11 mM imidazole, 92% of the adsorbed FII was eluted, with only a 3% PC loss. At 40 mM, 97% of the adsorbed PC was recovered. For the protein mixture, very similar adsorption and elution results were obtained, but slightly greater PC loss (16%) during elution at 11 mM imidazole. This result shows that there is a high potential for the PC purification from CFIV-1 by appropriately adjusting the imidazole concentration and pH in the IMAC process.

11.1 Introduction

Protein C (PC) is an anticoagulant, antithrombotic, anti-inflammatory [1,2], and anti-apoptotic [3,4]. Therefore, PC can be a valuable therapeutic for patients with PC deficiency, various thrombo-embolisms, advanced sepsis, and stroke [5]. Currently known PC sources are transgenic animal milk, recombinant mammalian cells, and blood plasma, all of which are very expensive and not always available. Cohn Fraction IV-1 (CFIV-1) is a by-product of the plasma fractionation process, and retains approximately 90% of the PC in plasma. CFIV-1 used to be discarded and, therefore, is a very inexpensive PC source. Purification of

[1]Department of Chemical Engineering, University of Louisville, Louisville, KY 40292.
[2]Department of Chemical and Biochemical Engineering, University of Maryland Baltimore County, Baltimore, MD 21250 and Synthesizer Inc., Ellicott City, MD 21043.

Table 11.1 Half-life and amount of the homologous proteins found in CFIV-1*

	PC	FII	FVII	FIX	FX
Half-life (hrs) [9,10]	8	85	5.5	22	24
Amount (μg/g CFIV-1)	100	1200	30	14	270

* Modified from Rezania, et al [8].

PC from CFIV-1 using usual bio-purification methods, such as ion-exchange chromatography, is difficult, if not impossible, because CFIV-1 contains various coagulants that are structurally homologous to PC (factors II, VII, IX, and X [6,7]; Table 11.1). Among these coagulants, factor II (FII) is the most problematic one because of its long half-life and abundance in the source [8]. Thus, our study focus has been the separation of FII from PC.

Histidine is a strong electron-donor and has a high affinity to metal ions (electron acceptor) [11]. The separation of FII from PC by immobilized metal affinity chromatography (IMAC) with Cu^{2+} and iminodiacetic acid (IDA) chelator has been studied by our group [12,13]. In IMAC operation, the imidazole concentration and pH are the two major parameters used for adjusting the affinity between the metal ions and the proteins [14]. Our previous study results showed that, when no imidazole was added in the buffer, the amounts of adsorbed PC and FII to the IMAC matrix were similar for pHs 6.0, 7.0, and 8.0. The elution study results showed that approximately 20% of the adsorbed FII was eluted at 15 mM imidazole, pH 6.0, without eluting PC. Here, the effect of both the pH and imidazole concentration on the adsorption and elution of PC and FII were characterized.

11.2 Instruments, Materials, and Methods

11.2.1 IMAC Matrix Preparation

Chelating Sepharose Fast Flow™ matrix with iminodiacetic acid (IDA) (Amersham Biosciences; Piscataway, NJ) was reacted with cupric sulfate pentahydrate (Sigma-Aldrich; St. Louis, MO) to immobilize Cu^{2+}, following the manufacturer's instruction. One mL of the settled Cu^{2+} immobilized matrix was measured in a graduated cylinder and then transferred into a 15 mL centrifuge tube (Fisher Scientific; Chicago, IL). Then the matrix was centrifuged (Marathon 3200R; Fisher Scientific) at 4000 rpm for one minute and the supernatant was discarded. Next, the Cu^{2+} immobilized matrix was equilibrated with 10 mL of the equilibration buffer (20 mM sodium phosphate buffer and 0.5 M NaCl) at a predetermined pH, centrifuged, and the supernatant was discarded. The pH equilibrated matrix was pre-equilibrated with 1 mL of the equilibrium buffer at the predetermined pH and imidazole concentration for ten minutes. The equilibrated Cu^{2+}-IDA-imidazole matrix was then divided into aliquots of 100 μL in 1.5 mL micro-centrifuge tubes (Fisher Scientific).

11.2.2 Adsorption and Elution of PC and/or FII

For the adsorption study, 100 μL of the equilibrated Cu^{2+}-IDA-imidazole matrix at predetermined conditions were reacted either with 20 μg of PC (Innovative Research; Southfield, MI), 20 μg of FII (Innovative Research), or the mixture of 20 μg-PC and 20 μg of FII in 200 μl of an equilibration buffer condition to be tested, at room temperature (22–24 °C). The vial was gently vortexed for 2 seconds and then was placed on a nutating mixer (Clay Adams® Nutator; Becton-Dickinson; Franklin Lakes, NJ) for ten minutes, and was then centrifuged at 4000 rpm for one minute. The protein in the supernatant after the adsorption reaction was quantified by ELISA. Following the protein adsorption, the matrix was washed with 400 μL of the equilibration buffer at pH 6.0 for ten minutes and then centrifuged. The protein in the supernatant was quantified by ELISA.

For the elution study, the protein adsorbed matrix was reacted with 400 μL of the phosphate equilibrium buffer at a predetermined imidazole concentration and pH 6.0 for ten minutes. The matrix was centrifuged and the protein in the supernatant was quantified by ELISA. For some cases, the elution process was repeated more than once.

11.2.3 ELISA for PC and FII

ELISA of PC and FII were performed using the procedure described by Lee et al [14]. EIA/RIA 96-well flat-bottom plate (Corning, NY) was first coated with a rabbit anti-human-PC Gig (Sigma; St. Louis, MO) for PC quantification, or goat anti-human-FII IgG (Boomed; Foster City, CA) for FII. For PC, the plate was washed and then blocked with BSA for 90 minutes. For FII, the plate was washed and was incubated for 90 minutes without BSA during the blocking step. After blocking, the plate was washed; the samples were applied to the wells vertically. The samples were diluted across the plate horizontally by serial dilution in the dilution buffer, and then incubated for 90 minutes. Then the plate was washed and incubated for 90 minutes with a goat anti-human-PC IgG (American Diagnostica Inc.; Hauppauge, NY), or a mouse anti-human-FII IgG (Enzyme Research; Southbend, IN). Next, the plate was washed and incubated for 20 minutes with horseradish peroxidase-conjugated to a rabbit anti-goat IgG (Sigma), or horseradish peroxidase-conjugated to a goat anti-mouse IgG (Sigma). O-phenylenediamine dichloride (OPD; Sigma) was added for color development, and the optical density was measured at 450 nm using an ELISA plate reader (Bio-Rad; Hercules, CA).

11.3 Result and Discussion

11.3.1 Effect of Imidazole and pH on the Adsorption and Elution of PC or FII

Our previous study results showed that the pH change (6∼8) alone did not affect the adsorption of either FII or PC [14]. As a next step, the adsorption behaviors of these proteins were studied with the changes in both the imidazole concentration between 0 and 11 mM and the pH between 6 and 8. For each protein, 20 μg was reacted with 100 μl of Cu^{2+}-IDA IMAC matrix and the amount of the adsorbed protein was quantified (Fig. 11.1).

FII adsorption decreased as imidazole concentration increased, at all pHs [Fig. 11.1 (a)]. Also, with the presence of imidazole, as the pH increased, FII adsorption decreased significantly. At 8 mM imidazole and pH 8.0, only 2.9 μg (14%) of FII was adsorbed. For PC [Fig. 11.1 (b)], most of the reacted PC was adsorbed, with little difference for the pHs and imidazole concentrations tested. At 11 mM imidazole, the amount of adsorbed PC decreased only slightly as the pH increased. A minimal FII adsorption with good PC adsorption was shown at 8 mM imidazole and pH 8.0.

Our previous elution study result at pH 6.0 showed that, 4.1 μg (∼20%) of FII and less than 0.5 μg (3%) of PC were eluted at both 15 and 20 mM imidazole [14]. This time, the elution behavior of FII was studied at an imidazole concentration range between 7 and 15 mM. For PC, a concentration between 20 and 40 mM was studied because of its higher affinity to Cu^{2+} ions. As described above, after the adsorption process at 8 mM imidazole and pH 8.0, the IMAC matrix

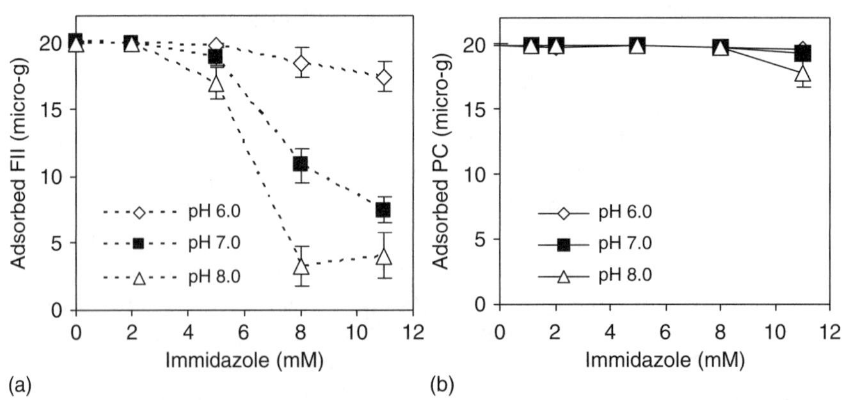

Fig. 11.1 Amounts of adsorbed (a) FII and (b) PC in the IMAC matrix with the changes in the imidazole concentration and the pH. 20 μg of FII or PC was reacted with 100 μl of Cu^{2+}-IDA matrix for 10 minutes.

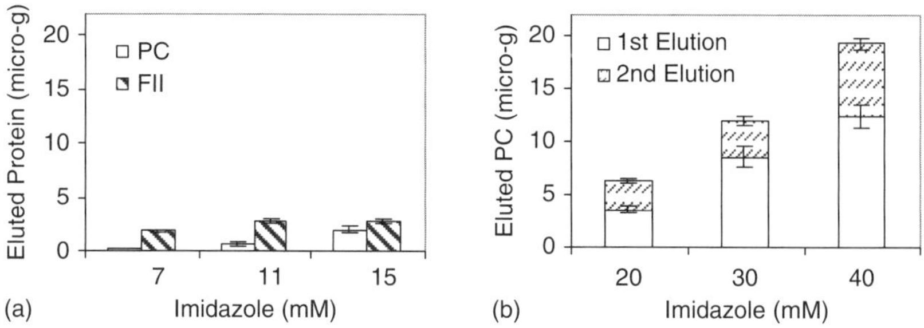

Fig. 11.2 Amounts of eluted (a) PC and FII at imidazole concentrations of 7, 11, and 15 mM and (b) PC at the imidazole concentrations of 20, 30, and 40 mM from the IMAC matrix with PC or FII. The matrix adsorbed with protein at 8 mM imidazole and pH 8.0 was reacted with 400 μl of the elution buffer for 10 minutes.

retained 2.9 μg of FII and 19.8 μg of PC. The amounts of PC and FII eluted by the buffer at various imidazole concentrations are shown in Fig. 11.2.

At 7 mM imidazole, 1.9 μg (66%) of the adsorbed FII was eluted with only 0.2 μg (1%) of PC [Fig. 11.2 (a)]. At 11 mM imidazole, the amount of eluted FII was slightly higher at 2.7 μg (93%), with only 0.7 μg (3%) of PC eluted. At 15 mM imidazole, a similar amount of FII was eluted but 2.0 μg (10%) of PC was also eluted. An imidazole concentration at 11 mM provided a maximal elution of FII with a minimal PC loss.

At 20 mM imidazole, 3.6 μg (18%) of PC was eluted during the 1st elution and 2.7 μg (14%) during the 2nd elution [Fig. 11.2 (b)], resulting in only 6.3 μg (32%) of the adsorbed PC (19.8 μg). At 30 mM imidazole, the total amount of PC eluted was 12.0 μg (60%). At 40 mM imidazole, the total PC recovered was 19.3 μg (97%).

11.3.2 Effect of Imidazole and pH on the Adsorption and Elution of the Mixture of PC and FII

With the information on the behavior of individual protein PC and FII in the IMAC process, a mixture of PC and FII at 1:1 ratio (20 μg of PC and 20 μg of FII) was used as the source material and the IMAC process was performed. Figure 11.3 shows the amounts of adsorbed PC and FII in the IMAC matrix at 8 mM imidazole, pH 8.0, when the protein mixture was reacted together with the matrix.

During the adsorption process, 5.3 μg (26%) of FII was adsorbed, slightly more than the amount adsorbed in the case with FII only (3.3 μg) [Fig. 11.3 (a)]. The amount of adsorbed PC [Fig. 11.3 (b)] was 19.3 μg (97%), similar to the amount for the adsorption with PC only (19.8 μg). The elution process was then performed using 11 mM and 40 mM imidazole (Fig. 11.4).

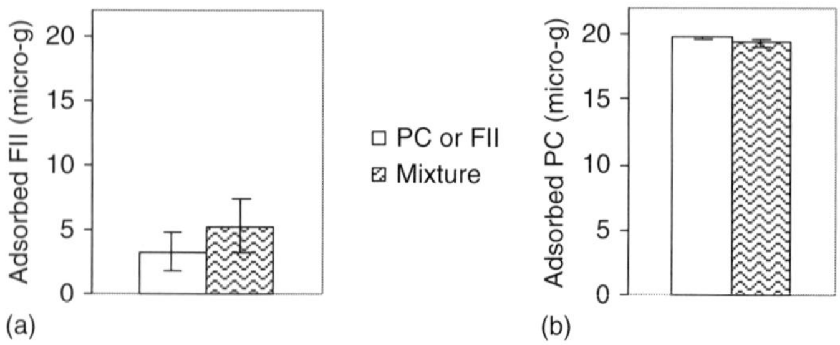

Fig. 11.3 Amounts of adsorbed (a) FII and (b) PC shown for each protein (20 μg PC or 20 μg FII) and for the mixture (20 μg PC with 20 μg FII). Proteins were reacted with 100 μl of Cu^{2+}-IDA matrix for 10 minutes at 8 mM imidazole and pH 8.0.

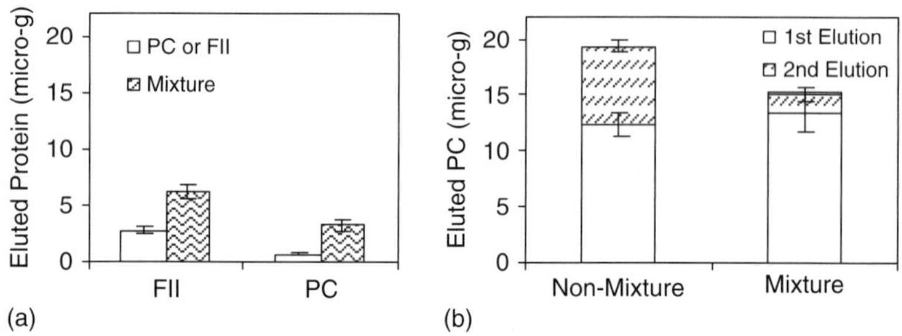

Fig. 11.4 Amounts of eluted (a) PC and FII at an imidazole concentration of 11 mM and (b) PC at an imidazole concentration of 40 mM.

At 11 mM imidazole [Fig. 11.4 (a)], nearly 100% of the adsorbed FII was eluted, but 3.2 μg (16%) of PC was also eluted, showing an increase from the case with only PC (0.6 μg). After this elution step, 16.1 μg of PC was expected to be in the matrix. The next two elutions at 40 mM imidazole removed 13.5 μg (84%) and 1.6 μg (10%) of PC for the 1st and 2nd elution steps, respectively, with the total of 15.1 μg (94%), slightly less than that for case with PC only (97%) [Fig. 11.4 (b)]. The adsorption at 8 mM imidazole and pH 8.0, followed by an elution at 11 mM imidazole for FII, and at 40 mM imidazole for PC elution has provided an effective separation of FII from a mixture with PC when the sample is a mixture of PC and FII at a ratio of 1:1.

11.4 Conclusions

The effect of imidazole concentration and pH on the adsorption and elution of PC and FII in IMAC operation was studied for an effective FII separation from PC. For the adsorption process, the imidazole concentration at 8 mM and pH

8.0 provided a minimal FII adsorption with a maximal PC adsorption, for both cases with FII or PC alone and with the PC/FII mixture. At a concentration of 40 mM imidazole, more than 94% of the adsorbed PC was eluted for both cases by eluting twice. In the IMAC process for PC purification from CFIV-1, the separation of FII from PC may be improved by selecting the suitable imidazole concentration and pH.

11.5 Future Study

For FII elution, the imidazole concentrations lower than 11 mM will be studied for less PC loss. A mixture of PC and FII at a ratio of 1:20, which is the ratio of PC and FII in the CFIV-1, will be studied for adsorption and elution of PC and FII.

Acknowledgment The authors thank the American Red Cross (Rockville, MD) for providing Cohn Fraction IV-1.

References

1. C. T. Esmon, The Anticoagulant and Anti-Inflammatory Roles of the Protein C Anticoagulant Pathway, *J. Autoimmun.* **15,** 113–116 (2000).
2. C. T. Esmon, Protein C anticoagulant pathway and its role in controlling microvascular thrombosis and inflammation, *Crit. Care Med.* **29**(7)**,** 48–51 (2001).
3. D. E. Joyce, L. Gelbert, A. Ciaccia, B. DeHoff and B. W. Grinnell, Gene expression profile of antithrombotic protein C defines new mechanisms modulating inflammation and apoptosis. *J. Biol. Chem.* **276,** 11199–11203 (2001).
4. T. Cheng, D. Liu, J. H. Griffin, J. A Fernandez, F. Castellino, E. D. Rosen, K. Fukudome and B. V. Zlokovic, Activated protein C blocks p53-mediated apoptosis in ischemic human brain endothelium and its neuroprotective, *Nat. Med.* **9**(3), 338–342 (2003).
5. D. Liu, T. Cheng, H. Guo, J. A. Fernandez, J. H. Griffin, X. Song and B. V. Zlokovic, Tissue plasminogen activator neurovascular toxicity is controlled by activated protein C, *Nat. Med.* **10**(12), 1379–1383 (2004).
6. R. M. Bertina, *Protein C and Related Proteins; Biochemical and Clinical Aspects*, (Churchill Livingstone, New York, 1998), pp. 1–54.
7. D. F. Bruley and W. N. Drohan, *Advances in Applied Biotechnology Series; Protein C and Related Anticoagulants*, **11** (Gulf Publications, Houston, TX 1990).
8. S. Rezania, D. G. Ahn and K. A. Kang, Separation of Protein C from Cohn Fraction IV-1 by Mini-Antibody, Proceedings of the 33rd Annual ISOTT meeting 2005. *Adv. Exp. Med. Biol.: Oxygen Transport to Tissue XXVIII,* **599**, (Maguire, D. J., Bruley, D. F., Harrison, D. K., eds.), p. 125–132, 2007.
9. L. H. Edmunds and E. W. Salzman, Hemostatic Problems, Transfusion Therapy, and Cardiopulmonary Bypass in Surgical Patients, *Hemostasis and Thrombosis*, 3rd ed., R. W. Colman, J. Hirsh, V. J. Marder and E. W. Salzman, (J. B. Lippincott Co, Philadelphia, 1994), p. 958.

10. G. J. Broze Jr. and J. P. Miletich, Biochemistry and Physiology of Protein C, Protein S, and Thrombomodulin, *Hemostasis and Thrombosis*, 3rd ed., R. W. Colman, J. Hirsh, V. J. Marder and E. W. Salzman, (J. B. Lippincott Co., Philadelphia, 1994), p. 262.
11. E. S. Hemdan, Y. J. Zhao, E. Sulkowski and J. Porath, Surface topography of hisitdine residues: A facile probe by immobilized metal ion affinity chromatography, *Proceedings of the National Academy of Science USA*, **86,** 1811–1815 (1989).
12. H. Wu and D. F. Bruley, Homologous human blood protein separation using immobilized metal affinity chromatography: protein C separation from prothrombin with application to the separation of factor IX and prothrombin, *Biotechnol. Progr.* **15,** 928 (1999).
13. H. Wu, D. F. Bruley, K. A. Kang, Protein C Separation form human plasma Cohn fraction IV-1 using immobilized metal affinity chromatography, *Adv. Exp. Med. & Bio.: Oxy. Trans. to Tis. XX,* **454,** (Plenum Press, New York, 1998), pp. 697–704.
14. J. J. Lee, D. F. Bruley, and K. A. Kang, Effect of pH and Imidazole on Protein C Purification from Cohn Fraction IV-1 by IMAC, Proceedings of the 33rd Annual ISOTT meeting 2005. *Adv. Exp. Med. Biol.: Oxygen Transport to Tissue XXVIII,* **599**, (Maguire, D. J., Bruley, D. F., Harrison, D. K., eds), p. 53–60, 2007.

Chapter 12
Separation of Factor V Leiden Molecule, a Mutated Form of Factor V, from Plasma of Homozygous Patient

Samim Rezania and Kyung A. Kang[1]

Abstract Factor V (FV) is a coagulant in plasma. The FV molecule consists of a heavy chain and a light chain, and Factor V Leiden (FVL) is mutated FV at a single amino acid in the heavy chain. FVL patients are in a dangerous hyper-coagulation state in their body. Current FVL diagnosis is done by DNA analysis, which is expensive and time consuming. Our group has been developing a real-time, cost effective immuno-optical biosensor for FVL diagnosis. For the sensor development, pure FVL, which is not currently available, is needed. Here, we have attempted FVL purification from FVL patient's plasma. Since plasma contains many proteins and some proteins are structurally homologous to FV, the purification must be done by a very specific method, such as immuno-affinity chromatography. However, an antibody that does not react with FV is not currently available. Because the mutation is in the heavy chain and the amino acid sequence of the light chain of FVL is identical to that of FV, antibodies generated against the light chain of FV were tested for purifying FVL. Plasma was obtained from a homozygous FVL patient. First, the plasma was pretreated by barium citrate and polyethylene glycol 6000, to remove the vitamin K-dependent proteins, alpha globulins, and other smaller than 6 kDa molecular weight proteins. The yield in the process was 54%. Immuno-affinity purification of FVL from patient plasma was then performed using an anti-FV light chain antibody immobilized CNBr-Sepharose, and the purification yield was 25%. In summary, the antibody against the light chain of FV was able to purify the single point mutated form of FV (FVL) from plasma with an overall yield of 14%. The same principle can probably be used for purification of the other single point mutated proteins.

[1]Samin Rezania and Kyung A. Kang, Department of Chemical Engineering, University of Louisville, Louisville, KY 40292.

K.A. Kang et al. (eds.), *Oxygen Transport to Tissue XXIX*

12.1 Introduction

Factor V (FV) is a coagulation factor in blood plasma. It accelerates clot formation initiated by factor X in the presence of phospholipid and calcium. It consists of a C-peptide region (Mw = 100,000), a heavy chain (Mw = 105,000), and a light chain (Mw = 74,000), and the two chains are non-covalently associated and its total molecular weight is approximately 300,000 [1,2,3]. Activated protein C (APC) inactivates activated FV (FVa), by cleaving the heavy chain of FVa [4] at positions Arg^{306}, Arg^{506}, and Arg^{679}. Factor V Leiden (FVL) has a single point mutation from the arginine at the position 506 to glutamine [5]. FVL has the same coagulant function as FV, but due to the lack of Arg^{506}, it is not deactivated by APC, leading the body to a dangerous hyper-coagulating state.

FVL is the most common blood coagulation disorder, present in 3–8% of the general US and European populations [5]. Nevertheless, FVL screening is not a routine clinical procedure. Currently, the diagnosis is made by DNA analysis, which is complex, expensive, and time consuming [6]. Therefore, the ultimate goal of our study is to develop a rapid, accurate, and cost-effective immuno-biosensor to diagnose FVL.

For this sensor development, both pure FVL molecule and antibodies against FVL are needed, but neither is currently available. The only source for FVL is FVL patient's plasma. Since plasma contains several proteins homologous to FVL, the purification of FVL from plasma should be done by a highly specific method, such as immuno-affinity chromatography. The single point mutation for FVL is in the heavy chain and the amino acid sequences of the light chains of FV and FVL are identical. Here, commercially available antibodies generated against the light chain of FV molecule were tested to purify FVL from homozygous FVL patient's plasma.

12.2 Materials, Methods, and Equipment

Unless otherwise specified, all the materials for different experiments were purchased from Sigma-Aldrich (St. Louis, MO).

12.2.1 Pretreatment of Plasma

Normal human plasma was obtained from Sigma-Aldrich. Plasma was obtained from a FVL homozygous patient (FVL plasma) by the plasmaphresis procedure, following the IRB approved by the University of Louisville Human Subjects Protection Program. The plasma was collected and was kept at –70 °C until the purification process.

Before pretreatment, normal plasma was reconstituted using 1 milliliter (mL) DI water and the FVL plasma was thawed at 4 °C. Then, protease inhibitors, 156.6 mg of benzamidine hydrochrolide, 50 mg of soybean trypsin inhibitor, and 25 μg of phenylmethanesulfonyl fluride per 1 liter of plasma, were added to both the normal and patient plasma.

The plasma was then pretreated following the procedure as described by Dahlback et al [3]. First, to remove Vitamin-K dependent proteins to reduce the risk formation of small amounts of thrombin or other active coagulation factors [7], 1 M barium chloride was added dropwise to plasma at a concentration of 80 mL/L-plasma, and the mixture was stirred for 1 hr. It was then centrifuged (TJ-MI refrigerated centrifuge; Beckman Coulter; Kansas City, MO) at 6,000 g for 10 min. 40 g solid polyethylene glycol 6000 (PEG; Alfa Aesar; Ward Hill, MA) was added to a liter of the supernatant, was stirred for 1 hr, and was centrifuged to remove alpha globulins, proteins smaller than 6 kDa, and reduce the sensitivity of plasma to proteolytic enzymes [8]. Then, PEG was added to the supernatant from the last step at a ratio of 80 g/L-supernatant. The solution was stirred for 1 hr and centrifuged as described above. After centrifugation, the supernatant was discarded and the precipitate was dissolved in Tris buffer (pH 7.4), to a final volume of 30 mL, and was kept at –70 °C before the immuno-affinity purification.

12.2.2 Selection of Antibody and Immuno-purification of FV/FVL

Four commercially available antibodies against the light chain of FV were purchased from Haematologic Inc. (HTI; Essex Junction, VT), QED Inc. (QED; San Diego, CA), Biodesign International (Biodesign; Saco, MA), and Fitzgerald Inc. (Fitzgerald; Concord, MA). The enzyme linked immunoassay (ELISA) for the antibody was performed as follows: Microtiter ELISA plates (Nalgen Nunc International; Roskilde, Denmark) were coated with 2 μg/mL of the antibodies. After blocking with 1% bovine serum albumin, FVL patient plasma was applied and a ½ serial dilution was performed. After washing the plate, horseradish peroxide conjugated antibody (HRP) was applied to the wells. Then, o-phenylenediamine dichloride (OPD) solution was added and optical density was measured at 450 nm using an ELISA plate reader (Bio-Rad; Hercules, CA).

The two affinity chromatography matrices tested for purification were Actigel ALD™ (Actigel; Sterogene; Carlsbad, CA) and CNBr-activated Sepharose™ 4B (CNBr; Amersham Biosciences; Piscataway, NJ). Following the manufactuer's instruction, 1 mg of antibody was immobilized on 1 mL of each matrix, and then the matrix was packed in a chromatography column (d = 0.7 cm; BioRad; Hercules, CA). The column was equilibrated with 10-column volume (CV) of 0.02 M imidazole, 5.0 mM $CaCl_2$, 0.15 M NaCl, pH 6 (Equilibrium/Washing buffer).

The method of immuno-affinity chromatography of FV/FVL is based on the protocol developed by Katzmman et al [9]. The regeneration protocol was modified as described by Kang et al [10]. Briefly, 6 mL of pretreated plasma was applied to the 1 mL anti-FV immobilized CNBr matrix, equilibrated with equilibration buffer as described above. The adsorption was allowed for 10 min and the column was washed with 10 CV of washing buffer. Adsorbed FV/FVL was eluted with elution buffer containing high salt concentration (0.02 M imidazole, 5.0 mM $CaCl_2$, 1.2 M NaCl, pH 6.5).

12.3 Results and Discussion

12.3.1 Selection of Antibody

To select the antibody among four commercially available antibodies generated against the light chain of FV, the relative affinities of the antibodies against FVL (in FVL homozygous patient plasma) were measured by ELISA (Fig. 12.1).

The antibody from Haematologic (HTI) showed the highest affinity and the one from Fitzgerald showed the lowest. Because, the affinity of ligand is usually significantly reduced during the immobilization process, HTI was selected to test first.

12.3.2 Selection of Affinity Chromatography Gel Matrix

Two commercially available affinity chromatography gel matrices, Actigel and CNBr were tested for their FV purification efficiency. Actigel and CNBr have a 5-atoms and 1-atom spacers, respectively. Our previous study results [11]

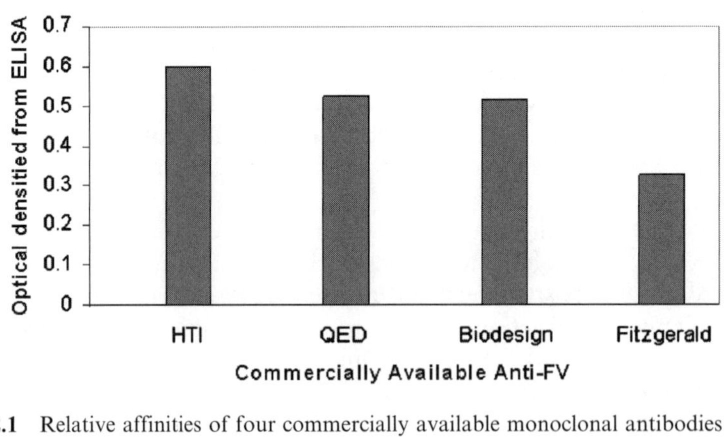

Fig. 12.1 Relative affinities of four commercially available monoclonal antibodies against the light chain of FV to FVL plasma.

Table 12.1 Performance of two commercially available immuno-affinity gel matrixes for FV purification

	Actigel		CNBr	
	HTI	Fitzgerald	HTI	Fitzgerald
Immobilization efficiency (%)	99.2 ± 0.28		99.9 ± 0.07	
Washing (%)	2.5 ± 0.28	3.8 ± 0.07	5.5 ± 0.1	6.1 ± 0.07
Elution (%)	1.2 ± 0.13	7.5 ± 0.63	2.7 ± 0.21	30 ± 19

showed that the spacer length between the antibody and the matrix can affect the affinity of the immobilized antibody to antigen. Actigel has shown a better performance than CNBr and, therefore, it was tested first.

The HTI antibody was immobilized on Actigel as described in the Methods section, and the immobilization efficiency was 99.2% (Table 12.1). FV purification performance of the matrix was then tested, using 40 μg pure FV and the amounts of FV in various fractions were quantified by ELISA (Table 12.1). Only 2.5% of FV was washed away during the washing step, but only 1% was eluted (Table 12.1), indicating a too high affinity between the antigen and the antibody. Therefore, the antibody showing the lowest affinity (Fitzgerald) was tested next. With the Fitzgerald antibody, 3.8% of the FV was washed away and 7.5% of FV was eluted, showing a slightly better recovery, but still very high affinity. We attempted to further reduce the affinity by using the CNBr matrix [10].

The antibody immobilization efficiency to the CNBr matrix was 99.9% (Table 12.1). The immuno-affinity chromatography for FV was again performed. During the washing step, 5.5% and 6.1% of the FV was washed away from the matrices immobilized with HTI and Fitzgerald, and the purification yields were 2.7 and 30% (Table 12.1), showing that the combination of the Fitzgerald antibody and the CNBr matrix provided the best performance.

12.3.3 FV Purification from Blood Plasma

As a next step, FV purification from normal plasma (5 mL) was studied. Human plasma was pretreated as described in the Methods section and the amounts of FV in each step were quantified by ELISA (Table 12.2).

Table 12.2 Pretreatment of normal plasma before affinity purification of FV

Fraction	Amount of FV (μg)	Yield (%)
Plasma (5 mL)	41.00	
Barium citrate supernatant	40.04	98 ± 1
PEG-6000 I supernatant	31.40	76 ± 3
PEG-6000 II precipitate	20.00	49 ± 8

Table 12.3 Immuno-affinity purification of FV from plasma using CNBr

Amount of FV in the source, μg	Washing, μg (%)	Elution, μg (%)
20	0.94 (4.7 ± 0.2)	5 (25 ± 4)

Then immuno-affinity chromatography of FV was performed using the pretreated plasma, and 1 ml of the anti-FV immobilized CNBr. The amount of FV in each process was determined (Table 12.3). Approximately 5% of FV was washed away during the washing step. A purification yield of 25% was achieved for the immuno-affinity chromatography of FV from plasma (Table 12.3). Therefore, the overall FV purification yield from plasma is approximately 13% combining the yield from the pretreatment of plasma (49%) and the yield (25%) for the immuno-affinity purification of FV. The maximum overall yield for FV purification from plasma reported in other references is 20% [3,4,9], confirming that our result is similar to others.

12.3.4 FVL Purification from Homozygous Patient Plasma Using FV Antibody

Since the performance of the anti-FV immobilized CNBr for FV purification was reasonable, the same procedure was used for FVL purification from the homozygous patient plasma. The plasma was pretreated as described previously. The recovery of the FVL from the barium citrate adsorption, PEG precipitation I, and PEG precipitation II was 93%, 89, and 54% (Table 12.4), which was similar to the yields for the pretreatment of normal plasma. The immuno-affinity purification of FVL showed a yield of 25%. The overall FVL purification yield from patient plasma was found to be 14% combining the yield from the pretreatment of homozygous plasma (54%) and the yield (25%) for immuno-affinity purification of FVL, which was similar to that of FV.

Table 12.4 Immuno-affinity purification of FVL from FVL homozygous patient plasma using anti-FV immobilized CNBr

Fraction	Amount of FVL (μg)	Yield (%)
Plasma	165.3	
Barium citrate supernatant	153.5	93 ± 2
PEG-6000 I supernatant	146.7	89 ± 1
PEG-6000 II precipitate	92	54 ± 5.5
Affinity chromatography	23.2	14 ± 2

12.4 Conclusions

FVL is the most common hereditary, abnormal blood-coagulation disorder. To develop a rapid and inexpensive biosensor for FVL diagnosis, FVL molecules were purified from homozygous patient plasma. The immuno-affinity chromatography of FVL was performed using antibodies against the light chain of FV. The antibody was able to purify the single point mutated form of FV (FVL) from plasma at an overall yield of 14%. The same principle can be used to purify the other single point mutated proteins.

Acknowledgment The authors thank the National Institutes of Health (5 R21EB003485-02) for the financial support, and Dr. Sharma at the Oncology Department of the University of Louisville for his help to obtain the plasma from a homozygous patient.

References

1. C. T. Esmon, The subunit structure of thrombin-activated factor V: Isolation of activated factor V, separation of subunits, and reconstitution of biological activity. *J. Biol. Chem.* **254**, 964–973 (1979).
2. W. H. Kane, P. W. Majerus, Purification and characterization of human coagulation factor V. *J. Biol. Chem.* **256**, 1002–1007 (1981).
3. B. Dahlback, Human coagulation factor V purification and thrombin-catalyzed activation. *J. Clin. Invest.* **66**, 583–591 (1980).
4. J. Rosing, and G. Tans, Factor V, *Int. J. Biochem.* **29**, 1123–1126 (1997).
5. E. Castoldi, J. M. Brugge, G. A. Nicolaes, D. Girelli, G. Tans, J. Rosing, Impaired APC cofactor activity of factor V plays a major role in the APC resistance associated with the factor V Leiden (R506Q) and R2 (H1299R) mutations, *Blood* **103**, 4173–4179 (2004).
6. M. Wilmer, C. Stocker, B. Buhler, B. Conell, A. Calatzis, Improved distinction of factor V wild-type and factor V Leiden using a novel prothrombin-based activated protein C resistance assay, *Am J Clin Pathol.* **122**, 836–842 (2004).
7. R. J. Alexander, T. C. Detwiler, Quantitative adsorption of platelet glycoprotein G (thrombin-sensitive protein, thrombospondin) to barium citrate. *Biochem. J.* **217**, 67–71 (1984).
8. H. A. Donald, K. C. Ingham, Mechanism of precipitation of proteins by Polyethylene Glycol. *J. Biol. Chem.* **256**, 12108–1211766 (1981).
9. J. A. Katzmann, M. E. Nesheim, L. S. Hibbard, K. G. Mann, Isolation of functional human coagulation factor V by using a hybridoma antibody. *Proc. Natl. Acad. Sci. USA.* **78**, 162–166 (1981).
10. K. A. Kang, D. Ryu, W. M. Drohan and C. L. Orthner, Effect of matrices on affinity purification of protein C. *Biotechnol Bioeng.* **39**, 1086–1096 (1992).
11. S. Rezania, D. G. Ahn, and K. A. Kang, Separation of protein C from Cohn fraction IV-1 by mini-antibody, Proceedings of the Annual 2005 ISOTT meeting. *Adv. Exp. Med. Biol.: Oxygen Transport to Tissue XXVIII,* **599**, (Maguire, D.J., Bruley, D.F., Harrison, D.K., eds.), p. 125–132, 2007.

Chapter 13
A Simple Volume Related Model of Arterial Blood Pressure Generation

Christopher B. Wolff[1], Benn S. Gooch[2], and James S. Douglas[3]

Abstract A single compartment model of the arterial circulation was used to generate an arterial blood pressure waveform from pre-determined stroke volume (SV) and arterial resistance (R). With fixed stroke volume and varying resistances blood pressure waveforms showed mean values proportional to resistance but amplitude lessening with higher pressure; the amplitude of the hypothetical volume waveform of the arterial system was the same for all resistance values. Where SV varied and R changed reciprocally, the waveform when analysed with the PulseCO^{TM} algorithm gave estimates slightly higher than the input stroke volumes (r 0.9998; y = 0.99x + 5.28 ml). Where SV varied with fixed R mean blood pressure varied with stroke volume; SV estimates were, again, slightly higher than the input stroke volumes (r 0.9994; y = 0.986x + 6.04 ml). Estimates of SV and R from Valsalva manoeuvre BP were used in the model to generate arterial blood pressure. SV estimates closely resembled the original model values (r 0.988; y = 1.0802x – 3.9251). The model appears capable of generating BP waveforms compatible with real BP waveforms since stroke volume estimates closely resemble the original stroke volumes used in the model.

13.1 Introduction

An algorithm, utilized commercially (PulseCO^{TM}) to derive stroke volume from arterial blood pressure includes, as part of the procedure, an equation for the generalized compliance properties of the arterial tree. It is assumed that the arterial volume, V, is related to the arterial blood pressure, P, according to the relationship:

[1]Clinical Pharmacology & Anaesthetics, William Harvey Research Institute, Barts and The Royal London, Charter House Square, London, UK, EC1M 6BQ.
[2]St. George's, University of London, Cranmer Terrace, London, UK, SW17 ORE.
[3]LiDCO Ltd, Unit M, South Cambridge Business Park, Babraham Road, Cambridge, UK, CB2 4JH.
Corresponding author: Christopher B. Wolff, e-mail: chriswolff@doctors.org.uk

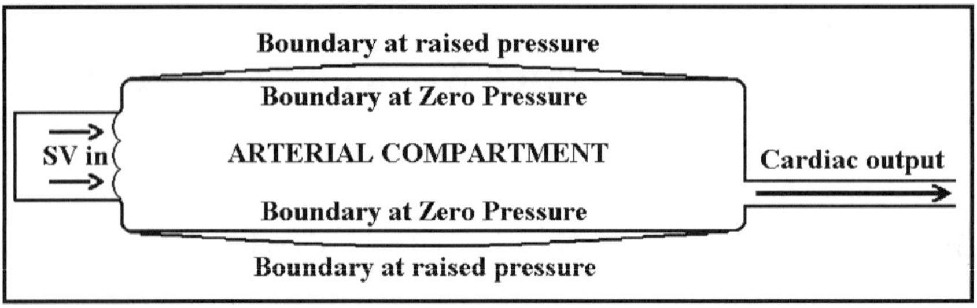

Fig. 13.1 The whole arterial compartment is represented as a single vessel with access from the heart and outflow to the periphery. 'SV in' represents the input, which will consist of increments adding up to the stroke volume during systole and a sequence of zeros during the time representing diastole. Cardiac output is continuous.

$$V = 250(1 - e^{(-0.0092P)}) \quad (13.1)$$

This allows calculation of a putative arterial volume waveform. The oscillation around its mean value is subjected to an autocorrelation manoeuvre. The derived auto-covariance gives an objective cardiac period and its amplitude is deemed related to the square of the stroke volume. Cardiac output values are displayed beat-by-beat. Since subjects vary in size and in their precise arterial wall properties a static, absolute, value of cardiac output is obtained intermittently by 'dye' dilution (Lithium chloride, LiDCO^{TM}) in order to obtain an individual scaling factor. In effect, the scaling factor (or calibration), simply adjusts the value of 250 in the equation above. The algorithm has been validated against Lithium dilution cardiac output measurement in a number of studies [1–3], and reviewed by Rhodes and Sunderland [4]. Further within-subject/patient validation is underway.

A single compartment model is presented here which includes the above assumption about the compliance properties of the arterial system. The input to the system consists of stroke volume increments fed in 1/100th second at a time and account is taken of the volume already present in the previous 1/100th second, outflow (in the previous 1/100th second) and arterial resistance. This allows generation of an arterial blood pressure waveform. Figure 13.1 shows the single arterial compartment diagrammatically, with a boundary at zero pressure and another shown for the larger volume which will accompany a finite arterial blood pressure. Details of the computations are given in the next section.

13.2 Model

The model arterial blood pressure is generated 1/100th second at a time. Increments of stroke volume (v, column D in Fig. 13.2) are added to a volume compartment (column E) to which the previous volume is added. The latest cardiac output

	B	C	D	E	F	G	H
	Equation line			E8+D9-G8	P = f (E9)	F9/H9	
	Function			V(n-1) +v - q(n-1)	P = -108.7xLN(1-V/250)	P/R	
			SV			Outflow	
5			increments	Arterial Vol	Pressure	(CO)	Resistance
6			vol IN/0.01s	excess > Vo	From V		(mmHg/(ml/.01s))
			v			q	
7		Time (s)	(ml/0.01s)	**V ml**	P (mm Hg)	(ml/0.01s)	R
8	Starting state	0	0	0	0	0	
9	**EQUATIONS**	0.01	6	**6**	**2.6**	**0.0**	100
10		0.02	8	**13.97359**	**6.3**	**0.1**	100
11		0.03	7	**20.91107**	**9.5**	**0.1**	100
12		0.04	5	**25.81612**	**11.8**	**0.1**	100
341		3.33	1	**166.97940**	**119.8**	**1.2**	100
342		3.34	1.5	**167.28112**	**120.2**	**1.2**	100
343		3.35	1	**167.07888**	**120.0**	**1.2**	100
344		3.36	1	**166.87930**	**119.7**	**1.2**	100
345		3.37	1	**166.68233**	**119.4**	**1.2**	100
346		3.38	1	**166.48794**	**119.2**	**1.2**	100
347		3.39	1	**166.29607**	**118.9**	**1.2**	100
348		3.4	1	**166.10670**	**118.7**	**1.2**	100
349		3.41	0	**164.91979**	**117.2**	**1.2**	100
350		3.42	0	**163.74815**	**115.7**	**1.2**	100
351		3.43	0	**162.59138**	**114.2**	**1.1**	100
352		3.44	0	**161.44909**	**112.8**	**1.1**	100
353		3.45	0	**160.32091**	**111.4**	**1.1**	100

Fig. 13.2 Starting conditions are given in an initial line (line 8 in this instance). Equations to calculate volume (V), arterial blood pressure (P) and blood flow (q) are present in line 9 in columns E, F and G respectively. V, P and q are calculated from stroke volume and resistance on the basis of several simple assumptions outlined in the text and embodied in the equations. The equations utilise stroke volumes in column D and values for arterial resistance in column H to generate values for the excess volume (V, above the volume at zero pressure), arterial blood pressure and cardiac output; they are applied to successive lines of the spreadsheet. Values soon reach a steady state where the arterial pressure oscillations and cardiac output are reproduced with every cycle (as in lines 341 to 353).

increment ('Outflow', column G) is subtracted from the total. The volume, V, here is the excess above the volume of the arterial system when blood pressure is zero (atmospheric).

Hence for the volume column:

$$Vn = v_n + V_{(n-1)} - q_{(n-1)} \tag{13.2}$$

To obtain the arterial blood pressure corresponding to this volume the equation for volume from blood pressure P (1) is reversed becoming:

$$P = 108.7 \times \ln(1 - V/250) \tag{13.3}$$

This is applied in column F in Fig. 13.2. For this model outflow, q, for each 1/100th second is derived in column G from P/R where R is arterial resistance (column H). The model includes starting conditions (all zeros here, in line 8) and, next, the first line of equations (line 9 in this example). Each row represents 1/100th second (column C).

Cardiac output is normally expressed as $l\ min^{-1}$ so the values in column G for outflow (ml per 0.01 s) are 1/6th (1000/(60 $\times$ 100)) of the usual units. Hence, cardiac output is obtained from column G using a factor 6 to scale it up from ml/0.01s to $l\ min^{-1}$.

A variety of different situations will be examined:

1. A single stroke volume of 100 ml will be entered into the model with different arterial resistance values to see the effect on the model arterial blood pressure and excess volume.
2. A range of stroke volumes (50, 80, 100, 120 and 150 ml will be entered into the model with the same value for arterial resistance (100 mm Hg per (ml/0.01s)); this will be expected to lead to mean blood pressures which increase as stroke volume increases.
3. The same range of stroke volumes will be entered into the model but with reciprocal values of arterial resistance; this is expected to give a constant value for the mean arterial blood pressure. For example, for 100 ml stroke volume the resistance entered will be 100 mm Hg per (ml/0.01s) and for 50 ml the resistance will be 200 mm Hg per (ml/0.01s).
4. The blood pressure records from situations 2. and 3. will be analysed with the PulseCO^{TM} algorithm to give estimates of stroke volume and these will be compared with the values originally entered into the model.
5. The arterial blood pressure recorded before during and after a Valsalva manoeuvre will be analysed to give estimates of stroke volume, cardiac interval and arterial resistance beat-by-beat. These values will be entered into the model to attempt regeneration of the original blood pressure waveform.
6. The blood pressure record generated by situation 5 will again be analysed to give beat-by-beat stroke volume estimates from the models blood pressure output. These will be compared with the values entered into the model.

Since the situation at any given time interval (of 100th s) depends solely on the values in the model for the present (SV) and immediately previous (100th s) time interval the starting conditions can be finite (they don't have to be zero); so one can enter known values from an existing steady state after which the outcome blood pressure will depend upon the stroke volume and resistance values entered into the model.

13.3 Results

13.3.1 Constant SV, Varying R

The overall arterial compliance curve is depicted in Fig. 13.3 (*left hand panel*). The arterial blood pressure and volume changes resulting from changing

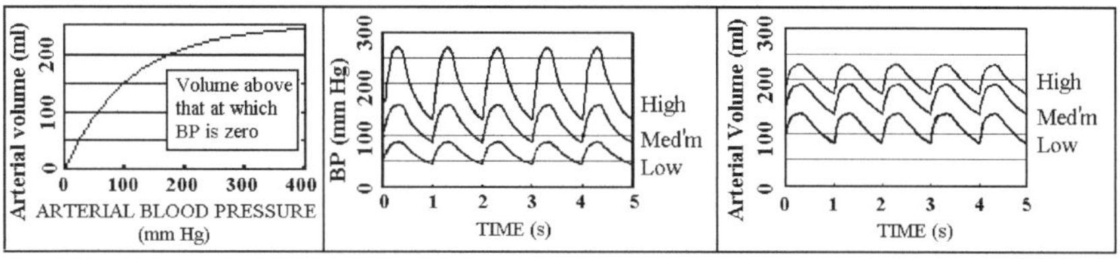

Fig. 13.3 Left panel: the compliance relationship between excess arterial volume (the volume above that where BP is zero) to the arterial blood pressure – a saturating exponential (1) above. Middle panel: blood pressure output from the model for constant stroke volume (SV, 100 ml), with three different arterial resistances (units, mm Hg per (ml. 0.01s $^{-1}$)). There is a progressive increase in mean blood pressure and in the arterial pressure oscillations as resistance increases. Right hand panel: the volume oscillations are of constant amplitude despite increases in the mean value; otherwise the algorithm to derive SV from blood pressure would give different stroke volumes.

peripheral resistance, with a constant stroke volume of 100 ml, are shown in the middle and right hand panels respectively.

13.3.2 Varying Stroke Volume

Arterial pressure waveforms were generated by the model with stroke volumes varying between 50 ml and 150 ml. They were then analysed by means of the PulseCO^{TM} algorithm to give estimates of stroke volume which were plotted against the original model values. The results are shown in Fig. 13.4.

Plot A shows where arterial resistance was constant (100 (mmHg/(ml/.01s)) with proportional changes in mean blood pressure; plot B shows the SV results where resistance was changed inversely with stroke volume so that the arterial mean pressure was constant.

Plots C and D show the differences between SV estimates from the model blood pressure and the original SV values; the differences in ml. are shown in C and then these are shown as a percentage of the original values in plot D.

Table 13.1 gives SV values entered in the model ('imposed SVs') and resistances used (fixed resistances, A; reciprocally varying resistances, B). SV estimates are also given.

13.3.3 A Valsalva Manoeuvre Blood Pressure Record – SV Estimates Used in the Model

Stroke volumes and cycle lengths were estimated from arterial blood pressure recorded before during and after a Valsalva manoeuvre (Fig. 13.5, *left hand panel*). Arterial Mean BP and cardiac output were then used to calculate beat-by-beat arterial resistance. These stroke volume and resistance values were then

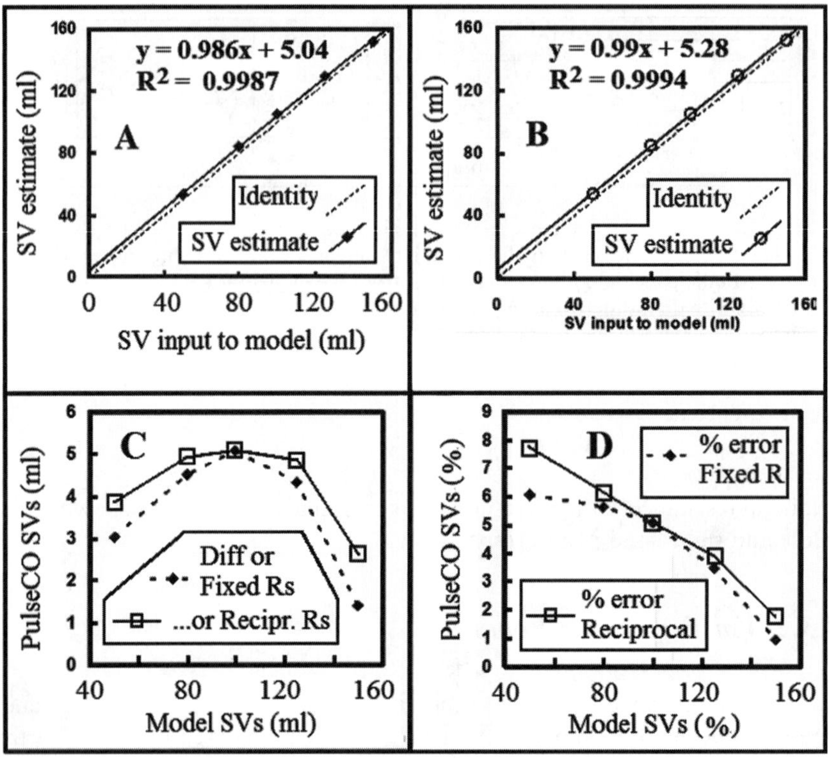

Fig. 13.4 This shows the results of estimating stroke volume by means of the PulseCO™ algorithm from arterial blood pressure waveforms generated by the model; the model stroke volumes were 50, 80, 100, 125 and 150 ml. A. Varying stroke volume with a fixed resistance ; B. Varying stroke volume with reciprocal resistance values (see Table 13.1). Errors are shown in C and D, expressed as differences in ml. and in D expressed in percentage terms.

Table 13.1 Stroke volumes and resistances imposed on the model and stroke volume values derived from the resulting model blood pressure records

	A		B	
Imposed SV (for model) (ml)	Fixed Resistance (mmHg/ (ml/.01s))	Derived SV (ml)	Reciprocal Resistance (inverse of SV) (mmHg/(ml/.01s))	Derived SV (ml)
50	100	53.0	200	53.3
80	100	84.5	120	84.4
100	100	105.1	100	104.6
125	100	129.3	80	129.1
150	100	151.4	66.7	152.7

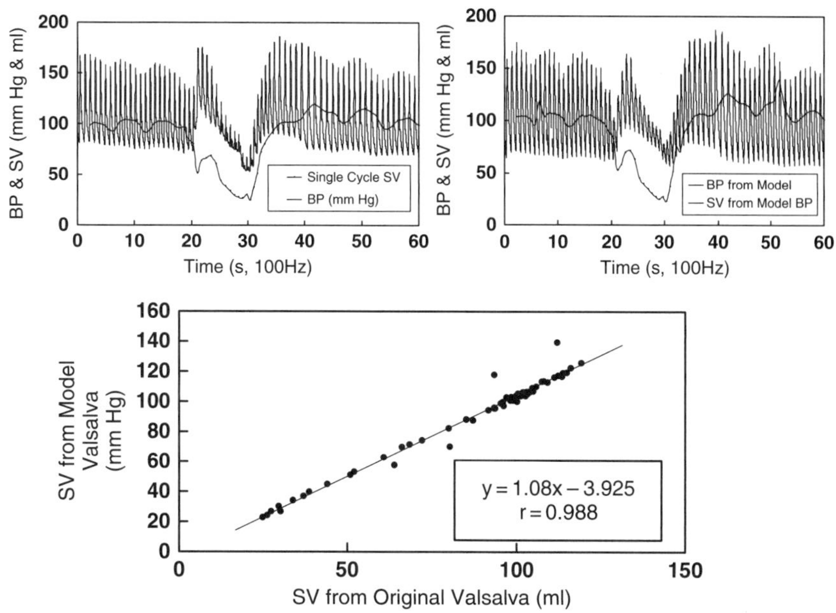

Fig. 13.5 Above left: Arterial blood pressure recorded before during and after a Valsalva manoeuvre (forced expiration against a closed airway for 10s) with stroke volumes beat by beat, derived using a true single cycle version of the PulseCO^{TM} algorithm (under development). Above right: Blood pressure waveform derived by the model from stroke volume estimates and arterial resistance calculated from the original Valsalva manoeuvre (*L panel*). Again, (*R panel*) stroke volume estimates are shown, this time derived from the model blood pressure. Below: The relationship between the stroke volume estimates from the model version of the blood pressure and the estimates from the original blood pressure (y = 1.08x – 3.925; r = 0.988).

used to run the model thereby generating a model blood pressure record. Stroke volumes were again estimated, this time from the blood pressure generated by the model.

Figure 13.5 shows the original arterial blood pressure record on the left with stroke volume estimates from it and, on the right, the model derived blood pressure and stroke volumes estimated from it. The lower panel shows the stroke volume estimates from the model blood pressure plotted against the stroke volume estimates from the original Valsalva BP record.

13.4 Discussion

The model generates a blood pressure waveform on the basis that stroke volume enters the model's arterial compartment intermittently. Expansion is less than stroke volume because of simultaneous outflow (cardiac output) at the periphery.[5] The outflow, in turn, depends upon the arterial blood pressure (BP). The blood pressure again depends upon the extent to which the arterial volume exceeds the volume which would be occupied at zero pressure (atmospheric). The curvilinear pressure dependency upon this 'excess' volume is represented by (1) and (its reverse) (3), the relationship illustrated in the left-hand panel of Fig. 13.3.

The PulseCO™ algorithm, working on blood pressure waveforms generated by the model from a wide range of stroke volumes, regenerates stroke volumes similar to those originally put into the model (imposed volumes). Derived volumes are around 3–4% greater than imposed volumes, presumably in someway due to the effect of the PulseCO™ algorithm's inclusion of auto-covariance of the volume converted blood pressure. The volume conversion can be seen to compensate for reduced arterial volume expansion at higher pressures (right hand panel of Fig. 13.3).

The similarity of derived stroke volumes to input stroke volumes underlies the close similarity of the original and model derived arterial blood pressure waveforms (Fig. 13.5). One obvious difference, however, is the lack of a reflected wave in the model derived blood pressure wave-form, as illustrated in Fig. 13.6 below. Here we see the original waveform with evidence of a reflected wave, the model only showing a smooth descent during the diastolic phase.

Validation of PulseCO™ [4] requires further support in the individual but the near agreement between the model and PulseCO™ (to the extent of PulseCO™ validation to date) supports the assumptions of the model. The assumptions made to derive the model are not new; they can be found in the papers of Remington, Hamilton and Dow [5] and Remington and Hamilton; [6] authors who, at that stage (1945), could not test their hypotheses readily with computers as illustrated here. It is hoped that the model will be of value in developing hypotheses concerning circulatory dynamics.

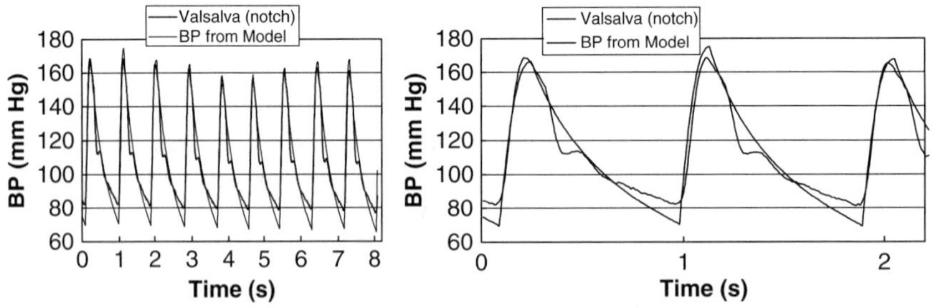

Fig. 13.6 A short section of the early part of original Valsalva manoeuvre blood pressure record, prior to the actual manoeuvre with the output from the model. On the left is an 8 second sequence with the presence of a dicrotic notch apparent in the original waveform (bold). The model blood pressure lines are thinner. In the right hand panel just over 2 cycles are shown and the dicrotic notch in the original record is more obvious. The model output shows a smooth decay.

13.5 Competing Interests

Christopher Wolff and James Douglas act as consultants to LiDCO who hold the patent for the commercial use of the PulseCO™ algorithm.

References

1. T. T. Hamilton, L. M. Huber and M. E. Jessen, PulseCO: a less-invasive method to monitor cardiac output from arterial pressure after cardiac surgery. *Ann. Thorac. Surg.* **74**, S1408–S1412 (2002).
2. M. M. Jonas and S. J. Tanzer, Lithium dilution measurement of cardiac output and arterial pulse waveform analysis: an indicator dilution calibrated beat-by-beat system for continuous estimation of cardiac output. *Curr. Opin. Crit. Care* **8**, 257–261 (2002).
3. R. M. Pearse, K. Ikram, and J. G. Barry, Equipment review: an appraisal of the LiDCO™*Plus* method of measuring cardiac output. *Crit. Care* **8**, 190–195 (2004).
4. A. Rhodes and R. Sunderland, Arterial pulse power analysis: the LiDCO™ system. In, *Functional Hemodynamic Monitoring Update in Intensive Care and Emergency Medicine 42,* edited by M. R. Pinsky and D. Payen (Springer-Verlag, Heidelberg, 2005), pp. 183–192.
5. J. W. Remington, W. F Hamilton and P. Dow, Some difficulties involvedin the prediction of the stroke voume from the pulse wave velocity. Am. J. Phyiol. **144**, 536–545 (1945).
6. J. W. Remington and W. F Hamilton, The construction of a theoretical cardiac ejection curve from the contour of the aortic pressure pulse. Am. J. Phyiol. **144**, 546–555 (1945).

Part IV
Tumor, Cancer and Oncology

Chapter 14
Strikingly High Respiratory Quotients: A Further Characteristic of the Tumor Pathophysiome

Peter Vaupel[1]

Abstract Conspicuously high respiratory quotients (RQs) are found in solid tumors in vivo. RQs in the range between 1.29 and 1.95 neither reflect the degree of substrate oxidation by tumor cells nor indicate the types of fuels involved in metabolic processes. Instead, such tumor RQs most probably are caused by (a) channeling of glycolytic end-products into lipogenesis, and by (b) CO_2 release from the tumor following extracellular buffering of H^+-ions by bicarbonate. H^+-ions exported from the intracellular space into the interstitial compartment titrate extracellular bicarbonate to CO_2 and H_2O with the aid of the ectoenzyme carbonic anhydrase IX, which is activated at low pH. Strikingly high (RQs) may thus be a further characteristic of the tumor microenvironment and of the tumor (patho-)physiome, the latter quantitatively describing the pathophysiologic characteristics of tumor cells and solid tumors.

14.1 Introduction

An indication of the types of fuels involved in metabolic processes is given by the respiratory quotient: RQ = CO_2 output/O_2 uptake. If the fuel is pure carbohydrate, the RQ = 1.0, for fat breakdown it is 0.7 [1,2]. In exceptional cases, the RQ can be outside the range of 0.7 – 1.0. It becomes greater than 1.0 when excess carbohydrates are consumed, so that fat stores are built up [1]. Moreover, extremely high RQ values are (temporarily) found in the early stages of a voluntary hyperventilation or when metabolic acidosis develops. In these cases, respiratory CO_2 output is greater than the metabolic CO_2 formation, so that the RQ values briefly rise above 1.0 (in some cases to as much as 1.5) [1].

[1]Institute of Physiology and Pathophysiology, University of Mainz, Duesbergweg 6, 55099 Mainz, Germany.

Enhanced lipogenesis has been described for tumor cells [3,4] and data have recently been communicated that glycolytic end-products can be channeled into lipogenesis in certain cancer models [5–7]. In this case, RQ values should greatly exceed the upper end of the reference range. To test this, a critical evaluation of in vivo data has been performed, and indeed excessively high RQ values have been calculated for experimental tumors in vivo.

14.2 Material and Methods

RQ values have been calculated from venous-arterial CO_2 concentration differences and arterio-venous O_2 concentration differences under steady state conditions. In vivo tumor models include experimental isotransplants in the rat [8] and xenografted human tumors in immunodeficient rnu/rnu-rats [9,10].

14.3 Results

Taking into account steady-state arterial and venous O_2 and CO_2 concentrations in tissue-isolated preparations, strikingly high RQ values of 1.91 have been calculated for experimental DS-sarcomas in SD-rats, and RQ values between 1.29 and 1.95 for xenografted human carcinomas in nude rats (see Table 14.1).

Table 14.1 Arterial (art.) and venous (ven.) O_2 and CO_2 concentrations (cO_2 and cCO_2, respectively), concentration differences (Δc), RQ values of experimental DS-sarcoma (DS-Sa) isotransplants in rats, and xenografted human breast and lung cancers in immunodeficient rnu/rnu-rats. Data are compared to normal (granulation/fat) tissue using a comparable tissue-isolated preparation. $RQ = \Delta cCO_2/\Delta cO_2$

Tumor	cO_2 (ml/dl)		ΔcO_2 (ml/dl)	cCO_2(ml/dl)		ΔcCO_2 (ml/dl)	RQ	Ref.
	art.	ven.		art.	ven.			
DS-Sa (rat)	16.40	10.85	5.55	42.8	53.4	10.6	1.91	[8]
Breast Ca. (human)	16.40	9.70	6.70	45.0	58.0	13.0	1.95	[9]
Breast Ca. (human)	16.60	9.27	7.31	47.0	57.7	10.7	1.46	[10]
Lung Ca. (human)	17.30	12.50	4.80	48.0	54.2	6.2	1.29	[9]
Granulation tissue	18.47	12.77	5.70	48.0	52.2	4.2	0.74	[10]

14.4 Discussion

In vitro studies described by Dickens and Simer [11] on tumor slices revealed RQ values distinctly below 1.0 (RQ = 0.82 – 0.91) which were associated with a high rate of glycolysis. These authors reasoned that in tumors, carbohydrate oxidation is limited by a defective mechanism for oxidation of glycolysis products, agreeing fundamentally with Warburg's idea of an impaired respiration in tumor cells [12,13]. However, there are several reasons why this hypothesis no longer seems tenable [6,7,14–16].

In contrast to Dickens and Simer [11], strikingly high RQ values are found in experimental tumors under in vivo conditions. These data may be explained by (at least) two mechanisms: (1) channeling of glycolytic endproducts into lipogenesis (i.e., conversion of carbohydrate to fat) which is known to yield RQ values above unity, and (2) hypoxia-induced upregulation of genes that can induce proteomic changes and that allow malignant cells to adapt to their O_2-deprived metabolic state (see Fig. 14.1).

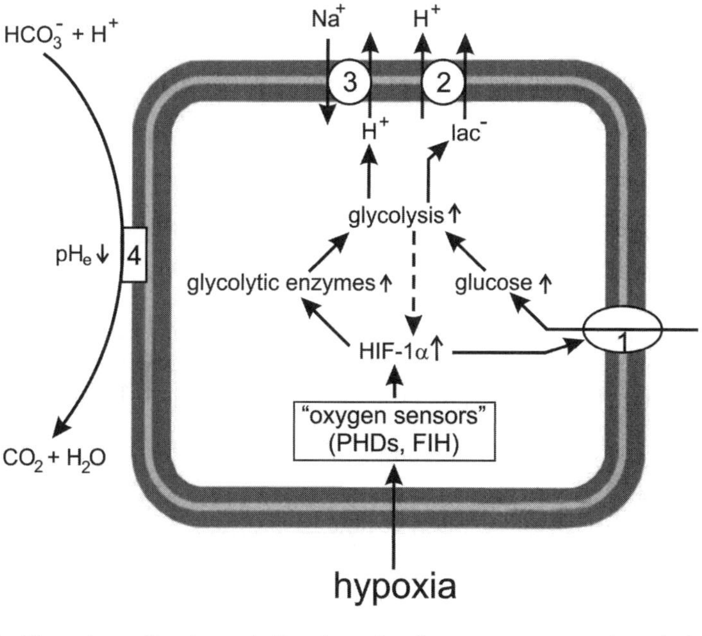

Fig. 14.1 Hypoxia-mediated metabolic adaptation for energy preservation. Activation of genes for glucose transporter-1 (GLUT-1 = 1) and glycolytic enzymes yields an increased glycolytic rate. H^+-ions produced are preferentially exported via a Na^+/H^+-antiporter (NHE-1 = 2) and a lactate$^-$/H^+-symporter (monocarboxylate transporter MCT-1 = 3) leading to a drop in extracellular pH (pH_e). Low extracellular pH activates the membrane-bound ectoenzyme carbonic anhydrase IX (CA IX = 4). HIF-1α = hypoxia-inducible factor 1α, PHDs = prolyl hydroxylases, FIH = asparagyl hydroxylase, lac$^-$ = lactic acid.

Over expression of GLUT-1 and of glycolytic enzymes in hypoxia facilitates cellular glucose uptake and enhances the capacity of tumor cells to catabolize glucose at even higher metabolic rates than those found under normoxic conditions. Overproduction of lactic acid is a mandatory consequence. To survive and proliferate, cells extrude H^+-ions via transporters to maintain a physiological intracellular pH, promoting extracellular acidification. The latter, together with a HIF-1α-induced upregulation of the membrane-bound ectoenzyme carbonic anhydrase (CA IX), finally leads to buffering of the exported protons by extracellular bicarbonate causing an intensified CO_2 release according to the following equation [17]

$$H^+ + HCO_3^- \rightarrow H_2O + CO_2 \qquad (14.1)$$

Total CO_2 output is thus greater than the metabolic formation of CO_2 from substrate oxidation. Conspicuously high RQs in solid tumors thus do not reflect the degree of substrate oxidation in tumor cells. Besides glycolysis, an intensified H^+-production results from substantial ATP hydrolysis, glutaminolysis and ketogenesis [18,19].

Experimental evidence for this latter notion are very high CO_2 partial pressures (79 mmHg) and low bicarbonate concentrations (19 mmol/l) found in the interstitial fluid of solid tumors compared to arterial blood (pCO_2 = 40 mmHg, bicarbonate concentration = 23–24 mmol/l) [20,21]

14.5 Conclusions

Strikingly high RQ values have been calculated for solid tumors in vivo. Most probably these are caused by (i) channeling of glycolytic end-products into lipogenesis, and (ii) buffering of exported H^+-ions from the intracellular space to the interstitial compartment. Acidosis-mediated activation of membrane-bound CA IX finally yields an intensified ("non-metabolic") CO_2 release from the tumor tissue. Conspicuously high RQs may thus be a further characteristic of the tumor pathophysiome.

In this context, the term "tumor pathophysiome" (in analogy to the physiome [22,23] and tumor metabolome [24]) is coined to quantitatively describe the (patho-)physiological dynamics and functional behavior of solid tumors in vivo, the quantitative evaluation and description of the tumor microenvironment included.

References

1. G. Thews and P. Vaupel, *Autonomic Functions in Human Physiology* (Springer, Berlin, Heidelberg, 1985).
2. W.F. Boron, Ventilation and perfusion of lungs, in: *Medical Physiology*, edited by W. F. Boron and E.L. Boulpaep (Saunders, Philadelphia, 2003).

3. M. Löffler and F. Schneider, Lipogenesis in Ehrlich ascites tumor cells under anaerobic culture conditions, *J. Cancer Res. Clin. Oncol.* **95,** 115–122 (1979).
4. F.P. Kuhajda, Fatty acid synthase and cancer: New application of an old pathway, *Cancer Res.* **66**(12), 5977–5980 (2006).
5. K. Uyeda and J.J. Repa, Carbohydrate response element binding protein, ChREBP, a transcription factor coupling hepatic glucose utilization and lipid synthesis, *Cell Metabol.* **4**(2), 107–110 (2006).
6. G. Hatzivassiliou, F. Zhao, D.E. Bauer, C. Andreadis, A.N. Shaw, D. Dhanak, S.R. Hingorani, D.A. Tuveson, and C.B. Thompson, ATP citrate lyase inhibition can suppress tumor cell growth, *Cancer Cell.* **8**(4), 311–321 (2005).
7. T. Bui and C.B. Thompson, Cancer's sweet tooth, *Cancer Cell.* **9**(6), 419–420 (2006).
8. P. Vaupel, Atemgaswechsel und Glucosestoffwechsel von Implantationstumoren (DS-Carcinosarkom) in vivo, *Funktionsanalyse biolog. Systeme* **1**, 1–138 (1974).
9. F. Kallinowski, K.H. Schlenger, S. Runkel, M. Kloes, M. Stohrer, P. Okunieff, and P.Vaupel, Tumor blood flow: The principal modulator of oxidative and glycolytic metabolism, and of the metabolic micromilieu of human tumor xenografts in vivo, *Int. J. Cancer* **44**, 266–272 (1989).
10. S. Runkel, Durchblutung, O_2-Verbrauch und Substrat-Umsatzraten xenotransplantierter menschlicher Mammakarzinome, Dr. med. Thesis, Faculty of Medicine, University of Mainz (1988).
11. F. Dickens and F. Simer, The metabolism of normal and tumour tissue. II. The respiratory quotient, and the relationship of respiration of glycolysis, *Biochem. J.* **24**, 1301–1326 (1930).
12. O. Warburg, *The Metabolism of Tumors* (Arnold Constable, London, 1930).
13. O. Warburg, On the origin of cancer cells, *Science* **123**, 309–315 (1956).
14. S. Weinhouse, On respiratory impairment in cancer, *Science* **124**, 267–268 (1956).
15. X.L. Zu and M. Guppy, Cancer metabolism: facts, fantasy, and fiction, *Biochem. Biophys. Res. Comm.* **313**, 459–465 (2004).
16. V.R. Fantin, J. St-Pierre, and P. Leder, Attenuation of LDH-A expression uncovers a link between glycolysis, mitochondrial physiology, and tumor maintenance, *Cancer Cell.* **9**(6), 425–434 (2006).
17. J. Piiper, Physiologie der Atmung, in: *Atmung, Physiologie des Menschen, Vol. 6*, edited by O.H. Gauer, K. Kramer, and R. Jung (Urban & Schwarzenberg, München, Berlin, Wien, 1975), pp. 1–159.
18. P. Vaupel, Tumor microenvironmental physiology and its implications for radiation oncology, *Semin. Radiat. Oncol.* **14**(3), 198–206 (2004).
19. P. Vaupel, Abnormal microvasculature and defective microcirculatory function in solid tumors, in: *Vascular-targeted Therapies in Oncology,* edited by D.W. Siemann (Wiley & Sons, Chichester, UK, 2006), pp. 9–29.
20. P.M. Gullino, Techniques for the study of tumor physiopathology, in: *Methods in Cancer Research,* edited by H. Busch (Academic Press, New York, 1970), pp. 45–91.
21. P.M. Gullino, Extracellular compartments of solid tumors, in: *Cancer Vol. 3,* edited by E.F. Becker (Plenum, New York, 1975), pp. 327–354.
22. P.J. Hunter and T.K. Borg, Integration from proteins to organs: the Physiome Project, *Nat. Rev. Mol. Cell Biol.* **4**(3), 237–243 (2003).
23. J.B. Bassingthwaighte, Strategies for the Physiome Project, *Ann. Biomed. Engin.* **28**(8), 1043–1058 (2000).
24. S. Mazurek and E. Eigenbrodt, The tumor metabolome, *Anticancer Res.* **23**, 1149–1154 (2003).

Chapter 15
Endogenous Hypoxia Markers: Case Not Proven!

Arnulf Mayer[1], **Michael Höckel**[2], **and Peter Vaupel**[3]

Abstract The pivotal role of hypoxia within the pathophysiological framework of solid malignant tumors is now considered to be indisputable. The fact that hypoxia can cause resistance to various cancer therapies and promote malignant progression is reflected in its adverse impact on prognosis which is repeatedly shown for various tumor entities. Knowledge in this area is based on direct assessment of the oxygenation status using O_2-sensitive microsensors. However, weaknesses of this standard method are its invasiveness and limitation to accessible tumor entities. Hypoxia-inducible factor (HIF)-1α, the master transcriptional regulator of the hypoxic response, as well as certain downstream genes, e.g., glucose transporter (GLUT)-1 and carbonic anhydrase (CA) IX, have been considered to be suitable as surrogate biomarkers of hypoxia due to their tight regulation by O_2 levels under certain, well-defined in vitro conditions. The fact that statistical correlations between the expression of these proteins and direct pO_2 measurements in the clinic have been sporadically reported seemed to support their role as "endogenous hypoxia markers". Remaining disparities were mainly attributed to the influence of tumor heterogeneity. In a series of studies, we have addressed this question by examining the expression of HIF-1α, GLUT-1 and CA IX in tissue microareas where direct O_2 measurements had previously been carried out, so that the influence of tumor heterogeneity could be reduced to a minimum. Using this methodology, no correlation between the expression of "endogenous hypoxia markers" and direct pO_2 measurements could be found. In conclusion, while there may be a stringent association between these markers and the oxygenation status under standardized in vitro conditions, this is not transferable to the clinical assessment of oxygenation status in patients. The term "endogenous hypoxia markers" should therefore be avoided, at least in the clinical oncology setting.

[1]Institute of Physiology and Pathophysiology, University of Mainz, 55099 Mainz, Germany.
[2]Department of Gynecology and Obstetrics, University of Leipzig, 04103 Leipzig, Germany.
[3]Institute of Physiology and Pathophysiology, University of Mainz, 55099 Mainz, Germany.

15.1 Introduction

The microenvironment of malignant tumors is characterized by extensive areas of local hypoxia leading to therapeutic resistance [1] and hypoxia-induced malignant progression [2,3]. The latter is multifactorial, being the common end result of hypoxia-driven gene expression, mutation and clonal selection. These processes lead to, e.g., enhanced invasive capacity [4,5], tendency to metastasize [6,7], radioresistance [8] and chemoresistance [9,10]. Since hypoxia is the consequence of an inadequate and malfunctional microvasculature, hypoxic regions may additionally indicate areas inaccessible to cytostatic agents and could therefore become "phenotypically" chemoresistant [11]. Interactions of hypoxia with other aspects of the microenvironment (e.g., acidic extracellular pH, elevated interstitial fluid pressure) significantly add to these phenomena. The clinical relevance of these mechanisms is reflected in a poor prognosis in patients with hypoxic tumors of various entities [2,12–19]. These data have been acquired using O_2-sensitive microsensors (Eppendorf pO_2 histography) in the clinical setting. However, important drawbacks of this method include its invasive nature and a limited applicability due to its restriction to tumors accessible to needle electrodes.

Hypoxia-induced gene expression is controlled primarily by a small set of heterodimeric hypoxia-inducible (transcription) factors (HIFs). HIF-1 was described first [20] and remains the most important protein in this group. In normoxia, HIF-1α is hydroxylated at certain conserved prolyl residues [21,22]. Conformational changes induced by this modification allow for recognition by the von Hippel-Lindau protein (pVHL), which is part of an E3 ubiquitin ligase complex [23]. Ubiquitin labeling results in proteasomal degradation and thus in low HIF-1α levels under normoxia. This degradation process is interrupted under hypoxic conditions, since the activity of the prolyl hydroxylases (PHDs) required for this process is oxygen-dependent.

Many tumor-relevant mechanisms are triggered by hypoxia through the action of HIF-1 target genes, of which approx. 70 are currently known [24]. In addition to its role in tumor pathophysiology, HIF-1α and its target genes GLUT-1 and CA IX have been proposed as "endogenous markers" of tumor hypoxia. Markers of this type would not only be able to circumvent problems associated with the microsensor technique, but would also permit oxygenation status assessment in archival paraffin material.

The association of expression levels of potential endogenous hypoxia markers HIF-1α, GLUT-1 and CA IX with the oxygenation status is well established in vitro. The aim of the present report was therefore to review the validity of these markers in the clinical setting.

15.2 Hypoxia-inducible Factor-1 (HIF-1)

HIF-1α protein levels in cultured cells are tightly regulated by hypoxia. Elevation of HIF-1α protein by hypoxic inhibition of PHD activity as well as degradation following reoxygenation are both rapid processes. In HeLa cells, detectable levels of HIF-1α are reached within 2 min of exposure to hypoxia and exhibit a half-life of approx. 8 min upon reoxygenation [25]. The O_2 concentrations at which accumulation starts are below 5%, with half-maximal induction at 1.5% and maximal at 0.5% in these cells [26]. The in vivo situation, however, is less clear. Immunohistochemical staining of HIF-1α is typically found to increase as a function of distance from microvessels within the tumor stroma, but a diffuse pattern has also been observed in some cases [27]. Reported correlations of immunohistochemical HIF-1α expression with data obtained from electrode measurements in patients were weak [28] or even very weak [29]. This lack of a stringent association between the two assays was attributed to tumor heterogeneity. To test this hypothesis, HIF-1α staining by immunohistochemistry was assessed in biopsy specimens of O_2 electrode tracks. Using this methodology, both analyses were performed in almost identical tissue microareas. No correlation between HIF-1α expression and the directly measured oxygenation status could be found [30].

Both methodological and biological factors may help to explain this finding. Immunohistochemical detection of HIF-1α is often performed with a biotinyl-tyramide-based, signal-amplifying detection system [28,30–32], which is more difficult to standardize than the conventional ABC-technique and may yield high background staining. Immediate fixation of specimens, a necessary step due to the rapid degradation kinetics of HIF-1α, is not always part of routine pathology procedures. Additionally, the choice of fixation media and fixation times has been reported to severely influence results [28].

The stabilization of HIF-1α can also be triggered by microenvironmental factors other than hypoxia. Extracellular acidosis, a further hallmark of the hostile tumor micromilieu, has been shown to result in inhibition of pVHL ubiquitin-ligase activity, leading to normoxic HIF-1α stabilization [33]. Pyruvate and lactate, end products of glycolytic metabolism, have also been shown to stabilize HIF-1α [34]. Both elevated [35] and lowered [36] tissue glucose concentrations can prevent hypoxic HIF-1α stabilization.

It has long been recognized that HIF-1α expression can occur secondary to the activation of signaling pathways by growth factors or by mutations within these cascades [37,38]. Subsequently, increased levels of HIF-1α protein "override" the proteasomal degradation process and activate the transcription of target genes. Since many growth factor genes are target genes of HIF-1α, autocrine feedback loops are likely to be established. Indeed, this has been shown to apply to insulin-like growth factor 2 [39]. Mutations of the VHL gene are known to result in normoxic stabilization of HIF-1α [23], leading to tumor formation in patients with VHL disease [40]. Activating mutations of HIF-1α have also been described [41], as well as gene amplification [42] and stabilization secondary to mutations of the tumor suppressor p53 [43]. Normoxic HIF-1α degradation is also prevented in cells carrying mutations in genes for the mitochondrial enzymes succinate dehydrogenase and fumarate hydratase [44].

This diversity of mechanisms for non-hypoxic HIF-1α induction and modulation of hypoxic HIF-1α induction is probably the reason for the findings of our study on cervix cancers [30] and for the weak correlations described by others [28,29]. The individual relevance of these mechanisms is very likely dependent not only on the tumor type but also on the genetic background of individual tumors and possibly even cell clones within tumors.

15.3 Target Genes of HIF-1α

The possibility that target genes of HIF-1α may be suitable as endogenous hypoxia markers whereas HIF-1α itself is not, may at first seem inconsequent. However, in theory this may be the case owing to the fact that HIF-1α transcriptional activity is a separate O_2-regulated step. The hydroxylation of a specific asparagine residue by factor inhibiting HIF (FIH) reduces the interaction of HIF-1α with its transcriptional co-activators (e.g., p300/CBP). FIH-1 is inhibited at a lower pO_2 than the prolyl hydroxylases [45].

15.3.1 Carbonic Anhydrase IX (CA IX)

The gene for CA IX has been shown to be the most strongly hypoxia-inducible of a panel of 24,504 unique transcripts on a gene array [38,46]. The induction of CA IX protein is delayed, with the highest levels being reached after 24 hours of hypoxia. Additionally, CA IX expression has been demonstrated to be stable for more than 96 hours after reoxygenation [38,47]. Therefore, while CA IX seems to be unsuitable as a marker of acute and "episodic" hypoxia, it may nevertheless prove to be useful in the assessment of chronic hypoxia. The immunohistochemical expression pattern of CA IX is similar to that described for HIF-1α.

Correlation of CA IX expression with oxygenation status has been reported for cancers of the uterine cervix [48] and for non–small cell lung cancers [49]. However, other investigations were not able to confirm this [50,51], including our own study in uterine cervix cancers [52]. As in our investigation of HIF-1α, both O_2 electrode measurements and immunohistochemistry for CA IX were performed in identical tissue microareas. The different types of hypoxia assessed by these methods (chronic vs. acute) may partially explain the differences seen between direct oxygenation measurements and CA IX expression. However, since we repeatedly found – analogous to our HIF-1α study –strongly hypoxic tumors completely devoid of CA IX staining and relatively well-oxygenated tumors exhibiting strong CA IX staining, the discrepancy between both methods may be of a more general nature.

Obviously, the deregulation of HIF-1α by non-hypoxic stimuli does play a role and the clear relevance of an additional O_2-regulated step as described above is not supported by our own findings. Since CA IX is involved in pH regulation, it is noteworthy that low extracellular pH has been shown to inhibit CA IX expression under hypoxia in SiHa cells [53]. In addition to this, other microenvironmental factors such as low bicarbonate [54] or glucose levels [47] have been shown to have a hypoxia-independent impact on CA IX expression. As with the HIF-1α data cited above, low glucose levels have been shown to impede CA IX expression [47]. CA IX expression has also previously been shown to be primarily dependent on cell density [55]

15.3.2 Glucose Transporter 1 (GLUT-1)

The immunohistochemical expression pattern of GLUT-1 is again similar to that of HIF-1α, with the strongest staining being found at the greatest distance from vessels within the tumor stroma and in the viable cell layers around necrotic areas. The timescale for GLUT-1 induction is comparable to that of CA IX, with maximum mRNA levels being reached only after 16 hours of hypoxia in one study [54]. The hypoxia-induced GLUT-1 fraction as such would therefore be expected to be predominantly a marker of chronic hypoxia. Although degradation of GLUT-1 upon reoxygenation is much more rapid than is the case with CA IX, it is nevertheless incomplete [54]. Hence, the expression of GLUT-1 tends to be more extensive than that of CA IX [56].

A correlation between electrode measurements and GLUT-1 expression in advanced carcinomas of the uterine cervix has been described [57]. However, this correlation was only weak ($r \approx 0.28$). Contrary to these data, in a recent analysis of GLUT-1 expression and the oxygenation status in identical microareas, no correlation between results obtained with these two methods could be found [58,59].

As was the case for HIF-1α and CA IX, the impact of hypoxia-independent factors on GLUT-1 expression may be the decisive factor for the absence of a direct correlation. Besides hypoxia, induction of GLUT-1 expression has been reported to occur as a result of glucose depletion [60,61], and this in turn has been linked to the induction of the unfolded protein response, [62] although the molecular mechanisms remain unclear. Induction of GLUT-1 expression through glucose depletion seems to be of pivotal importance in cervical cancer since glucose levels have been shown to vary considerably in this tumor entity [63, 64]. Other examples of non-hypoxic GLUT-1 induction are (c-MYC) oncogene activation [65], osmotic stress [66,67] and the hormone triiodothyronine [68].

15.4 Conclusions

A role for HIF-1α, GLUT-1 or CA IX as endogenous markers of tumor hypoxia is currently not supported by the available clinical data. The same is true for a variety of other markers which could not be discussed within the scope of this review (e.g., VEGF [69], osteopontin [59,70], thymidine phosphorylase [71]). It currently appears likely that direct O_2 measurements and "endogenous markers" (HIF-1α and its respective target genes GLUT-1 and CA IX) assess different aspects of the tumor microenvironment.

References

1. P. Vaupel, O. Thews, M. Höckel, Treatment resistance of solid tumors: role of hypoxia and anemia, *Med. Oncol.* **18**(4), 243–259 (2001).
2. M. Höckel, K. Schlenger, B. Aral, M. Mitze, U. Schäffer, P. Vaupel, Association between tumor hypoxia and malignant progression in advanced cancer of the uterine cervix, *Cancer Res.* **56**(19), 4509–4515 (1996).
3. M. Höckel, P. Vaupel, Tumor hypoxia: definitions and current clinical, biologic, and molecular aspects, *J. Natl. Cancer Inst.* **93**(4), 266–276 (2001).
4. B. Krishnamachary, S. Berg-Dixon, B. Kelly, F. Agani, D. Feldser, G. Ferreira, N. Iyer, J. LaRusch, B. Pak, P. Taghavi, G. L. Semenza, Regulation of colon carcinoma cell invasion by hypoxia-inducible factor 1, *Cancer Res.* **63**(5), 1138–1143 (2003).
5. S. Pennacchietti, P. Michieli, M. Galluzzo, M. Mazzone, S. Giordano, P. M. Comoglio, Hypoxia promotes invasive growth by transcriptional activation of the met protooncogene, *Cancer Cell* **3**(4), 347–361 (2003).
6. R. A. Cairns, T. Kalliomaki, R. P. Hill, Acute (cyclic) hypoxia enhances spontaneous metastasis of KHT murine tumors, *Cancer Res.* **61**(24), 8903–8908 (2001).
7. E. K. Rofstad, H. Rasmussen, K. Galappathi, B. Mathiesen, K. Nilsen, B. A. Graff, Hypoxia promotes lymph node metastasis in human melanoma xenografts by up-regulating the urokinase-type plasminogen activator receptor, *Cancer Res.* **62**(6), 1847–1853 (2002).
8. L. H. Gray, A. D. Conger, M. Ebert, S. Hornsey, O. C. A. Scott, The concentration of oxygen dissolved in tissues at the time of irradiation as a factor in radiotherapy, *Br. J. Radiol.* **26**, 638–648 (1953).
9. A. Kondo, R. Safaei, M. Mishima, H. Niedner, X. Lin, S. B. Howell, Hypoxia-induced enrichment and mutagenesis of cells that have lost DNA mismatch repair, *Cancer Res.* **61**(20), 7603–7607 (2001).
10. K. M. Comerford, T. J. Wallace, J. Karhausen, N. A. Louis, M. C. Montalto, S. P. Colgan, Hypoxia-inducible factor-1-dependent regulation of the multidrug resistance (MDR1) gene, *Cancer Res.* **62**(12), 3387–3394 (2002).
11. A. I. Minchinton, I. F. Tannock, Drug penetration in solid tumours, *Nat. Rev. Cancer* **6**(8), 583–592 (2006).
12. M. Höckel, C. Knoop, K. Schlenger, B. Vorndran, E. Baussmann, M. Mitze, P. G. Knapstein, P. Vaupel, Intratumoral pO_2 predicts survival in advanced cancer of the uterine cervix, *Radiother. Oncol.* **26**(1), 45–50 (1993).
13. D. M. Brizel, G. S. Sibley, L. R. Prosnitz, R. L. Scher, M. W. Dewhirst, Tumor hypoxia adversely affects the prognosis of carcinoma of the head and neck, *Int. J. Radiat. Oncol. Biol. Phys.* **38**(2), 285–289 (1997).
14. A. W. Fyles, M. Milosevic, R. Wong, M. C. Kavanagh, M. Pintilie, A. Sun, W. Chapman, W. Levin, L. Manchul, T. J. Keane, R. P. Hill, Oxygenation predicts radiation response and survival in patients with cervix cancer, *Radiother. Oncol.* **48**(2), 149–156 (1998).
15. K. Sundfor, H. Lyng, E. K. Rofstad, Tumour hypoxia and vascular density as predictors of metastasis in squamous cell carcinoma of the uterine cervix, *Br. J. Cancer* **78**(6), 822–827 (1998).

16. M. Höckel, K. Schlenger, S. Höckel, P. Vaupel, Hypoxic cervical cancers with low apoptotic index are highly aggressive, *Cancer Res.* **59**(18), 4525–4528 (1999).
17. M. Nordsmark, J. Alsner, J. Keller, O. S. Nielsen, O. M. Jensen, M. R. Horsman, J. Overgaard, Hypoxia in human soft tissue sarcomas: adverse impact on survival and no association with p53 mutations, *Br. J. Cancer* **84**(8), 1070–1075 (2001).
18. M. Nordsmark, J. Overgaard, Tumor hypoxia is independent of hemoglobin and prognostic for loco-regional tumor control after primary radiotherapy in advanced head and neck cancer, *Acta Oncol.* **43**(4), 396–403 (2004).
19. M. Nordsmark, S. M. Bentzen, V. Rudat, D. Brizel, E. Lartigau, P. Stadler, A. Becker, M. Adam, M. Molls, J. Dunst, D. J. Terris, J. Overgaard, Prognostic value of tumor oxygenation in 397 head and neck tumors after primary radiation therapy. An international multi-center study, *Radiother. Oncol.* 77(1), 18–24 (2005).
20. G. L. Wang, G. L. Semenza, Purification and characterization of hypoxia-inducible factor 1, *J. Biol. Chem.* **270**(3), 1230–1237 (1995).
21. M. Ivan, K. Kondo, H. Yang, W. Kim, J. Valiando, M. Ohh, A. Salic, J. M. Asara, W. S. Lane, W. G. Kaelin, Jr., HIFα targeted for VHL-mediated destruction by proline hydroxylation: implications for O_2 sensing, *Science* **292**(5516), 464–468 (2001).
22. P. Jaakkola, D. R. Mole, Y. M. Tian, M. I. Wilson, J. Gielbert, S. J. Gaskell, A. Kriegsheim, H. F. Hebestreit, M. Mukherji, C. J. Schofield, P. H. Maxwell, C. W. Pugh, P. J. Ratcliffe, Targeting of HIF-α to the von Hippel-Lindau ubiquitylation complex by O_2-regulated prolyl hydroxylation, *Science* **292**(5516), 468–472 (2001).
23. P. H. Maxwell, M. S. Wiesener, G. W. Chang, S. C. Clifford, E. C. Vaux, M. E. Cockman, C. C. Wykoff, C. W. Pugh, E. R. Maher, P. J. Ratcliffe, The tumour suppressor protein VHL targets hypoxia-inducible factors for oxygen-dependent proteolysis, *Nature* **399**(6733), 271–275 (1999).
24. G. L. Semenza, Targeting HIF-1 for cancer therapy, *Nat. Rev. Cancer* **3**(10), 721–732 (2003).
25. U. R. Jewell, I. Kvietikova, A. Scheid, C. Bauer, R. H. Wenger, M. Gassmann, Induction of HIF-1α in response to hypoxia is instantaneous, *FASEB J.* **15**(7), 1312–1314 (2001).
26. B. H. Jiang, G. L. Semenza, C. Bauer, H. H. Marti, Hypoxia-inducible factor 1 levels vary exponentially over a physiologically relevant range of O_2 tension, *Am. J. Physiol.* **271**(4 Pt 1), C1172–1180 (1996).
27. G. Gruber, R. H. Greiner, R. Hlushchuk, D. M. Aebersold, H. J. Altermatt, G. Berclaz, V. Djonov, Hypoxia-inducible factor 1 alpha in high-risk breast cancer: an independent prognostic parameter?, *Breast Cancer Res.* **6**(3), R191–198 (2004).
28. H. K. Haugland, V. Vukovic, M. Pintilie, A. W. Fyles, M. Milosevic, R. P. Hill, D. W. Hedley, Expression of hypoxia-inducible factor-1α in cervical carcinomas: correlation with tumor oxygenation, *Int. J. Radiat. Oncol. Biol. Phys.* **53**(4), 854–861 (2002).
29. G. J. Hutchison, H. R. Valentine, J. A. Loncaster, S. E. Davidson, R. D. Hunter, S. A. Roberts, A. L. Harris, I. J. Stratford, P. M. Price, C. M. West, Hypoxia-inducible factor 1α expression as an intrinsic marker of hypoxia: correlation with tumor oxygen, pimonidazole measurements, and outcome in locally advanced carcinoma of the cervix, *Clin. Cancer Res.* **10**(24), 8405–8412 (2004).
30. A. Mayer, A. Wree, M. Höckel, C. Leo, H. Pilch, P. Vaupel, Lack of correlation between expression of HIF-1α protein and oxygenation status in identical tissue areas of squamous cell carcinomas of the uterine cervix, *Cancer Res.* **64**(16), 5876–5881 (2004).
31. H. Zhong, A. M. De Marzo, E. Laughner, M. Lim, D. A. Hilton, D. Zagzag, P. Buechler, W. B. Isaacs, G. L. Semenza, J. W. Simons, Overexpression of hypoxia-inducible factor 1α in common human cancers and their metastases, *Cancer Res.* **59**(22), 5830–5835 (1999).
32. D. M. Aebersold, P. Burri, K. T. Beer, J. Laissue, V. Djonov, R. H. Greiner, G. L. Semenza, Expression of hypoxia-inducible factor-1α: a novel predictive and prognostic parameter in the radiotherapy of oropharyngeal cancer, *Cancer Res.* **61**(7), 2911–2916 (2001).

33. K. Mekhail, L. Gunaratnam, M. E. Bonicalzi, S. Lee, HIF activation by pH-dependent nucleolar sequestration of VHL, *Nat. Cell Biol.* **6**(7), 642–647 (2004).
34. H. Lu, R. A. Forbes, A. Verma, Hypoxia-inducible factor 1 activation by aerobic glycolysis implicates the Warburg effect in carcinogenesis, *J. Biol. Chem.* **277**(26), 23111–23115 (2002).
35. S. B. Catrina, K. Okamoto, T. Pereira, K. Brismar, L. Poellinger, Hyperglycemia regulates hypoxia-inducible factor-1α protein stability and function, *Diabetes* **53**(12), 3226–3232 (2004).
36. S. J. Kwon, Y. J. Lee, Effect of low glutamine/glucose on hypoxia-induced elevation of hypoxia-inducible factor-1α in human pancreatic cancer MiaPaCa-2 and human prostatic cancer DU-145 cells, *Clin. Cancer Res.* **11**(13), 4694–4700 (2005).
37. G. Höpfl, O. Ogunshola, M. Gassmann, HIFs and tumors – causes and consequences, *Am. J. Physiol. Regul. Integr. Comp. Physiol.* **286**(4), R608–623 (2004).
38. A. Mayer, M. Höckel, P. Vaupel, Endogenous hypoxia markers in locally advanced cancers of the uterine cervix: reality or wishful thinking? *Strahlenther. Onkol.* **182**(9), 501–510 (2006).
39. D. Feldser, F. Agani, N. V. Iyer, B. Pak, G. Ferreira, G. L. Semenza, Reciprocal positive regulation of hypoxia-inducible factor 1α and insulin-like growth factor 2, *Cancer Res.* **59**(16), 3915–3918 (1999).
40. S. J. Mandriota, K. J. Turner, D. R. Davies, P. G. Murray, N. V. Morgan, H. M. Sowter, C. C. Wykoff, E. R. Maher, A. L. Harris, P. J. Ratcliffe, P. H. Maxwell, HIF activation identifies early lesions in VHL kidneys: evidence for site-specific tumor suppressor function in the nephron, *Cancer Cell.* **1**(5), 459–468 (2002).
41. X. S. Fu, E. Choi, G. J. Bubley, S. P. Balk, Identification of hypoxia-inducible factor-1α (HIF-1α) polymorphism as a mutation in prostate cancer that prevents normoxia-induced degradation, *Prostate* **63**(3), 215–221 (2005).
42. O. R. Saramaki, K. J. Savinainen, N. N. Nupponen, O. Bratt, T. Visakorpi, Amplification of hypoxia-inducible factor 1α gene in prostate cancer, *Cancer Genet. Cytogenet.* **128**(1), 31–34 (2001).
43. R. Ravi, B. Mookerjee, Z. M. Bhujwalla, C. H. Sutter, D. Artemov, Q. Zeng, L. E. Dillehay, A. Madan, G. L. Semenza, A. Bedi, Regulation of tumor angiogenesis by p53-induced degradation of hypoxia-inducible factor 1α, *Genes Dev.* **14**(1), 34–44 (2000).
44. A. King, M. A. Selak, E. Gottlieb, Succinate dehydrogenase and fumarate hydratase: linking mitochondrial dysfunction and cancer, *Oncogene* **25**(34), 4675–4682 (2006).
45. P. Koivunen, M. Hirsila, V. Gunzler, K. I. Kivirikko, J. Myllyharju, Catalytic properties of the asparaginyl hydroxylase (FIH) in the oxygen sensing pathway are distinct from those of its prolyl 4-hydroxylases, *J. Biol. Chem.* **279**(11), 9899–9904 (2004).
46. A. Lal, H. Peters, B. St Croix, Z. A. Haroon, M. W. Dewhirst, R. L. Strausberg, J. H. Kaanders, A. J. van der Kogel, G. J. Riggins, Transcriptional response to hypoxia in human tumors, *J. Natl. Cancer Inst.* **93**(17), 1337–1343 (2001).
47. D. Vordermark, A. Kaffer, S. Riedl, A. Katzer, M. Flentje, Characterization of carbonic anhydrase IX (CA IX) as an endogenous marker of chronic hypoxia in live human tumor cells, *Int. J. Radiat. Oncol. Biol. Phys.* **61**(4), 1197–1207 (2005).
48. J. A. Loncaster, A. L. Harris, S. E. Davidson, J. P. Logue, R. D. Hunter, C. C. Wycoff, J. Pastorek, P. J. Ratcliffe, I. J. Stratford, C. M. West, Carbonic anhydrase (CA IX) expression, a potential new intrinsic marker of hypoxia: correlations with tumor oxygen measurements and prognosis in locally advanced carcinoma of the cervix, *Cancer Res.* **61**(17), 6394–6399 (2001).
49. Q. T. Le, E. Chen, A. Salim, H. Cao, C. S. Kong, R. Whyte, J. Donington, W. Cannon, H. Wakelee, R. Tibshirani, J. D. Mitchell, D. Richardson, K. J. O'Byrne, A. C. Koong, A. J. Giaccia, An evaluation of tumor oxygenation and gene expression in patients with early stage non-small cell lung cancers, *Clin. Cancer Res.* **12**(5), 1507–1514 (2006).

50. D. Hedley, M. Pintilie, J. Woo, A. Morrison, D. Birle, A. Fyles, M. Milosevic, R. Hill, Carbonic anhydrase IX expression, hypoxia, and prognosis in patients with uterine cervical carcinomas, *Clin. Cancer Res.* **9**(15), 5666–5674 (2003).
51. B. Jankovic, C. Aquino-Parsons, J. A. Raleigh, E. J. Stanbridge, R. E. Durand, J. P. Banath, S. H. Macphail, P. L. Olive, Comparison between pimonidazole binding, oxygen electrode measurements, and expression of endogenous hypoxia markers in cancer of the uterine cervix, *Cytometry B Clin. Cytom.* **70**(2), 45–55 (2006).
52. A. Mayer, M. Höckel, P. Vaupel, Carbonic anhydrase IX expression and tumor oxygenation status do not correlate at the microregional level in locally advanced cancers of the uterine cervix, *Clin. Cancer Res.* **11**(20), 7220–7225 (2005).
53. B. S. Sorensen, J. Hao, J. Overgaard, H. Vorum, B. Honore, J. Alsner, M. R. Horsman, Influence of oxygen concentration and pH on expression of hypoxia induced genes, *Radiother. Oncol.* **76**(2), 187–193 (2005).
54. M. Rafajova, M. Zatovicova, R. Kettmann, J. Pastorek, S. Pastorekova, Induction by hypoxia combined with low glucose or low bicarbonate and high posttranslational stability upon reoxygenation contribute to carbonic anhydrase IX expression in cancer cells, *Int. J. Oncol.* **24**(4), 995–1004 (2004).
55. J. Pastorek, S. Pastorekova, I. Callebaut, J. P. Mornon, V. Zelnik, R. Opavsky, M. Zat'ovicova, S. Liao, D. Portetelle, E. J. Stanbridge, Cloning and characterization of MN, a human tumor-associated protein with a domain homologous to carbonic anhydrase and a putative helix-loop-helix DNA binding segment, *Oncogene* **9**(10), 2877–2888 (1994).
56. P. J. Hoskin, A. Sibtain, F. M. Daley, G. D. Wilson, GLUT1 and CAIX as intrinsic markers of hypoxia in bladder cancer: relationship with vascularity and proliferation as predictors of outcome of ARCON, *Br. J. Cancer* **89**(7), 1290–1297 (2003).
57. R. Airley, J. Loncaster, S. Davidson, M. Bromley, S. Roberts, A. Patterson, R. Hunter, I. Stratford, C. West, Glucose transporter Glut-1 expression correlates with tumor hypoxia and predicts metastasis-free survival in advanced carcinoma of the cervix, *Clin. Cancer Res.* **7**(4), 928–934 (2001).
58. A. Mayer, M. Höckel, A. Wree, P. Vaupel, Microregional expression of glucose transporter-1 and oxygenation status: lack of correlation in locally advanced cervical cancers, *Clin. Cancer Res.* **11**(7), 2768–2773 (2005).
59. K. I. Sakata, M. Someya, H. Nagakura, K. Nakata, A. Oouchi, M. Hareyama, M. Satoh, A clinical study of hypoxia using endogenous hypoxic markers and polarographic oxygen electrodes, *Strahlenther. Onkol.* **182**(9), 511–517 (2006).
60. I. Stein, M. Neeman, D. Shweiki, A. Itin, E. Keshet, Stabilization of vascular endothelial growth factor mRNA by hypoxia and hypoglycemia and coregulation with other ischemia-induced genes, *Mol. Cell Biol.* **15**(10), 5363–5368 (1995).
61. R. J. Boado, W. M. Pardridge, Glucose deprivation and hypoxia increase the expression of the GLUT1 glucose transporter via a specific mRNA cis-acting regulatory element, *J. Neurochem.* **80**(3), 552–554 (2002).
62. E. Wertheimer, S. Sasson, E. Cerasi, Y. Ben-Neriah, The ubiquitous glucose transporter GLUT-1 belongs to the glucose-regulated protein family of stress-inducible proteins, *Proc. Natl. Acad. Sci. U. S. A.* **88**(6), 2525–2529 (1991).
63. S. Walenta, M. Wetterling, M. Lehrke, G. Schwickert, K. Sundfor, E. K. Rofstad, W. Mueller-Klieser, High lactate levels predict likelihood of metastases, tumor recurrence, and restricted patient survival in human cervical cancers, *Cancer Res.* **60**(4), 916–921 (2000).
64. P. Vaupel, F. Kallinowski, P. Okunieff, Blood flow, oxygen and nutrient supply, and metabolic microenvironment of human tumors: a review, *Cancer Res.* **49**(23), 6449–6465 (1989).
65. R. C. Osthus, H. Shim, S. Kim, Q. Li, R. Reddy, M. Mukherjee, Y. Xu, D. Wonsey, L. A. Lee, C. V. Dang, Deregulation of glucose transporter 1 and glycolytic gene expression by c-Myc, *J. Biol. Chem.* **275**(29), 21797–21800 (2000).

66. K. Barnes, J. C. Ingram, O. H. Porras, L. F. Barros, E. R. Hudson, L. G. Fryer, F. Foufelle, D. Carling, D. G. Hardie, S. A. Baldwin, Activation of GLUT1 by metabolic and osmotic stress: potential involvement of AMP-activated protein kinase (AMPK), *J. Cell. Sci.* **115**(Pt 11), 2433–2442 (2002).
67. D. Y. Hwang, F. Ismail-Beigi, Stimulation of GLUT-1 glucose transporter expression in response to hyperosmolarity, *Am. J. Physiol. Cell Physiol.* **281**(4), C1365–1372 (2001).
68. L. C. Moeller, A. M. Dumitrescu, S. Refetoff, Cytosolic action of thyroid hormone leads to induction of HIF-1α and glycolytic genes, *Mol. Endocrinol.* **90**(2), 936–943 (2005).
69. C. M. West, R. A. Cooper, J. A. Loncaster, D. P. Wilks, M. Bromley, Tumor vascularity: a histological measure of angiogenesis and hypoxia, *Cancer Res.* **61**(7), 2907–2910 (2001).
70. S. Lukacova, J. Overgaard, J. Alsner, M. R. Horsman, Strain and tumour specific variations in the effect of hypoxia on osteopontin levels in experimental models, *Radiother. Oncol.* **80**(2), 165–171 (2006).
71. P. Kabuubi, J. A. Loncaster, S. E. Davidson, R. D. Hunter, C. Kobylecki, I. J. Stratford, C. M. West, No relationship between thymidine phosphorylase (TP, PD-ECGF) expression and hypoxia in carcinoma of the cervix, *Br. J. Cancer* **94**(1), 115–120 (2006).

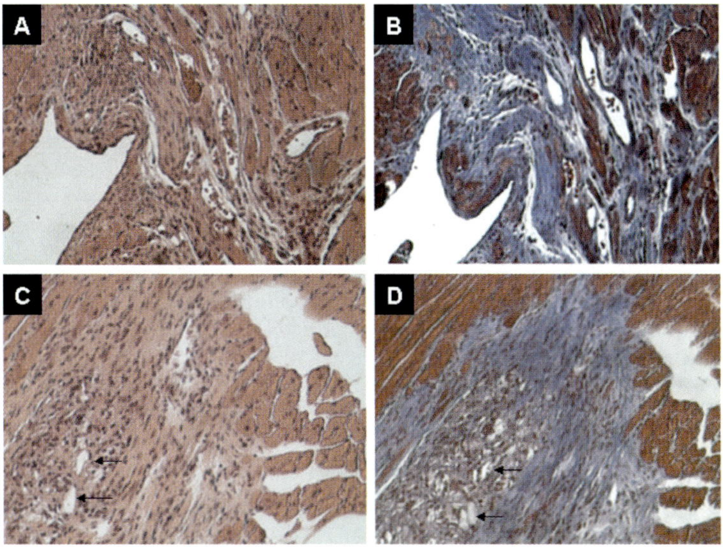

Fig. 6.1 Histology showing needle-track injury in mouse hearts at 3 weeks after implantation (200x). (**A**) Hematoxylin and Eosin (H & E) staining, (**B**) Masson Trichrome (MT) staining showing collagen deposition and intense fibrosis (blue), (**C**) H & E staining after OxySpin implantation, and (**D**) MT staining after OxySpin implantation showing fibrosis. The arrows in (**C**) and (**D**) show the OxySpin probes in the tissue.

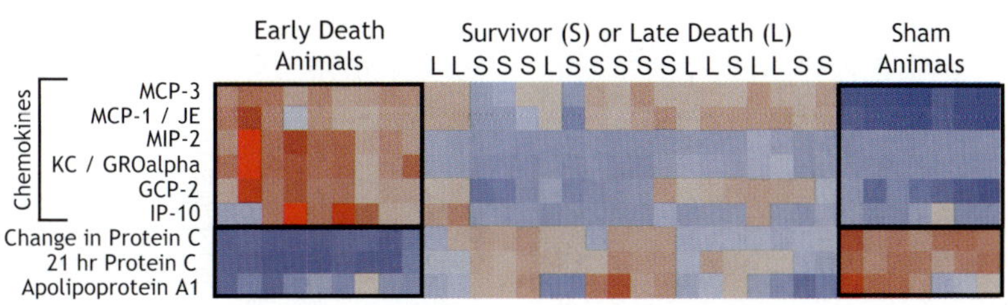

Fig. 10.1 Clustering of plasma markers as a function of outcome in the CLP model. The levels of various plasma markers from the Rodent Multi-Analyte Profile were clustered in JMP 5.1 using the Ward method. Dark blue to dark red change indicates low to high relative levels, respectively, for each marker.

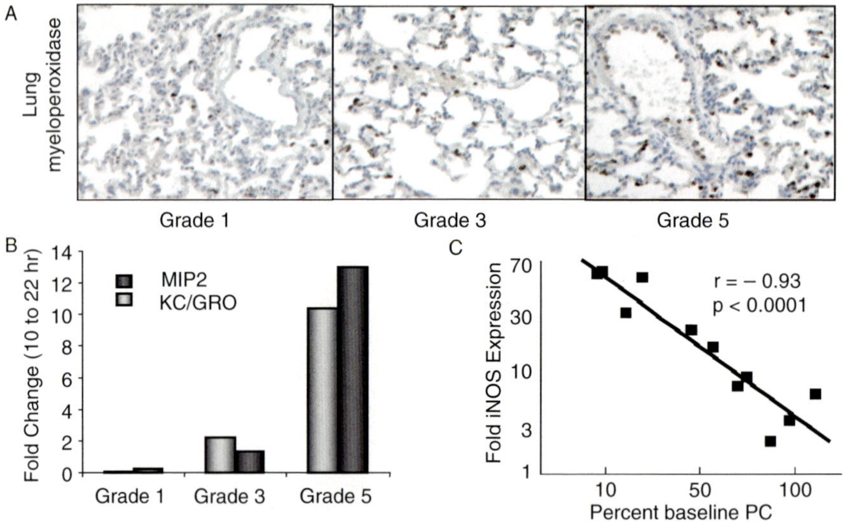

Fig. 10.2 Tissue injury, inflammatory markers and protein C suppression in the rat CLP. (A) Example of pathology in the lung by severity grade. (B) Analysis of change in protein C plasma level as a function of MIP2 and KC/Gro. Arteriole blood draws were performed at approximately 10 and 22h and plasma was analyzed for various analytes as described previously.[5] (C) Relationship of the change in PC as a function of lung iNOS expression, analyzed by quantitative real-time PCR (TaqMan®) n = 11 animals.

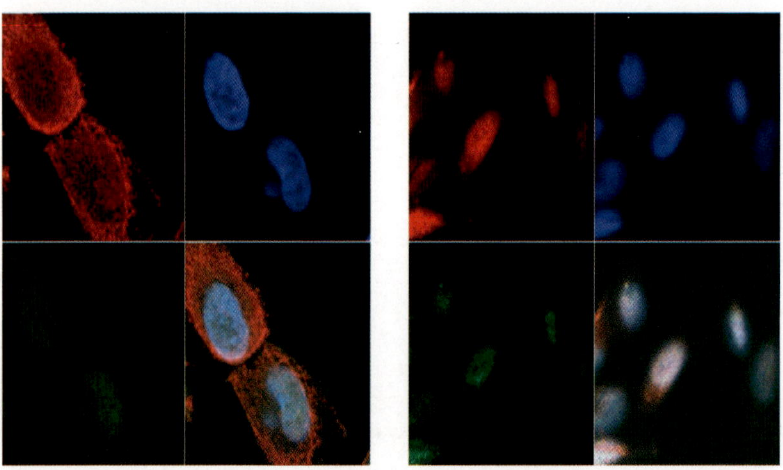

Fig. 16.2 Cells were irradiated in G_1 (left panel) or S (right panel) and stained 2 h later. The cells were stained for PCNA (red), DNA (blue), RAD18 (green), and examined by confocal microscopy. The lower right quadrant of each panel is the merged image.

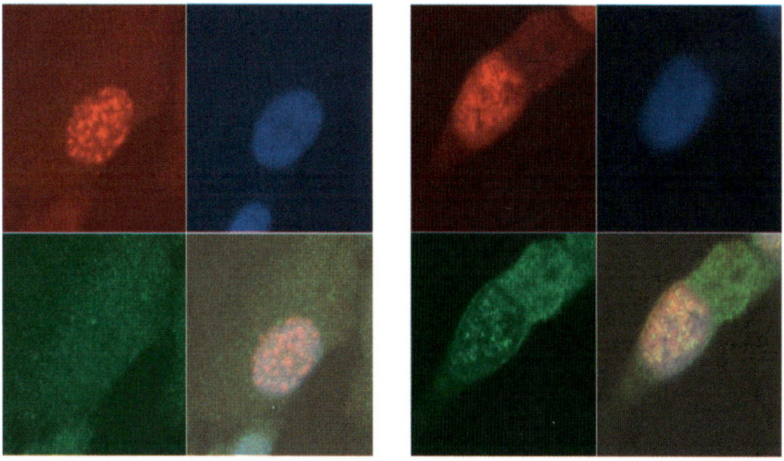

Fig. 16.3 Left panel, cells irradiated in G_1. Pcna (red), DNA (blue), pol iota (green), lower right quadrant, merged images. Right panel, cells irradiated in S-phase.

Chapter 16
RAD18 Signals DNA Polymerase IOTA to Stalled Replication Forks in Cells Entering S-phase with DNA Damage

Shelly Kakar, Nicholas B. Watson, and W. Glenn McGregor[1]

Abstract Endogenously generated reactive oxygen species and genotoxic carcinogens can covalently modify bases in cellular DNA. If not recognized and removed prior to S-phase of the cell cycle, such modifications can block DNA replication fork progression. If blocked forks are not are not resolved, they result in double strand breaks and cell death. Recent data indicate that the process of translesion DNA synthesis (TLS) is a highly conserved mechanism for bypassing lesions in template DNA. Although not fully understood, in yeast a ubiquitin ligase (RAD18) signals error-prone Y family polymerases to the blocked fork to bypass the damage with potentially mutagenic consequences. Homologs of the yeast proteins are found in higher eukaryotic cells, including human. We are examining the hypothesis that RAD18 acts as a proximal signal to Y-family polymerases to bypass damage, in a manner analogous to yeast but with additional layers of complexity. Here we report that RAD18 accumulates in nuclear foci after UV irradiation only in cells entering S-phase with DNA damage. These foci co-localize with proliferating cell nuclear antigen (PCNA). In addition, a newly described DNA polymerase, pol iota, also forms nuclear foci in a damage- and S-phase dependent manner. These data support our overall hypothesis that RAD18 accumulates at blocked forks and initiates the signal to recruit TLS polymerases.

16.1 Introduction

Cellular DNA is continuously under assault by the byproducts of endogenous metabolic processes, particularly reactive oxygen species (ROS), and from exogenous genotoxic agents. Principal among the latter is ultraviolet radiation, which induces DNA damage by the production of ROS as well as photoaddition products between adjacent pyrimidines. These damages distort the DNA

[1]Shelly Kakar, Nicholas B. Watson, and W. Glenn McGregor, Department of Pharmacology and Toxicology and Department of Medicine, James Graham Brown Cancer Center, University of Louisville, Louisville, KY 40202. SK and NBW contributed equally to this work.

helical structure to varying degrees, and prokaryotic and eukaryotic organisms have developed remarkably conserved mechanisms to recognize and remove such lesions. In general, oxidized bases are removed by the base excision repair system and larger bulky adducts are removed by the nucleotide excision repair system with some redundancy between the two.

It was recognized several years ago that most mutations are induced in DNA when the repair systems fail to remove DNA damage before the cells enter S-phase [1]. It was known that the damaged bases block the highly processive and accurate replicative DNA polymerase complexes, but the mechanisms that result in incorrect copying of the template were largely unknown. Within the last several years, dramatic advances have afforded significant new insights into the molecular mechanisms of mutagenesis. Replication forks that are blocked by bulky adducts in the DNA are resolved by error-free recombination and error-prone translesion synthesis (TLS). The latter process is responsible for the majority of base substitutions induced in the DNA. TLS is defined as the incorporation of a nucleotide across from DNA damage followed by extension of the potentially mispaired primer-template. Data indicate that this process is undertaken by accessory DNA polymerases. Several such polymerases have been discovered in higher eukaryotic cells, and several have been purified. Based on structural homology, these polymerases fit into one of three families: the Y-family (REV1, pol η, ι, and κ), the B-family (pol ζ) or the X-family (pol μ, λ, σ, and Trf5). The cellular roles of this universe of polymerases are not known [2]. In particular, the extent to which each of these polymerases participates in translesion replication most likely depends on the structure of a particular adduct and on the sequence context. It has been suggested that pol η,ι, and/or κ inserts a base directly across from a lesion, and that pol ζ extends the mispair to form a template-primer that can be extended by pol δ [3].

The unrestrained activity of error-prone polymerases would lead to widespread mutagenesis and genomic instability, so there are signaling mechanisms that tightly control polymerase switching events. We are studying the cell biology of a key regulator of this process, a ubiquitin ligase encoded by the RAD18 gene [4]. The present studies were undertaken to examine the hypothesis that RAD18 is required to signal the error-prone DNA polymerase iota to sites of DNA replication forks stalled by damage in the DNA template. The ultimate goal of these studies is to examine the suitability of this gene product as a molecular target to prevent carcinogen-induced mutagenesis, and to translate these data into a practical way to reduce the frequency of mutations and incidence of carcinogen-induced cancers, such as those of the skin and lung.

16.2 Instruments, Materials, and Methods

16.2.1 Cells and Cell Culture

The primary fibroblast cell strain GM1604 (Coriell Institute) was originally derived from human fetal lung tissue, and was telomerase immortalized. The immortalized cells (NF1604) were provided by Dr. Lisa McDaniels (University of Texas Southwestern Medical Center, Dallas) under the terms of MTA 3025 between WGM and Geron Corporation. Cells were kept in exponential growth in DMEM medium supplemented with 10% supplemented calf serum (Hyclone), 2 mM glutamine, penicillin (100 U/ml) and streptomycin (100 μg/ml).

16.2.2 Cell Synchronization

Cells were synchronized with 500 μM mimosine for 24 h. Under these conditions, inhibition of ribonucleotide reductase causes the cells to arrest in late G_1 by depleting the pool of available deoxynucleotides. For studies of cells in S-phase, the medium was replaced with fresh medium without mimosine, and the cells were irradiated 6 h later. The percentage of cells in each phase of the cell cycle was determined by propidium iodide staining and determination of DNA content using a FACScan flow cytometer and the ModFit program.

16.2.3 UV Irradiation

Cells were plated at a density of 2.5×10^4 cells/cm^2 per well of a 24 well plate 24 h prior to treatment. The wells also contained lysine coated coverslips (Becton-Dickinson). Cells were irradiated in G_1-phase, or in S-phase, 6 h after removal of the mimosine block. The UV source was a Spectroline germicidal lamp, and the flux was measured at 254 nm using a research radiometer fitted with a SED240 photodetector and a W diffuser (International Light, Newburyport, MA, USA). Cells were washed two times with sterile phosphate buffered saline pH 7.4 (PBS), irradiated, then fresh complete cell culture medium was added. The cells on coverslips were fixed in methanol 2 h after irradiation.

16.2.4 Localization of Proteins

Following fixation, the cells were washed with 1x PBS three times for five min. each. After the final wash, the cells were permeabilized with 0.2% PBST for 10 min. Cells were then blocked for 30 min. on parafilm squares in 0.2% PBST containing 10% goat serum (GPBST) in a humid chamber at room temperature. While the cells incubated, the primary antibody was conjugated with the corresponding Zenon fluorescent secondary (Molecular Probes). 2 μg of mouse anti-PCNA (PC-10, Dako) was incubated with 20 μL of goat anti-mouse Alexa-Fluor 594 for five minutes at room temp. Similarly, 2 μg of mouse anti-RAD18 or anti-pol iota (Abcam) were incubated with 20 μL of goat anti-mouse Alexa-Fluor 488 for five minutes at room temp. After this incubation, 20 μL of non-specific IgG was added to both tubes to remove any non-bound fluorescent probe. Both solutions were then added together into a 10% GPBST solution. Cells were incubated in the dark for

1 hour on fresh parafilm in a humid chamber. After fluorescently labeling the cells, a second fixation is performed with 1x PBS containing 4% formaldehyde for 15 min. at room temperature to prevent disassociation of the secondary from the primary. The cells were then washed 3x with PBS and mounted on slides using Slowfade Light Antifade Kit with DAPI (Molecular Probes) for the mounting medium. The slides were sealed with clear fingernail polish. Slides were imaged using a Zeiss Axiovert 100M confocal microscope with 100x oil-immersion objective. The images were analyzed with LSM 510 software.

16.3 Results and Discussion

16.3.1 Cell Synchrony

Mimosine is thought to act by inhibiting ribonucletide reductase [5]. Therefore, the available pool of deoxynucleotide precursors is greatly reduced and the cells arrest in G_1. Note that all four deoxynucleotides are affected equally, so there are no potentially mutagenic pool imbalances. To confirm the degree of synchrony, we examined the percentage of cells in all four phases of the cell cycle of various populations using flow cytometry. In Fig. 16.1A, the cells were exposed to mimosine for 24 h, then harvested and stained with propidium iodide. The great majority of cells (81%) had 2N DNA content, indicating that they were in G_1 (or G_0). In experiments that examined cells in G_1, the cells were irradiated after 24 h in mimosine, held for 2 h in medium containing mimosine, then fixed for immuno-histochemistry. Figure 16.1B shows the effect of removing the mimosine then

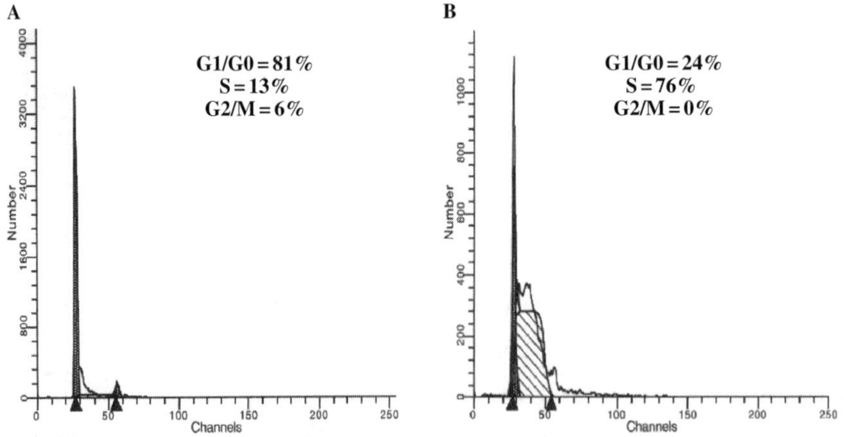

Fig. 16.1 A. Cells were synchronized with mimosine for 24 h, then harvested and examined for DNA content with flow cytometry. 81% of the cells were in G_1-phase. B. Mimosine was removed and the cells were harvested 6 h later. Under these conditions, 76% of the cells were in S-phase.

harvesting the cells after 6 h in fresh medium. Under these conditions, the great majority (76%) of cells had entered S-phase. The S-phase experiments were done by irradiating at this point then harvesting the cells 2 h later. These data confirm the effectiveness and lack of toxicity of mimosine under the conditions examined.

16.3.2 Subcellular Localization of RAD18

Proliferating cell nuclear antigen (PCNA) is a homotrimeric protein that acts as a "clamp" that tethers DNA polymerase δ and ϵ to the template. It functions in a variety of DNA transactions in addition to replication, notably in mismatch repair and in the gap filling step of nucleotide excision repair. We showed that PCNA staining increases in the cytoplasm throughout G_1-phase, and is also found within the nucleus, as expected, in cells that do not have DNA damage [6]. RAD18, on the other hand is restricted to the nucleus in a diffuse pattern in such cells [7].

This pattern is reflected in cells irradiated in late G_1 (Fig. 16.2, left panel). The PCNA (red) is present most intensely in a perinuclear distribution. Nuclear PCNA is present in small focal areas. These have been found to colocalize with cyclobutane dimers [7], and are presumed to represent PCNA involvement in the gap filling step of nucleotide excision repair. In contrast, RAD18 was present in the nucleus in a diffuse pattern in cells irradiated in G_1 (Fig. 16.2, green staining).

In cells irradiated in S phase (right panel), PCNA is principally in the nucleus, and is primarily in a focal pattern. The distribution of RAD18 is also principally in a focal pattern, which colocalizes with PCNA in the merged image. Several Y-family DNA polymerases are known to form

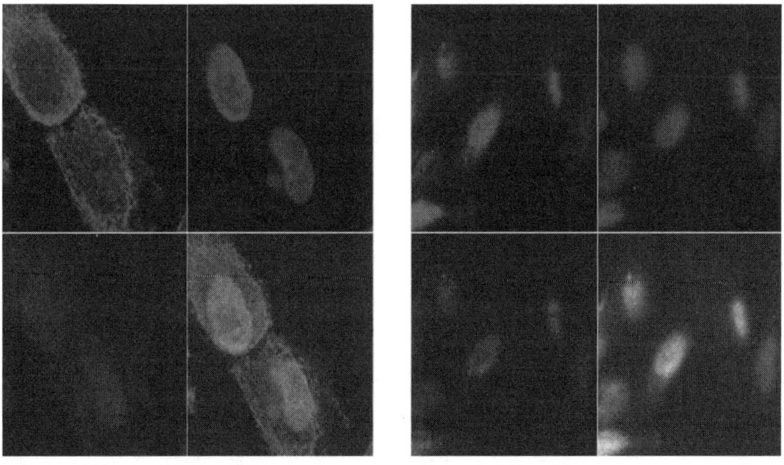

Fig. 16.2 Cells were irradiated in G_1 (left panel) or S (right panel) and stained 2 h later. The cells were stained for PCNA (red), DNA (blue), RAD18 (green), and examined by confocal microscopy. The lower right quadrant of each panel is the merged image. (See also color insert.)

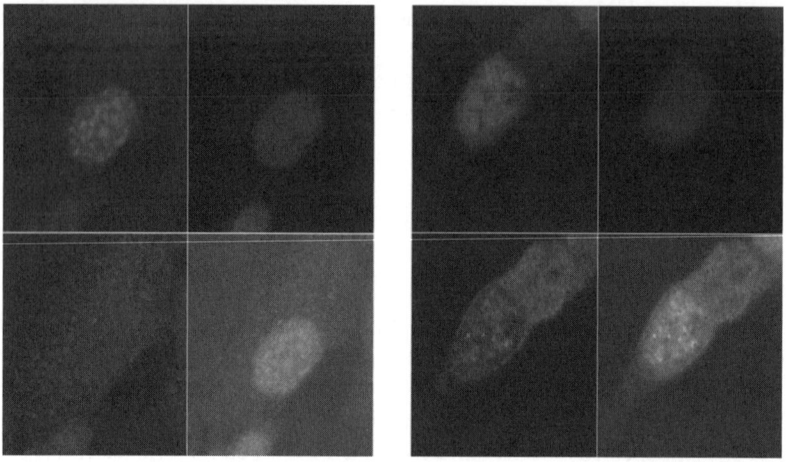

Fig. 16.3 Left panel, cells irradiated in G_1. Pcna (red), DNA (blue), pol iota (green), lower right quadrant, merged images. Right panel, cells irradiated in S-phase. (See also color insert.)

foci in the nuclei of cells entering S-phase with DNA damage [7,8]. These foci are thought to represent stalled DNA replication forks, to which TLS proteins have been recruited. These data are consistent with the proposed role of RAD18, which is to signal translesion polymerases to sites of stalled replication forks.

Figure 16.3 shows staining for cells treated exactly as those in Fig. 16.2, but stained for pol iota instead of RAD18. In the G_1 cells (left panel) the PCNA pattern is similar to that shown in Fig. 16.2, but a cell with a more prominent focal pattern of PCNA is illustrated. In these cells, little or no pol iota could be demonstrated. In the S-phase cells (right panel) pol iota was clearly demonstrated to be in nuclear foci that colocalize with PCNA. It was organized in a focal pattern only in cells irradiated in S-phase, and these foci colocalized with PCNA.

These results support the hypothesis that RAD18 is recruited to sites of DNA replication forks that are stalled by UV damage, and the protein is involved in the subsequent recruitment of newly discovered error-prone DNA polymerases that can complete the replication of DNA containing UV damage. In separate studies, we have shown that reducing the level of RAD18 using antisense technology prevents the formation of foci of pol iota and other Y-family polymerases.

16.4 Conclusions

RAD18 forms nuclear foci that colocalize with PCNA only in cells entering S phase with DNA damage. These foci represent stalled replication forks, to which translesion polymerases have been recruited. In support of this idea, pol iota, a newly discovered translesion polymerase, is recruited to these sites in a RAD18-dependent manner. Recent data support a model in which RAD18 recruits a ubiquitin ligase (RAD6) to the site of the stalled replication complex. At least one target of RAD6 is PCNA, which is ubiquitylated in a unique linkage that serves a signaling function. A working hypothesis is that the Y family polymerases have a higher affinity for PCNA when it is ubiquitylated in this way, resulting in recruitment to the stalled replication complexes.

16.5 Future Study

We have found that reducing the level of RAD18 using antisense technology greatly reduces the frequency of mutations induced by UV and chemical carcinogens. We are investigating the possibility that such cells may become genetically unstable by virtue of illegitimate recombination induced in cells that cannot perform TLS. If the cells resolve the blocked forks by error-free homologous recombination, and are not genetically unstable, then RAD18 may present and attractive target for cancer chemoprevention.

Acknowledgment This work was supported by a NASA-Ames NAG2-1647 and NIH/USPHS grants CA112197 and CA112664.

References

1. McGregor, W.G., Chen, R.H., Lukash, L., Maher, V.M. & McCormick, J.J. Cell cycle-dependent strand bias for UV-induced mutations in the transcribed strand of excision repair-proficient human fibroblasts but not in repair-deficient cells. *Mol. Cell Biol.* **11**, 1927–1934 (1991).
2. Friedberg, E.C., Lehmann, A.R. & Fuchs, R.P. Trading places: how do DNA polymerases switch during translesion DNA synthesis? *Mol. Cell.* **18**, 499–505 (2005).
3. Prakash, S. & Prakash, L. Translesion DNA synthesis in eukaryotes: a one-or two-polymerase affair. *Genes Dev.* **16**, 1872–1883 (2002).
4. Watson, N.B., Mukhopadhyay, S. & McGregor, W.G. Translesion DNA replication proteins as molecular targets for cancer prevention. *Cancer Lett.* **241**, 13–22 (2006).
5. Krude, T. Mimosine arrests proliferating human cells before onset of DNA replication in a dose-dependent manner. *Exp. Cell Res.* **247**, 148–159 (1999).
6. Mukhopadhyay, S., Clark, D.R., Watson, N.B., Zacharias, W. & McGregor, W.G. REV1 accumulates in DNA damage-induced nuclear foci in human cells and is implicated in mutagenesis by benzo[a]pyrenediolepoxide. *Nucleic Acids Res.* **32**, 5820–5826 (2004).
7. Tissier, A. et al. Co-localization in replication foci and interaction of human Y-family members, DNA polymerase pol eta and REV1 protein. *DNA Repair (Amst)* **3**, 1503–1514 (2004).
8. Bi, X., et al. RAD18 regulates DNA polymerase kappa and is required for recovery from S-phase checkpoint-mediated arrest. *Mol. Cell Biol.* **26**, 3527–3540 (2006).

Chapter 17
Alanine in HI: A Silent Mutation Cries Out!

J. H. Shah[1], D. J. Maguire[1], T. B. Munce[2], and A. Cotterill[3]

Abstract It is a widely held paradigm in molecular biology that a change in the third base of a codon is silent in terms of expression. In this investigation, results are presented that challenge that paradigm, at least in terms of one polymorphism in KCNJ11, which is one of five genes that have been implicated in the disorder Hyperinsulinism of Infancy. In two cohorts of Australian patients, an uneven distribution of KCNJ11 SNP's was observed. A silent polymorphism at codon 190 was over-represented in the patients who responded well to medical treatment and under-represented in those that required radical surgical intervention. In an attempt to investigate this polymorphism, it was expressed in vitro and western blot analysis showed that there were virtually no bands from the homozygous variant samples, while strong bands were seen in normal controls. The human genome is highly redundant in terms of tRNA species for each amino acids but enigmatically under-represents a number of specific codons. The polymorphism in question occurs within one such codon. We propose that the presence of a base change at the third position of codon that is not represented by a corresponding anti-codon within the human nuclear tRNA leads to a decreased rate of expression of the protein.

17.1 Introduction

Hyperinsulinism of Infancy (HI) is a metabolic syndrome associated with the unregulated release of insulin from the pancreatic beta cells, accompanied by hypoglycaemia that is hard to control. HI is the most common form of permanent hyperinsulinism of infancy. Patients usually present in the first 72 hours of

[1]School of Biomolecular and Biophysical Science, Griffith University, Nathan, Q4111, Australia
[2]Pathology, Mater Health Services, South Brisbane, Queensland, 4101–Metabolic Medicine, Mater Health Services, South Brisbane, Queensland, 4101
[3]Paediatric Endocrinology and Diabetes, Mater Children's Hospital, South Brisbane, Queensland, 4101, Australia

life with symptoms of hypoglycaemia including jitteriness, nausea, vomiting and seizures. In severe cases coma or death can occur [1]. Due to the difficulty in controlling hypoglycaemic seizures, up to 50% of all patients suffer from long-term neurological impairment. Management can be difficult, with treatment aimed at maintaining euglycemia [2].

The known causes of HI are linked to the expression, structure or function of the ATP dependent potassium channel of the pancreatic β cells. The K-ATP channel is an octameric complex consisting of 4 subunits each of the sulfonyl urea receptor (SUR1) and the inwardly rectifying subunit Kir6.2. SUR1 and Kir6.2 are encoded by the genes ABCC8 and KCNJ11 respectively, which lie adjacent to each other at 11p15.1 [1]. Mutations in ABCC8 and KCNJ11 have been implicated in HI, with more than a hundred mutations reported in ABCC8. The genes encoding the enzymes Glucokinase [3] (GCK), glutatmate dehyrogenase [4] (GLUD1) and 3-Hydroxyacyl-CoA dehydrogenase [5] (SCHAD) have also been implicated in HI. The genetic cause, however, remains unknown in up to 50% of HI patients [6].

In an effort to further understand of the genetic aetiology of HI and the genotype-phenotype correlations of the disease, we genotyped a cohort of Australian HI patients. The patients were segregated by the severity of their disease and their response to treatment. Mainstream medical therapy involves the use of glucose and glucagon infusions coupled with diazoxide and octreotide. Those who failed to respond to this treatment at maximal doses required a partial or subtotal pancreatectomy to maintain euglycemia. They were assumed to have the more severe form of the disease and constituted the surgical cohort. Those patients who responded well to this medical treatment were assumed to have a milder form of the disease and were classified as the medical cohort [7]. Our genetic findings are described in Shah, J.H et al [8]. Upon sequencing KCNJ11 we identified a differential distribution of 5 polymorphisms between our two cohorts ($p<0.01$), also described in Shah et al [8]. One of these polymorphisms was the A190A [9]. a GCC to GCT at codon 190 that did not alter the transcribed amino acid. The polymorphism was identified in 30.5% of medical patients and 74% of surgical patients. The allele frequency in 100 control subjects was 27.6%.

Nucleotide base changes are classified as silent in cases where an alteration does not lead to a change in the amino acid sequence of a protein. These are common, but are usually considered to have no phenotypic effect. Some studies have shown that silent polymorphisms can have an effect on post translational modification as well as alternate splicing [10]. In the case of the A190A polymorphism, we have investigated the translation process to identify a possible link between the silent polymorphism and the disease phenotype

17.2 Hypothesis and Aims

We suggest that in some cases, such as with A190A in HI, the effect of a silent mutation may be enhanced by the relative abundance of the tRNA species for the codon.

Table 17.1 Population of alanine tRNA in the Human genome (The genomic tRNA database: http://lowelab.ucsc.edu/GtRNAdb/)

Anti Codon.	No. of tRNA in humans
AGC	28*
GGC	0#
CGC	5
TGC	9
Total	42

* Wildtype in HI; #polymorphic in HI

There are 42 tRNA molecules that carry anticodons specific to those that transcribe alanine [11]. Alanine has a redundancy of 4 and is transcribed by the codons GCC, GCG, GCT and GCA. However, there is an uneven distribution of tRNA molecules for each codon. An assessment of the alanine tRNA population in humans (Table 17.1), revealed that there is no tRNA molecule carrying the anticodon GGC which is specific for the GCC codon that is polymorphic in HI patients.

We hypothesise that the relative abundance of specific tRNA molecules has an impact on the efficiency of translation, thus implying that the wobble hypothesis in translation is not always true. We propose that the efficiency of translation is reduced in those individuals carrying the variant form of the alanine codon at amino acid 190 in KCNJ11. Using Kir6.2 as a model, we studied the efficiency of translation and investigated the possible role that a reduced translational efficiency would have in the disease progression in HI.

17.3 Methodology

Six patients and one normal control were chosen for our investigations. The patients were selected based on their genotype such that we had one homozygous variant, one heterozygous and one homozygous wildtype from each cohort. The normal control was a homozygous wildtype and was not known to have PHHI or any other endocrine disease. Other polymorphisms were identified in the KCNJ11 gene, however these were identical in all participants in this study.

Genotyping of KCNJ11 was carried out by PCR followed by sequencing as described in Shah et al [8]. Initial experiments for measuring in vitro translational efficiency were undertaken using the TNT® T7 Coupled Reticulocyte Lysate System (Promega).

The T7 promoter sequence was ligated 5' to the KCNJ11 promoter using PCR. This was confirmed by sequencing. The PCR products were then purified using the High Pure PCR purification kit (Roche) and a low melting point agarose gel. 1 mg of the purified PCR product was introduced into the TNT rabbit reticulocyte system as described by the manufacturer. *S35-Met was used as described by the manufacturer. Reactions were stopped by heating to 98 $^{\circ}$ C after predetermined time intervals.

Protein production in the in vitro reactions was analysed using denaturing SDS-PAGE, autoradiography and by Western blot analysis using an antibody raised against the N terminal of the protein (Santa Cruz antibodies). *S35-Met incorporation was measured by using a beta scintillation counter.

17.4 Results

Protein production at the end of in vitro trancription-translation reactions were analysed without specific purification. Initial results obtained from SDS-PAGE – autoradiography showed that there was a slight difference in Kir6.2 protein synthesis between the patients analysed (Fig. 17.1a). Through western blotting however, significant differences were noted. A strong band representing the Kir6.2 protein was observed for the homozygous wildtype samples whereas there was virtually no signal for the homozygous variants (Fig. 17.1b).

Quantitative scintillation counting for studying incorporation of *S-Met in the patients was found to follow the same trend as the western blots. In vitro translation using wildtype DNA showed a significantly higher rate of *S-Met incorporation and had more protein produced compared to incorporation and protein production in experiments using homozygous variant DNA (Fig. 17.2).

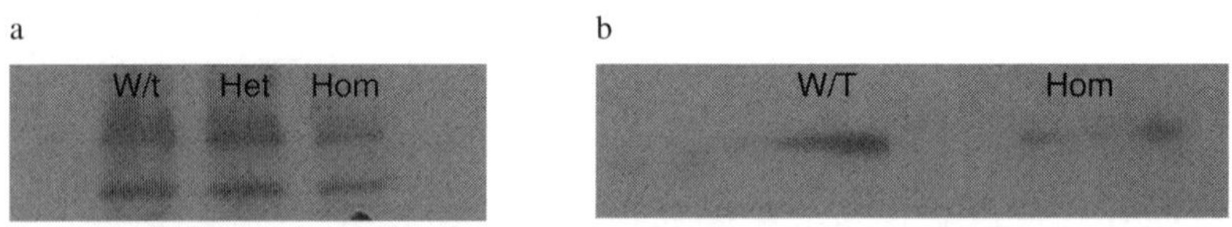

Fig. 17.1 SDS PAGE autoradiography (a) and Western Blotting (b) of in vitro translation products. W/t –wildtype, Het – heterozygous, Hom- homozygous.

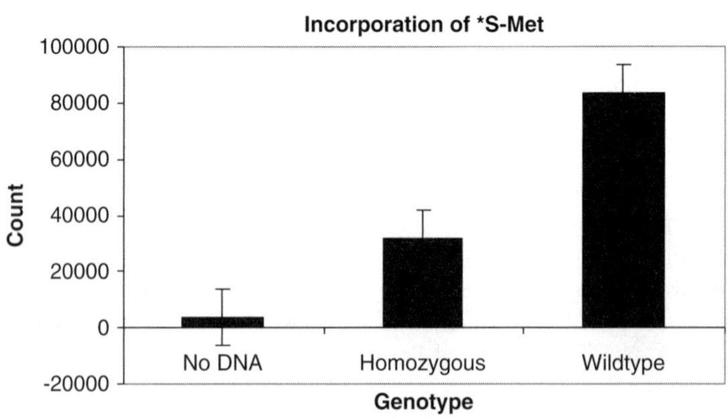

Fig. 17.2 Measure of *S-Met incorporation in patients after an in vitro translation experiment.

17.5 Discussion

Our initial results indicate an apparent reduced amount of Kir6.2 protein produced after 90 min in vitro transcription and translation. In some experiments, a fragmented product was observed. Western blots and *S-Met incorporation studies suggest an even more striking depletion of Kir6.2 in vitro expression studies than autoradiography. There are several possible explanations for this observation, including a decreased rate of protein production due to the lack of directly binding tRNA species, the slower rate of Kir6.2 protein translation could lead to of improper folding or altered pre mRNA editing.

With reference to HI, we can speculate that the presence of the polymorphism reduces the rate of Kir6.2 production, and thus channel assembly. It is however interesting to note that the GCC codon for alanine that is abnormal in KCNJ11 seems to be the most commonly used alanine codon in the human genome [12]. Further work will be undertaken to investigate the impact of translational efficiency in Kir6.2 and other proteins.

Acknowledgment The authors would like to acknowledge members of the Mater Children's Hospital PHHI research group Dr. D Cowley, Dr. F Bowling, Dr M Harris, Dr R Greer and others.

References

1. Dunne MJ, Cosgrove KE, Shepherd RM, Aynsley-Green A, Lindley KJ. *Hyperinsulinism in Infancy: from Basic Science to Clinical Disease*. Physiol Rev, 2004. **84(1),** 239–275.
2. Aynsley-Green A, Hussain K, Hall J, Saudubray JM, Nihoul-Fekete C, De Lonlay-Debeney P, Brunelle F, Otonkoski T, Thornton P, Lindley KJ. *Practical management of hyperinsulinism in infancy*. Arch Dis Child Fetal Neonatal Ed. Arch Dis Child, 2000. **82**(2), F98–F107.
3. Davis EA, Cuesta-Munoz A, Raoul M, Buettger C, Sweet I, Moates M, Magnuson MA, Matschinsky FM. *Mutants of glucokinase cause hypoglycaemia- and hyperglycaemia syndromes and their analysis illuminates fundamental quantitative concepts of glucose homeostasis*. Diabetologia, 1999. **42**(10), 1175–1186.
4. Stanley CA, Lieu YK, Hsu BY, Burlina AB, Greenberg CR, Hopwood NJ, Perlman K, Rich BH, Zammarchi E, Poncz M. *Hyperinsulinism and hyperammonemia in infants with regulatory mutations of the glutamate dehydrogenase gene*. N Engl J Med, 1998. **338**(19), 1352–1357.
5. Molven A, Matre GE, Duran M, Wanders RJ, Rishaug U, Njolstad PR, Jellum E, Sovik O. *Familial Hyperinsulinemic Hypoglycemia caused by a defect in the SCHAD enzyme of mitochondrial fatty acid oxidation*. Diabetes, 2004. **53(**1), 221–227.
6. Molven A, Rishaug U, Matre GE, Njolstad PR, Sovik O. *Hunting for a hypoglycemia gene: Severe neonatal hypoglycemia in a consanguineous family*. Am J Med Genet, 2002. **113(**1), 40–46.
7. Jack MM, Greer RM, Thomsett MJ, Walker RM, Bell JR, Choong C, Cowley DM, Herington AC, Cotterill AM. *The outcome in Australian children with hyperinsulinism of infancy: Early extensive surgery in severe cases lowers risk of diabetes*. Clin Endocrinol (Oxf), 2003. **58**(3), 355–364.

8. Shah JH, Maguire DJ, Brown D, Cotterill AM. *The role of ATP sensitive channels in insulin secretion and the implications in Persistent Hyperinsulinemic Hypoglycaemia of Infancy (PHHI)*, In Press.
9. Database of Single Nucleotide Polymorphisms (dbSNP). Bethesda (MD): National Center for Biotechnology Information, National Library of Medicine. dbSNP accession:{rs5218}. Available from: http://www.ncbi.nlm.nih.gov/SNP/
10. Nielsen KB, Sorensen S, Cartegni L, Corydon TJ, Doktor TK, Schroeder LD, Reinert LS, Elpeleg O, Krainer AR, Gregersen N, Kjems J, Andresen BS. *Seemingly neutral polymorphic variants may confer immunity to splicing-inactivating mutations: A synonymous SNP in exon 5 of MCAD protects from deleterious mutations in a flanking exonic splicing enhancer*. Am J Hum Genet. 2007 **80**(3), 416–432.
11. Lowe TM, Eddy SR. *tRNAscan-SE: A program for improved detection of transfer RNA genes in genomic sequence* Nucl. Acids Res. 1997. **25**, 955–964. (The genomic tRNA database. (http://lowelab.ucsc.edu/GtRNAdb/)
12. Nakamura, Y. Codon Usage Datebase. (http://www.kazusa.or.jp/codon/) Accessed 12th June 2006.

Chapter 18
Biomathematics in Cancer Detection: Simulation of Lipogenesis in Cancer

Ping Huang, and Britton Chance[1]

Abstract The usual mechanisms for biochemical events are steady-state systems without dynamic simulation. Our study is to simulate lipogenesis from the breakdown of glucose coupled with oxidative phosphorylation in mitochondria by using JSim (for Java Simulator) as software development environment, which enables non-linear differential equations to be used in a simulation giving a time course through a variety of non-steady-state conditions. Glycolysis and lipogenesis coupled with oxidative phosphorylation in mitochondria non-linear differential model is built in this paper. Simulation and discussion on lipogenesis by carbohydrate responsive element-binding protein (ChREBP) are given. Our model provides a potential way to analyze the experimental databank.

18.1 Introduction

The usual mechanisms for biochemical events are steady-state systems without dynamic simulation. We present here a mechanism based upon JSim (for Java Simulator), which is developed by National Simulation Resource. JSim is a software environment for scientific modeling that provides tools for development of models, for their run-time control, and for analysis of their behavior. JSim enables non-linear differential equations to be used in a simulation giving a time course through a variety of non-steady-state conditions. We learn from this of the very efficient transfer of reducing equivalents from glucose to lipid, a pathway of some importance for lipid biosynthesis, particularly in cancer.

[1]Ping Huang and Britton Chance, Department of Biophysics and Biochemistry, University of Pennsylvania, Philadelphia, Pennsylvania 19104. Corresponding authors: Ping Huang and Britton Chance, e-mail: huangping2001@hotmail.com, chance@mail.med.upenn.edu.

K.A. Kang et al. (eds.), *Oxygen Transport to Tissue XXIX* 151

18.2 Method

The energy released by the breakdown of glucose can be used to phosphorylate ADP, forming ATP; we see that glucose by itself generates a much smaller amount of ATP. However, during the breakdown of glucose, a large amount of mitochondrial NADH is produced; it is this reducing agent that dramatically increases the amount of ATP produced. We simulate glucose metabolism to increase ATP production.

Glycolysis as well as the lipogenic process could be easily incorporated into our program as long as a simple quantitative kinetic description of this process in brain and muscle is appropriate.

The pathway from glucose to pyruvate to acetyl Co-A to citrate is in the cytoplasm and α-ketoglutarate activity is in the mitochondria, the pathway from citrate to MalonylCoA to Palmitate to lipids can be simulated as in Fig. 18.1. Pyruvate reduces NAD to NADH and citrate. The definitions and units of the symbols used in the model are listed in Table 18.1. Other equations of oxidative phosphorylation in mitochondria are in a related papers [1–5].

Glycolysis and lipogenesis non-linear differential equations are given as following (18.1)–(18.7):

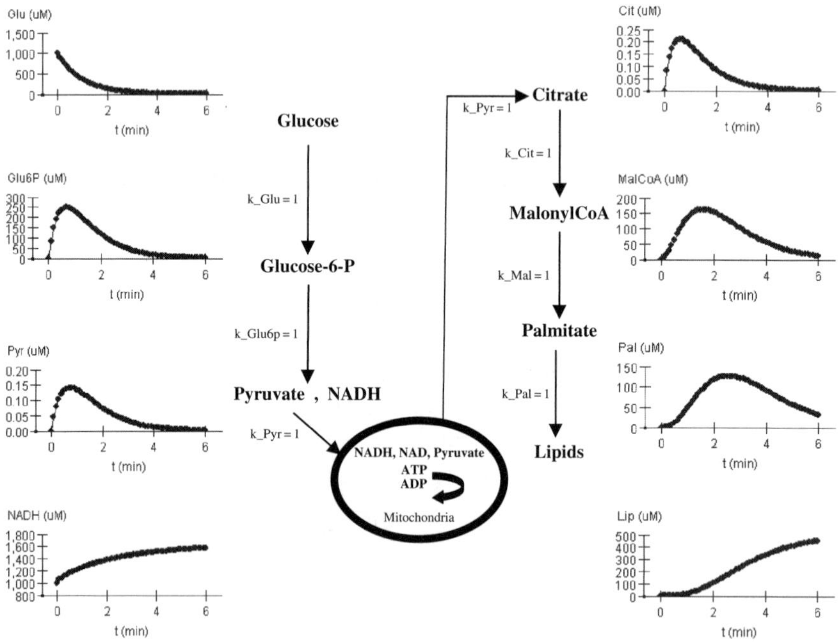

Fig. 18.1 Pathway of lipogenesis from glucose to lipids and simulation results.

Table 18.1 Definitions and units of the symbols used in the model

Symbol	Definition	Unit
Glu	concentration of Glucose	uM
Glu6P	concentration of Glucose-6-P	uM
Pyr	concentration of Pyruvate	uM
Cit	concentration of Citrate	uM
MalCoA	concentration of MalonylCoA	uM
Pal	concentration of Palmitate	uM
Lip	concentration of Lipids	uM
t	time	Min
k_Glu	the rate of reaction : Glu → Glu6P	min^{-1}
k_Glu6P	the rate of reaction: Glu6P → Pyruvate	min^{-1}
k_Pyr	the rate of reaction : Pyruvate+NAD → NADH+Citrate	uM^{-1}. min^{-1}
k_Cit	the rate of reaction : Citrate+NADH → MalonylCoA	uM^{-1}. min^{-1}
k_MalCoA	the rate of reaction: MalonylCoA → Palmitate	min^{-1}
k_Pal	the rate of reaction: Palmitate → Lipids	min^{-1}
k_Glu6P_GPGLc	the rate of reaction: Glu6P → GPGLc	min^{-1}

$$dGlu/dt = -k\_Glu \cdot Glu; \tag{18.1}$$

$$dGlu6P/dt = k\_Glu \cdot Glu - k\_Glu6P \cdot Glu6P - k\_Glu6P\_GPGLc \cdot Glu6P; \tag{18.2}$$

$$dPyr/dt = k\_Glu6P \cdot Glu6P - k\_Pyr \cdot Pyr \cdot NAD; \tag{18.3}$$

$$dCit/dt = k\_Pyr \cdot Pyr \cdot NAD - k\_Cit \cdot Cit \cdot NADH; \tag{18.4}$$

$$dMalCoA/dt = k\_Cit \cdot Cit \cdot NADH - k\_MalCoA \cdot MalCoA; \tag{18.5}$$

$$dPal/dt = k\_MalCoA \cdot MalCoA - k\_Pal \cdot Pal; \tag{18.6}$$

$$dLip/dt = k\_Pal \cdot Pal; \tag{18.7}$$

18.3 Simulation Results and Discussion

ATP hydrolysis is 700 uM/min (rest state). In the lipid biosystem, all reaction rates are set to 1 (units are shown in Table 18.1). The initial values are shown in Table 18.2.

The 6-minute simulation results are shown in Fig. 18.1. The breakdown of 1000uM glucose can eventually produce 450 uM lipids in 6 minutes. If we gave

Table 18.2 Initial values used in the simulation

Name	Value	Unit
Glucose	1000	uM
Glucose-6-P	0	uM
Pyruvate	0	uM
NADH	1000	uM
Citrate	0	uM
MalonylCoA	0	uM
Palmitate	0	uM
Lipids	0	uM

2000 uM glucose as initial value, we can obtain around 900uM lipids in 6 minutes. Therefore in the case of all reaction rates being set to 1, the efficiency of transfer of reducing equivalents from glucose to lipid is about 45%.

Also we figure out the rate of glucose-6-P breakdown to pyruvate can affect the amount of product lipids. If the rate of glucose-6-P breakdown to pyruvate is increased to 5/min, the breakdown of 1000 uM glucose can eventually produce 800 uM lipids in 6 minutes, and 2000 uM glucose can eventually produce 1600 uM lipids in 6 minutes, so the efficiency of transfer of reducing equivalents from glucose to lipid is about 80%. The changes of other rates in the pathway did not affect the produced lipid amount. Therefore in this mechanism the rate of glucose-6-P breakdown to pyruvate can affect the efficient transfer of reducing equivalents from glucose to lipid, helping us to investigate lipid biosynthesis in cancer.

18.4 Simulation and Discussion on Lipogenesis by ChREBP

Dr. Kosaku Uyeda from University of Texas Southwestern Medical Center has indicated that the activation of lipogenesis by Carbohydrate responsive element-binding protein (ChREBP) requires many complicated processes including the activation of inactive ChREBP, localized in cytosol, by dephosphorylation by PP2A, then ChREBP is imported into the nucleus. mRNA synthesis requires formation of an initiation complex and binding to DNA followed by RNA polymerase reaction. To get mRNA synthesis at high glucose takes at least 24 hrs and an initial lag period may take at least 3–4 hrs. So during this initiation period there are several steps involved, the activation of ChREBP, formation of the initiation complex, and many enzyme steps in RNA polymerase reaction. But neither the rate limiting step of the overall reaction nor each major step are known.

We simulated glucose going from Glucose-6-P to xylulose 5-phosphate (Xu-5-P) pathway. We changed the rate value of ribulose-5-phosphate (Ribu-5-P) to Xu-5-P from 3/min to 12/min (4-fold increase), which stimulated ChREBP from inactivated state (in Cytosol) to activated state (in nucleus).

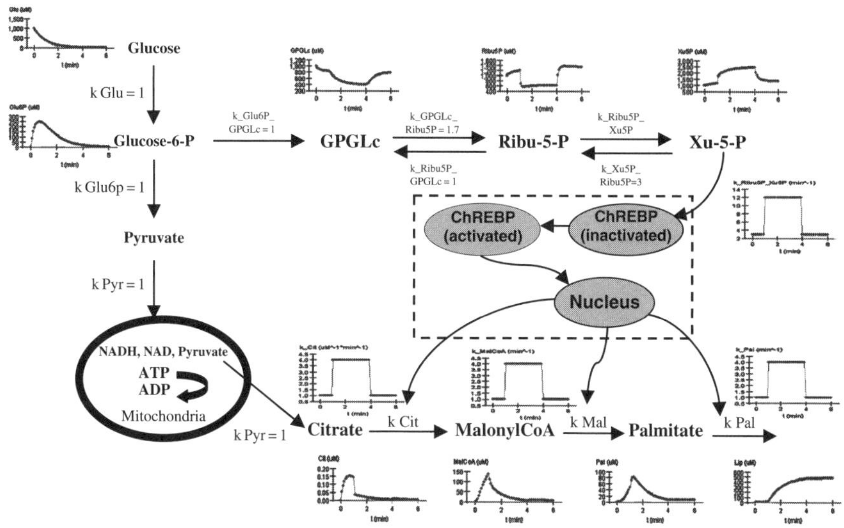

Fig. 18.2 Pathway of lipogenesis from glucose to lipids by ChREBP and simulation results.

Then the activated ChREBP signaled the enzymes of the lipogenesis pathway.The rates on the pathway from citrate to lipids is increased accordingly (4 fold). The lipogenesis by the ChREBP pathway and simulation results are shown in Fig. 18.2, as are the dynamic flux changes. So Xu-5-P is activated by glycolysis, and it activates the lipid biosynthesis pathway by ChREBP so that the pyruvate which we get from activated glycolysis goes into the CAC, makes citrate, from citrate. It makes triglyceride with activation from Xu-5-P.

18.5 Conclusions

Our model provides a potential way to analyze the experimental databank by fitting the data to our model to deduce the parameters of lipogenesis from the breakdown of glucose coupled with oxidative phosphorylation in mitochondria glycolysis and lipogenesis in the cancer state.

References

1. Ping Huang and Britton Chance, Simulation of Mitochondrial Function in Brain and Muscle Tissues, *Biomedical Optics 2006 OSA (Optical Society of America)*, ME39, ISBN 1-55752-807-1. Fort Lauderdale, Florida, USA. March 19–22 (2006).
2. Zheng Li, Tada Yipintsoi and James B. Bassingthwaighte, "Nonlinear Model for Capillary-Tissue Oxygen Transport and Metabolism," *ABE* **25**, 604–619 (1997).

3. Bernard Korzeniewski, Jerzy A. Zoladz, "A model of oxidative phosphorylation in mammalian skeletal muscle," *JBC* **92,** 17–34 (2001).
4. Chandan K. Sen, Lester Packer and Osmo O.P. Hanninen, *Handbook of oxidants and antioxidants in exercise* (Elsevier Science B.V., 2000).
5. "Modeling and imaging, the national simulation resource in circulatory mass-transport & exchange," *NSR Simulation Analysis Workshop*, University of Washington, Sept. (2001).

Chapter 19
Activity of Drug Efflux Transporters in Tumor Cells Under Hypoxic Conditions

Oliver Thews[1], Birgit Gassner[2], Debra K. Kelleher[1], and Michael Gekle[2]

Abstract Tumor cells exhibit mechanisms by which chemotherapeutic drugs can be actively pumped out of the cell (e.g., p-glycoprotein pGP, MRP1), resulting in a multidrug resistant phenotype. Many human tumors show pronounced hypoxia which can result in a local ATP depletion which in turn may compromise the efficacy of these transporters. The aim of this study was therefore to assess the transport activity and expression of drug transporters under hypoxic conditions. Prostate carcinoma cells (R3327-AT1) were exposed to hypoxia (pO_2<0.5 mmHg) for up to 24h and pump activity was determined by an efflux assay. The results showed that exposing cells to hypoxia for 3–6 h led to a moderate increase in pGP activity. After 24 h pGP activity was reduced by 44% compared to control levels. Hypoxia reduced the MRP1 activity to a lesser extent (by 25%). However, the expression of pGP and MRP1 was almost independent of the medium pO_2. In conclusion, pronounced hypoxia had only minor effects on the activity of drug transporters with the activity decreasing only after 12–24 h under hypoxia, possibly as a result of ATP depletion. Instead, indirect effects of hypoxia leading to extracellular acidosis seem to have a much more pronounced effect on pGP activity.

19.1 Introduction

Many tumors exhibit mechanisms by which chemotherapeutic agents can be actively transported out of the cell, leading to a chemoresistant phenotype (MDR = multidrug resistance) [1]. Up until now, a large number of these transporters belonging to the ABC (ATP-binding cassette) family have been identified. Clinically important representatives of this family are the p-glycoprotein (pGP), the product of the MDR1 gene, and the multidrug resistance-related protein 1 (MRP1 [1,2]. These membrane proteins are able to

[1]Institute of Physiology and Pathophysiology, University of Mainz, 55099 Mainz, Germany.
[2]Institute of Physiology, University of Würzburg, 97070 Würzburg, Germany.

actively pump various drugs (e.g., doxorubicin, vinblastine, paclitaxel) out of the cell and in this way can reduce the cytotoxic efficacy of these drugs [1]. At the same time, various inhibitors of pGP have been identified (e.g. verapamil, probenecid), which are capable of reducing the chemoresistance of pGP- or MRP1-expressing tumor cells, at least in cell culture experiments [1,3,4].

The tumor vasculature exhibits numerous structural and functional abnormalities, leading to an inadequate, chaotic perfusion thus resulting in an insufficient O_2 supply to the tissue [5]. As a consequence, the mean oxygen partial pressure (pO_2) in tumors is often considerably lower than in the surrounding normal tissue, with areas of severe hypoxia or even anoxia in vital tumor tissue in approx. 60% of human tumors [6]. Correspondingly, tumor cells switch to anaerobic glycolysis which is – from an energetic point of view – less effective and result in a lack of ATP [5].

Since ABC-transporters use ATP hydrolysis as an energy source, the question arises of whether tumor hypoxia can affect the transport efficacy of pGP or MRP1. The aim of this study was to analyze the transport activity and expression of ABC-transporters under hypoxic conditions in vitro.

19.2 Material and Methods

19.2.1 Cell Line

The subline AT1 of the R-3327 Dunning-prostate carcinoma of the rat was used in all experiments. Cells were grown in RPMI 1640 medium supplemented with 10% fetal calf serum (FCS) at 37°C under a humidified 5% CO_2 atmosphere and passaged once per week. Twenty-four hours prior to the experiments, cells were transferred to RPMI medium without FCS supplementation. The medium was buffered with 20mM HEPES adjusted to a pH of 7.4. In hypoxia experiments, cells were gassed with a mixture containing 95% N_2 and 5% CO_2 for up to 24 h.

19.2.2 MDR1 and pGP Transport Activity

In order to assess the activity of the MRP1 or the pGP, the efflux rate of a fluorescent substrate of the respective transporter in the presence or absence of a specific inhibitor was measured. For pGP measurements, cells were incubated with rhodamine (0.5 μM, dissolved in Ringer solution) for 30 min at 37°C. Subsequently, the rhodamine-containing solution was removed, the cells rapidly washed with PBS at 4°C and then incubated with fresh rhodamine-free Ringer solution. Samples (100 μL) of the supernatant were taken at 0, 5, 15, and 45 min after medium exchange. The rhodamine-123 efflux rate could then be calculated from the increase in the rhodamine-123 concentration in the

Ringer solution as determined by fluorimetric measurements using a fluorescence microplate reader (Victor [2], Wallac, Turku, Finland) with excitation/emission wavelengths of 485/535 nm. In order to determine the pGP-mediated efflux rate, a second set of cells were primarily incubated with rhodamine-123 as described above, but thereafter the solution was replaced by a rhodamine-free Ringer solution containing verapamil (10 μM, dissolved in EtOH). Due to the inhibition of pGP by verapamil, the rhodamine-efflux rate was slower. The ratio of the efflux rate without and with verapamil was used as a measure of the activity of the pGP-mediated efflux [7]. Measurements of the MRP1 activity were performed in the same manner except that Calcein-AM (Invitrogen, Carlsbad, USA; 0.1 μM, dissolved in Ringer solution) was used as the fluorescent substrate and probenecid (2 mM dissolved in Ringer solution) as an MRP1 inhibitor.

All values obtained were normalized relative to the protein content in each Petri dish determined with the bicinchoninic acid (BCA) assay (Pierce, KMF Laborchemie, Sankt Augustin, Germany).

19.2.3 MRP1 and pGP Expression

The cellular expression of MRP1 and pGP was determined in a whole cell ELISA, as described previously [8]. In brief, after cell fixation with 4% paraformaldehyde for 60 min, cells were washed with a permeabilizing buffer containing 0.1% Triton X-100 and then incubated for 20 min with this buffer to which 0.6% H_2O_2 had been added. After incubation with the primary anti-pGP (C219, Signet Laboratories, Dedham, USA) or anti-MRP1 (N19, Santa Cruz Biotechnology, Heidelberg, Germany) antibody (diluted 1:1000), respectively, at 4°C overnight, cells were washed and incubated with a secondary anti-mouse peroxidase antibody (diluted 1:1000) for 1 h. Thereafter, cells were incubated with a HRP-substrate (containing 0.5 mg/mL o-phenylenediamine, 11.8 mg/mL Na_2HPO_4 $2H_2O$, 7.3 mg/mL citric acid and 0.015% H_2O_2) for 15 min and measured photometrically at a wavelength of 490 nm using a microplate reader (Victor2, Wallac, Turku, Finland). In order to normalize pGP-expression for the number of cells in each well, the permeabilized cells were subsequently incubated with 0.2% trypan blue solution for 5 min, washed with PBS, dissolved in 1% SDS and the trypan blue concentration as a measure of cell number was determined photometrically.

19.2.4 Metabolic Parameters

The glucose and lactate concentrations of the medium were measured enzymatically using standard test kits (#1447521 and #1822837; Roche-Diagnostics, Indianapolis, USA).

19.2.5 Statistical Analysis

Results are expressed as means ± standard error of the mean (SEM). Differences between the groups were assessed using the two-tailed Wilcoxon test for unpaired samples. The significance level was set at $\alpha = 5\%$.

19.3 Results

Maintaining AT1 cells under pronounced hypoxia ($pO_2 < 0.5$ mmHg) for up to 24 h leads to forced glycolysis as indicated by a substantial increase in the lactate concentration of the medium up to 14.3±0.2 mmol/L (Fig. 19.1B). In parallel, the glucose level decreased to 1.1±0.4 mmol/L after 24 h under hypoxia (Fig. 19.1A). Since our intention was to solely study the effect of hypoxia on the transporter activity, the medium was buffered with 20 mM HEPES. For this reason, the pH decreased only slightly to 7.18±0.01 after 24 h (Fig. 19.1C).

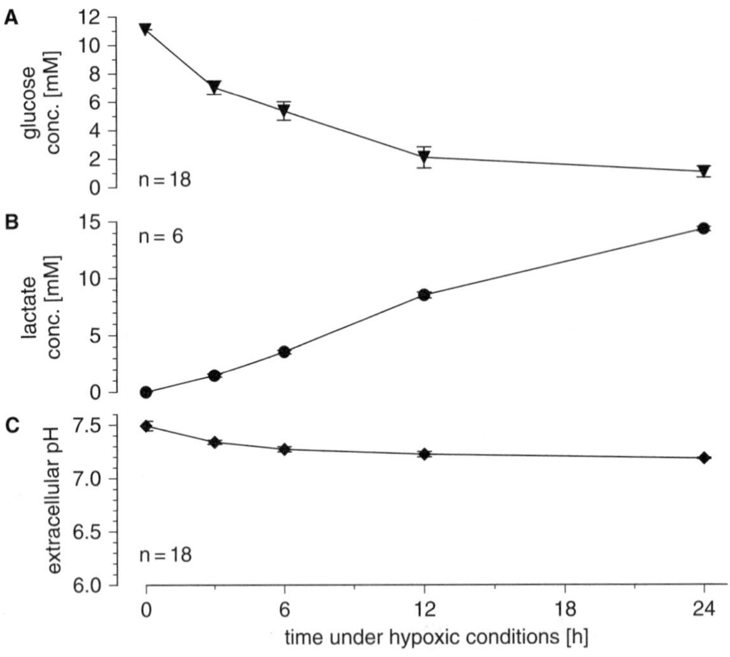

Fig. 19.1 (A) Glucose and (B) lactate concentrations as well as (C) the extracellular pH in the medium under hypoxic conditions ($pO_2 < 0.5$ mmHg) for up to 24 h. Data are expressed as mean ± SEM; n: number of experiments. All values were statistically significantly different from the values at t = 0 h.

The transport activity of pGP was measured by determining the rhodamine efflux rate in the absence and presence of an inhibitor. The ratio of the efflux without and with verapamil was used as a measure of the activity of pGP [7]. All values were normalized with respect to the pump activity under control conditions at t = 0 h. Exposing the cells to hypoxia for 3–6 h led to a moderate (but not statistically significant) increase in pGP activity by 19–24% (Fig. 19.2A). After 12 h under hypoxic conditions the rhodamine efflux rate returned to control levels whereas after 24 h pGP activity was reduced by 44±15% below control levels (Fig. 19.2A). These differences in activity could be either the result of an increase in cellular pGP-expression or a higher activity of the pre-existing transporters. The analysis of expression showed that hypoxic conditions over 3-6 h caused almost no change in pGP protein content (Fig. 19.2B). After 12 h, a small (however not statistically significant) increase by 19±8% in expression was seen. After 24 h the expression decreased 19±4% below the control level.

The analysis of the multidrug resistance-related protein 1 (MRP1) showed that hypoxic conditions for up to 6 h had almost no impact on the transporter activity (Fig. 19.3A). Thereafter the MRP1 transport rate decreased by 24 to 28%. MRP1 expression did not change appreciably throughout the observation period (Fig. 19.3B).

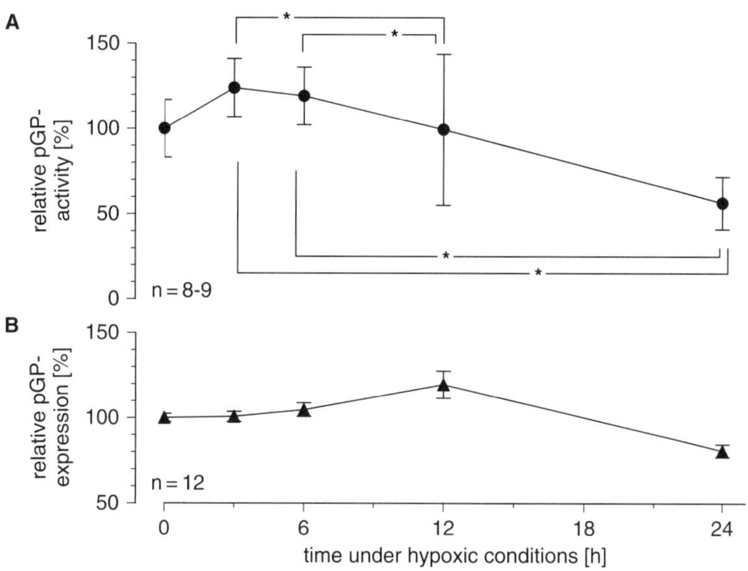

Fig. 19.2 (A) Activity and (B) expression of pGP (p-glycoprotein) in AT1 cells under hypoxic conditions ($pO_2<0.5$ mmHg) for up to 24 h. Data are normalized to control values at t = 0 h and expressed as mean ± SEM; n: number of experiments; (*) $p<0.05$.

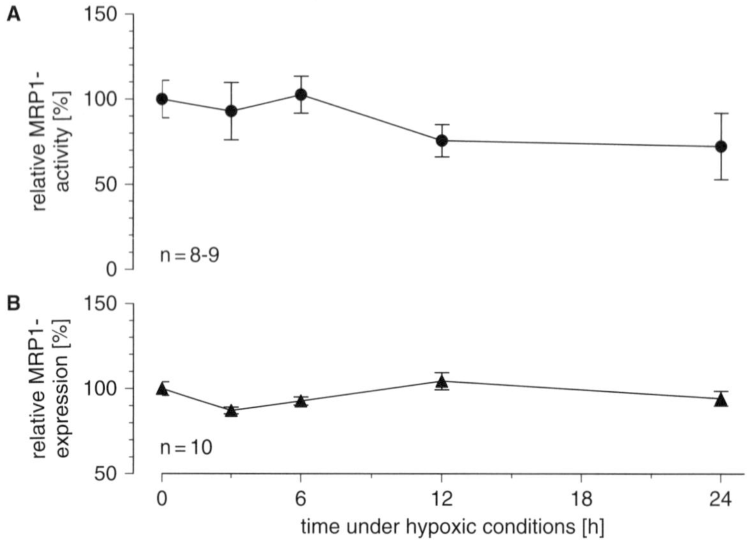

Fig. 19.3 Relative (A) activity and (B) expression of MRP1 (multidrug resistance related protein 1) in AT1 cells under hypoxic conditions (pO_2<0.5 mmHg) for up to 24 h. Data are normalized with respect to control values at t = 0 h and expressed as mean ± SEM; n: number of experiments.

19.4 Discussion

In the present study, the impact of hypoxia – a common phenomenon in human tumors [6] – on the activity and expression of two drug transporting proteins were analyzed in cell culture experiments. For this, tumor cells were exposed to a hypoxic atmosphere (pO_2<0.5 mmHg) for up to 24 h forcing glycolytic metabolism as indicated by a rapid increase in the extracellular lactate concentration (Fig. 19.1B). Since the aim of this study was to solely analyze the impact of hypoxia and not changes due to a concomitant extracellular acidosis, the medium was buffered so that the medium pH remained practically constant (Fig. 19.1C). This experimental design was necessary since previous results showed a medium pH of 6.6 (without hypoxia) to have a strong impact on the pGP-mediated transport rate [9].

The R-3327 AT-1 cell line used is known to functionally express the p-glycoprotein [4]. In the present study, the daunorubicin efflux from the cells was 1.75±0.37 times slower in the presence of the pGP-inhibitor verapamil. The cells also functionally express MRP1 as indicated by a 2.27±0.28 times lower calcein-AM efflux rate when cells were simultaneously incubated with the known MRP1-inhibitor probenecid. Since in more than 60% of human tumors the oxygen supply to the tissue is insufficient [6], solid tumors often show a pronounced glycolytic metabolism which may result in an ATP deficiency [5,10]

For this reason, it might be expected that ATP-consuming cellular processes such as active drug transport might be compromised. However, compared to control conditions, pGP-activity was even higher (approx. 20%, not statistically significant) despite the increased glycolytic metabolism (Fig. 19.2A). A reduction in the daunorubicin efflux of 44% was observed only after exposure to hypoxia for 24 h. After this period the glucose level in the medium decreased to 1.1±0.4 mmol/L which may in turn have reduced cellular ATP formation resulting in a reduction in pGP activity. This slight decrease in the transport rate after 24 h was also seen for the MRP1-mediated efflux (Fig. 19.3A). However, further measurements are necessary to analyze the cellular ATP level after 24 h of severe hypoxia. A decrease in ATP could be responsible for a reduction in drug transporter activity. It has been proposed that inhibition of aerobic metabolism (leading to a dramatic decrease in ATP levels) could be a possible therapeutic strategy for overcoming drug resistance. Xu and coworkers [11] showed that inhibition of mitochondrial respiration leads to an increase in the cytotoxic efficacy of various chemotherapeutic drugs, probably as a result of a reduced pGP-activity which was attributed to a depletion of ATP.

Despite the slight increase in pGP-activity 3 to 6 h after exposure to hypoxic conditions, the environmental hypoxia did not markedly change pGP-expression (Fig. 19.2B) over this time interval. These results are in accordance with previous studies showing that hypoxia/anoxia had almost no impact on pGP expression [12,13]. Only Comerford and colleagues [14] found an induction of MDR1 after 48 h in an O_2-deprived medium. However, in the latter study, hypoxia was defined as a pO_2 in the culture medium of 20 mmHg which is not comparable to the pO_2 used in the present study and which is much higher than the pO_2 found in solid-growing tumors [5,6]. The discrepancy between the studies may be the result of differences in the cell lines, levels of hypoxia and durations of hypoxia exposure used.

In conclusion, the results of the present study show pronounced hypoxia (comparable to that found in solid-growing tumors) to have only minor effects on the activity of drug transporters. The small increase in pGP activity after 3–6 h was not the result of a change in the number of transport molecules (as indicated by the expression measurements) but instead due to a functional modulation of the pump. However, previous studies using the same tumor cell model clearly demonstrated that indirect effects of hypoxia leading to extracellular acidosis have a much stronger effect on the p-glycoprotein activity. When cells were exposed to an extracellular pH of 6.6, the pGP-mediated transport rate was more than doubled resulting in a significant reduction of cytotoxicity of chemotherapeutic drugs known to be substrates for the p-glycoprotein [9]. Insufficient oxygen supply to tumor tissue seems to play a role in the chemosensitivity of tumor cells although this impact seems to be indirectly mediated through the acidic environment which develops due to tumor hypoxia.

Acknowledgment This study was supported by the Deutsche Krebshilfe (grants 106774 and 106906).

References

1. A.H. Schinkel and J.W. Jonker. Mammalian drug efflux transporters of the ATP binding cassette (ABC) family: an overview. *Adv. Drug Deliv. Rev.* **55**, 3–29 (2003).
2. S.V. Ambudkar, C. Kimchi-Sarfaty, Z.E. Sauna, and M.M. Gottesman. P-glycoprotein: from genomics to mechanism. *Oncogene* **22**, 7468–7485 (2003).
3. T. Fojo and S. Bates. Strategies for reversing drug resistance. *Oncogene* **22**, 7512–7523 (2003).
4. M.J. Siegsmund, C. Kreukler, A. Steidler, T. Nebe, K.U. Kohrmann, and P. Alken. Multidrug resistance in androgen-independent growing rat prostate carcinoma cells is mediated by P-glycoprotein. *Urol. Res.* **25**, 35–41 (1997).
5. P. Vaupel, F. Kallinowski, and P. Okunieff. Blood flow, oxygen and nutrient supply, and metabolic microenvironment of human tumors: a review. *Cancer Res.* **49**, 6449–6465 (1989).
6. M. Höckel and P. Vaupel. Tumor hypoxia: definitions and current clinical, biological, and molecular aspects. *J. Natl. Cancer Inst.* **93**, 266–276 (2001).
7. G. Lee and M. Piquette-Miller. Cytokines alter the expression and activity of the multidrug resistance transporters in human hepatoma cell lines; analysis using RT-PCR and cDNA microarrays. *J. Pharm. Sci.* **92**, 2152–2163 (2003).
8. H.H. Versteeg, E. Nijhuis, G.R. van den Brink, M. Evertzen, G.N. Pynaert, S.J. van Deventer, P.J. Coffer, and M.P. Peppelenbosch. A new phosphospecific cell-based ELISA for p42/p44 mitogen-activated protein kinase (MAPK), p38 MAPK, protein kinase B and cAMP-response-element-binding protein. *Biochem. J.* **350**, 717–722 (2000).
9. O. Thews, B. Gassner, D.K. Kelleher, G. Schwerdt, and M. Gekle. Impact of extracellular acidity on the activity of p-glycoprotein and the cytotoxicity of chemotherapeutic drugs. *Neoplasia* **8**, 143–152 (2006).
10. P. Vaupel, C. Schaefer, and P. Okunieff. Intracellular acidosis in murine fibrosarcomas coincides with ATP depletion, hypoxia, and high levels of lactate and total P_i. *NMR Biomed.* **7**, 128–136 (1994).
11. R.H. Xu, H. Pelicano, Y. Zhou, J.S. Carew, L. Feng, K.N. Bhalla, M.J. Keating, and P. Huang. Inhibition of glycolysis in cancer cells: a novel strategy to overcome drug resistance associated with mitochondrial respiratory defect and hypoxia. *Cancer Res.* **65**, 613–621 (2005).
12. B.C. Liang. Effects of hypoxia on drug resistance phenotype and genotype in human glioma cell lines. *J. Neurooncol.* **29**, 149–155 (1996).
13. K. Sakata, T.T. Kwok, B.J. Murphy, K.R. Laderoute, G.R. Gordon, and R.M. Sutherland. Hypoxia-induced drug resistance: comparison to P-glycoprotein-associated drug resistance. *Br. J. Cancer* **64**, 809–814 (1991).
14. K.M. Comerford, T.J. Wallace, J. Karhausen, N.A. Louis, M.C. Montalto, and S.P. Colgan. Hypoxia-inducible factor-1-dependent regulation of the multidrug resistance (MDR1) gene. *Cancer Res.* **62**, 3387–3394 (2002).

Chapter 20
Antioxidants Reduce Consequences of Radiation Exposure

Paul Okunieff, Steven Swarts, Peter Keng, Weimin Sun, Wei Wang, Jung Kim, Shanmin Yang, Hengshan Zhang, Chaomei Liu, Jacqueline P. Williams, Amy K. Huser, and Lurong Zhang[1]

Abstract Antioxidants have been studied for their capacity to reduce the cytotoxic effects of radiation in normal tissues for at least 50 years. Early research identified sulfur-containing antioxidants as those with the most beneficial therapeutic ratio, even though these compounds have substantial toxicity when given in-vivo. Other antioxidant molecules (small molecules and enzymatic) have been studied for their capacity to prevent radiation toxicity both with regard to reduction of radiation-related cytotoxicity and for reduction of indirect radiation effects including long-term oxidative damage. Finally, categories of radiation protectors that are not primarily antioxidants, including those that act through acceleration of cell proliferation (e.g. growth factors), prevention of apoptosis, other cellular signaling effects (e.g. cytokine signal modifiers), or augmentation of DNA repair, all have direct or indirect effects on cellular redox state and levels of endogenous antioxidants. In this review we discuss what is known about the radioprotective properties of antioxidants, and what those properties tell us about the DNA and other cellular targets of radiation.

20.1 Introduction

There are many types of radiation damage to normal tissues. The types of damage depend on the cells and organs being irradiated, the dose and dose rate of the exposure, and the time after exposure that is being assayed for a radiation effect. Many of the types of damage seen after irradiation can be ameliorated by antioxidants. This review will outline a number of radiation-related toxicological processes and discuss the role antioxidants might play in affecting these processes in terms of the likely cellular types or compartments in

[1]Paul Okunieff, Steven Swarts, Peter Keng, Weimin Sun, Wei Wang, Jung Kim, Shanmin Yang, Hengshan Zhang, Chaomei Liu, Jacqueline P. Williams, Amy K. Huser, and Lurong Zhang, Department of Radiation Oncology, University of Rochester Medical Center, Rochester, NY 14642.

which an antioxidant is employed. The role that different combinations of antioxidants might play in preventing each of these individual effects will also be explored.

20.2 Cell Components

Exposure of a cell to ionizing radiation results in the formation of free radicals within the cell, leading to damage of cellular components. Here we will provide some examples of how antioxidants reduce or prevent the damaging effects of radiation at three sensitive targets in the cell, the nucleus, cellular membranes and mitochondria.

20.2.1 Nucleus

20.2.1.1 Immediate Effects by Antioxidants

Radiation-induced DNA damage is the best studied effect of radiation. An oxygen enhancement ratio (OER) of 2.5 to 3 in the yield of DNA damage is observed in the presence of oxygen tensions of 5 mmHg or higher compared to maximally hypoxic conditions (<1 mmHg). In accordance with this difference in DNA damage, there is a 3-fold difference in cell reproductive survival measured by clonogenic assays in the presence of oxygen which is generally independent of the phase of the cell cycle [1]. Prevention of immediate radiation-induced genotoxicity requires that an antioxidant be present at the time of irradiation [2]. To be maximally effective the antioxidant must be present near the DNA and thus must have access to the nucleus. It must be able to either, 1) react with all the oxygen-related free radicals and detoxify them to radicals that are not themselves genotoxic and/or 2) effectively compete with oxygen to repair damage to the DNA chemically through reactions with free radicals on the DNA. Thiol-based compounds are especially good antioxidants because these compounds are capable of both scavenging oxygen radicals and affecting chemical repair of some forms of DNA damage with the subsequent formation of sulfur-based radicals, which are not reactive with DNA [3]. Incorporating one or more positive charges on the thiol-based antioxidant has the effect of changing the proximity of the compound to the DNA [4,5]. The resulting counter-ion condensation between the positive charge of the thiol and the negatively charged sugar-phosphate backbone of the DNA binds the thiol close to the DNA, facilitating the competition of the thiol with oxygen in reactions with DNA radicals, thereby, reducing DNA damage and increasing cell survival [5,6].

Like the synthetic antioxidants (e.g., amifostine, captopril, and NAC), antioxidants derived from natural sources also exhibit dose-modifying effects on DNA damage and cell survival when present at the time of irradiation. This immediate protection is mediated by the scavenging of radicals. For example, there are a number of antioxidants, including caffeine, melatonin, flavonoids, polyphenols, and other phytochemicals (e.g., albana), which are shown to decrease radiation-induced damage in either plasmid or cellular DNA through the scavenging of oxygen radicals and/or peroxides [7–12].

Uptake and distribution of antioxidants also plays a role in their dose-modifying effects. With amifostine, there is differential uptake of the compound in tumors and normal tissues. In tumors, the uptake is predominantly through passive diffusion, which is slow due to the hydrophilicity of the compound [13]. This is in contrast to the dephosphorylated form of the compound, WR-1065, which is less hydrophilic and readily crosses the tumor cell membrane. In fact, Brown et al [14]. suggested that the hydrophilicity of the compound could be useful for designing or selecting better differential radioprotectors. This is supported by their work that showed increases in the therapeutic gain (ratio of the dose reduction factors for the hematopoietic system and tumor) for 6 hydrophilic thiols, ranging from 1.59 to 2.29, in comparison to values ranging from 0.88 to 1.59 for 5 lipophilic thiols. In normal tissues, there is active uptake of amifostine through the polyamine transport system [15]. This active transport results in a preferential uptake of the antioxidant into normal tissues as compared to tumors [13]. Another factor that aids in the differential uptake of amifostine into normal tissues is the higher concentration of alkaline phophatases in these tissues as compared to tumors, converting amifostine to WR-1065, which is then readily taken up by normal tissues [16]. However, the levels and distributions of the aminothiol can vary between and within tissues, leading to variations in the dose-modifying effects of this compound in irradiated tissues [17]. These variations can be attributed to differing degrees of negative feedback on the polyamine transport as a consequence of variable polyamine concentrations with tissues, thereby reducing or inhibiting the uptake of amifostine, and to differences in oxygen concentration within tissues [17,18]. Also, there is a limit to which cells can take up and accumulate thiol-based antioxidants before the compounds become cytotoxic. For example, in several tumor lines, concentrations of WR-1065 greater than 25–30 nmole/10^6 cells induced significant cytotoxicity in unirradiated cells [19]. In-vivo these agents cause peripheral neuropathies and hypotension [20]. Clinically speaking, although the impact of amifostine and similar thiol-based antioxidants can theoretically be as great as a factor of 3 in dose modification, no antioxidant reaches this potential and few if any alter the in-vivo tolerance to irradiation by more than a factor of 1.3 when administered at concentrations below those that elicit hemodynamic- or cyto-toxicities in unirradiated controls [2,11,14].

20.2.1.2 Chronic Radioprotective Effects by Antioxidants

For many antioxidants, the impact that these compounds have on radiation-induced damage and the biological consequences of the damage within cells and tissues can extend to their direct or indirect interaction with other cellular targets. For example, melatonin has been shown to augment the activity of glutathione peroxidase in addition to stimulating the activity of glutathione reductase and increasing the synthesis of glutathione (GSH); all of which are important in reducing levels of oxygen radicals and peroxides in cells [8]. In addition, WR-1065 has been shown to induce a delayed radioprotective effect through the activation of the redox-sensitive nuclear transcription factor, NFκB, and subsequent expression of the antioxidant enzyme, manganese superoxide dismutase (MnSOD). Other thiols, such as captopril ([S]-1-[3-mercapto-2-methyl-1-oxo-propyl]-L-proline), mesna (sodium-2-mercapto-ethane-sulfonate), and NAC demonstrated similar effects to those observed for WR-1065 with respect to increasing cell survival [21]. The involvement of NFκB in the induced expression of MnSOD was shown in experiments where pretreatment of human microvascular endothelial cells with Helenalin, an inhibitor of NFκB, prior to treatment with WR-1065, prevented the thiol-induced activation of NFκB and subsequent elevation in MnSOD levels [22].

20.2.2 Membranes

The irradiation of lipid membranes is known to cause an increase in the formation of lipid radicals and peroxides that can result in damage or release of membrane proteins [23], in addition to the liberation of products formed from the peroxidation of lipids that subsequently react with and alter cellular components [24]. Various natural and synthetic antioxidants are known to decrease the peroxidation of membrane lipids. For example, pretreatment of mice with diethyldithiocarbamate (DDTC) prior to whole body dose resulted in a two-fold decrease in lipid peroxidation in isolated liver microsomes, as compared with irradiated control mice [25]. In a recent study, disulfiram, a drug that is used in treating alcohol abuse, inhibited lipid peroxidation in microsomes and reduced lipid peroxides in whole-body-irradiated mice by 65% compared with unirradiated controls [26]. The flavonoid, luteolin, reduced lipid peroxidation by almost 4-fold 48 h post-radiation in comparison with radiation controls in mouse bone marrow cells when mice were pre-treated with the flavonoid for 2 hours before irradiation [27]. In a variation on structural design, the antioxidant, tocopherol-monoglucoside (TMG), is a water soluble derivative of the lipophilic parent compound, α-tocopherol. This structural modification allows TMG to scavenge oxygen radicals, such as peroxides, superoxides, and hydroxyl radicals, in both water and lipid phase [28]. Additionally, the compound also increases the levels of glutathione peroxidase (GPx) in treated cells.

It is known that the active form of vitamin E in membranes is maintained through reactions with ascorbic acid [29]. Without this regenerative mechanism, the active form of vitamin E would be rapidly exhausted in membranes. Therefore, the optimal properties of antioxidants designed to protect cellular membranes are,

1) an ability to scavenge lipid radicals and react with lipid peroxides in membranes at concentrations that will not alter the structure or properties of the membrane, and 2) provide for the maximum interaction of the compound with cytosolic-reducing agents (ascorbic acid or GSH) to regenerate the antioxidant. This strategy also necessitates the use of multiple antioxidant therapy, for example the combination of vitamin E and vitamin C, which provide both an effective protection of membranes and increased radioresistance in cells [30, 31].

20.2.3 Mitochondria

The mitochondrion is the cellular organelle responsible for energy generation in the cell through the production of ATP [23]. Mitochondria, like the nucleus, contain DNA and this DNA is required for proper mitochondrial function and for mitochondrial replication. Replication of mitochondria occurs naturally in non-dividing cells. The impact of radiation on mitochondrial DNA likely does not result in changes in reproductive integrity and thus clonogenic survival, which is perhaps why it is rarely studied. Long-term cellular health however clearly requires cells to have a continuous supply of mitochondria for normal functioning. Mitochondrial DNA has the advantage over nuclear DNA in that it is present in many replicates (instead of just duplicate), can increase the number of DNA copies in response to radiation exposure, and the mitochondria is naturally high in antioxidant capacity [23,32]. In comparison, however, to the nuclear DNA, nucleotide excision repair of mitochondrial DNA is lacking [33] and repair is not efficient for specific classes of DNA damage, such as bulky lesions, and some types of alkaline-labile sites and single strand breaks [34,35] (Table 20.1). Also, although not yet shown, the fidelity of the repair of radiation-induced damage at clustered sites in mitochondrial DNA is likely to be adversely impacted in a similar fashion to clustered lesions in nuclear DNA [36–38].

A consequence of mitochondrial energy generation (ATP synthesis) is the evolution of heat (entropy) and the production of ROS. Mitochondria have an inherent antioxidant capacity (e.g., the interaction between GSH, GPx, glutathione reductase [GRd], and MnSOD) to counteract much of the ROS. Stressors, such as ionizing radiation, damage the mitochondrial function, likely leading to additional ROS production which can overwhelm the antioxidant capacity of the organelle. The unscavenged ROS may produce further damage to mitochondrial components, including mitochondrial DNA, leading to additional mitochondrial damage and ROS formation. Providing additional antioxidant capacity to mitochondria, either through uptake of additional antioxidant agents like vitamin E, or through increasing the levels of GSH and mitochondrial antioxidant enzymes, can provide the necessary antioxidant buffer to scavenge additional ROS produced as a consequence of exposure to radiation and thereby minimize damage to mitochondria and its DNA.

Table 20.1 Characteristic differences between DNA in the nucleus and mitochondria

Parameter	Nucleus	Mitochondria	Advantage
Target Size	Under 30,000 genes	37 genes	*Mitochondria*
DNA/Gene Ratio	High	Low	*Nucleus*
Oxygen Tension	Normoxic	Potentially Hypoxic	*Mitochondria*
Repair Capacity	>99.9% SSB and 98% DSB repaired	Low repair	*Nucleus*
Gene Copies	One duplicate copy per cell	High number of replicates per cell	*Mitochondria*
Radical Levels	Low radical environment	High radical environment	*Nucleus*
Antioxidant Level	Moderate antioxidant environment	High antioxidant environment	*Mitochondria*

The DNA in the nucleus and mitochondria have different oxidative environments and mechanisms for repair of oxidative damage. This leads to different temporal and functional DNA damage responses following irradiation. Mitochondrial DNA has an advantage in the case of radiation due to its small mass, its large number of replicates, and its naturally high antioxidant capacity. Nuclear DNA enjoys a powerful set of enzymatically mediated DNA repair pathways; mitochondrial DNA instead relies more on the presence of antioxidants. Due to the lower degree of repair capability and fidelity of direct damage in mitochondrial DNA, continuous low dose rate radiation and very late manifestation of radiation damage might be a relative disadvantage to mitochondrial DNA compared with nuclear DNA. After therapeutic radiation or other high dose or high dose rate exposure, early cytotoxicity is probably not due to DNA damage of the mitochondria. There are no comprehensive studies of late radiation toxicity to the mitochondria, so the degree to which this organelle impacts certain radiation scenarios months or years after exposure remains unknown.

The antioxidant, melatonin, is particularly effective at protecting mitochondria by increasing the efficiency of oxidative phosphorylation, thereby reducing the leakage of electrons from the electron transport chain [8]. The reduction of electron leakage decreases the formation of ROS from these electrons and, therefore, damage to mitochondria. Additionally, melatonin induces the levels of antioxidant enzymes, such as GPx and, more importantly, also increases GSH levels within the cell. This latter effect can reduce the levels of radiation-induced oxygen radicals and peroxides in mitochondria through the increased availability of glutathione for GSH/GSSG cycling that is used in regenerating GPx [8]. A similar redox cycle has been proposed for WR-1065 to explain the regeneration of the thiol after it is converted to the disulfide form following reactions with lipid peroxides in the mitochondrial membrane. In this case, the disulfide form of WR-1065 is recycled to the reduced state through the oxidation of GSH to the disulfide, GSSG. The GSSG is then reduced to GSH by GRd [39].

Protection of the mitochondria can be further facilitated through the development of antioxidants that are designed either for increased uptake into mitochondria, or to increase the activity of antioxidant enzymes. Linking the positively-charged functional group, alkyl-triphenyl-phosphonium ion, to vitamin E or ubiquinone (CoQ) increased the uptake of these antioxidants into the mitochondrial matrix [23]. However, studies to determine how this structural modification might influence the radioprotection of mitochondria have not yet been performed. Increasing the levels of antioxidant enzyme activity in mitochondria has been shown to occur with the administration of SOD mimetics or through over-expression of MnSOD by transfection of a transgene [40]. Another approach to increasing mitochondrial content of an antioxidant is to take advantage of the low pH outside of the inner membrane of the mitochrondrion whereby functional groups on the compound undergo protonation to change the charge on the molecule and, thereby, prevent the elimination of the compound from the mitochondrion.

20.3 Apoptosis

Reactive oxygen species play a pivotal role in the initiation of apoptosis, and antioxidants have been shown to have the ability to inhibit apoptosis. This inhibitory effect appears to occur through a number of pathways but has as a common result, the preservation of mitochondrial membrane integrity and the electrochemical gradient (ΔP) across the membrane. It is suggested that scavenging of ROS by antioxidants interferes with the initiation of apoptosis by depleting ROS levels in cells and maintaining membrane integrity [41]. Also, antioxidants like the water soluble vitamin E derivative, trolox, reduce both lipid membrane peroxidation and the post-irradiation uptake of calcium, thereby inhibiting apoptosis [42]. Reduction in lipid peroxides and decreased apoptotic indices were also found in irradiated mice treated either with SOD or, more effectively, with the combination of catalase and trolox [43].

Antioxidants also have the ability to affect apoptosis through inhibiting proteins in the apoptotic cascade or modification of gene expression. By inhibiting the cleavage of caspase-3 and its substrate, poly(ADP-ribose) polymerase, the green tea polyphenol, (-)-epigallocatechin, was found to prevent apoptosis in HaCaT human keratinocytes when pretreated 16 h before irradiation [44]. Pretreatment of human microvascular epithelial cells with WR-1065 thirty min prior to irradiation was found to down-regulate a host of genes associated with apoptosis [45]. A greater than two-fold reduction in the expression of 12 genes was observed, including the caspases 2, 4, and 9; the cyclins A, G1, G2, and D3; the DNA check damage/checkpoint proteins, ATM, DNA-PK, and RAD 23B; TNF receptor 1; and FAST kinase. Also, treatment with WR-1065 significantly reduced the accumulation of cells in an apoptotic sub-G1 population 1–2 days following irradiation to levels that were not statistically distinguishable ($p<0.05$) from non-irradiated cells. What is not clear from these results are the relative contributions of the radical scavenging properties and the modifying effect that amifostine has on gene expression to the observed reduction in apoptosis.

20.4 Tissue-based Radiation Effects

Late fibrovascular effects of radiation include vascular dysfunction. Ischemia itself injures tissue, and ischemia followed by reperfusion is thought to further the injury through the production of a rapid burst of ROS. This rapid production of ROS can overwhelm the antioxidant capacity of the tissue and lead to free radical-mediated damage to all intracellular and tissue compartments [46]. This can be especially problematic under conditions of chronic ischemia where recurrent injury to tissues is expected to occur. The involvement of chronic oxidative stress, and concomitant production of ROS, has been suggested as a driving force in the amplification of late radiation effects such as fibrosis, chronic inflammation, and oncogenesis in irradiated tissues [47,48]. Consequently, increasing the antioxidant capacity of the involved tissues is expected to reduce tissue injury due to ROS-mediated late radiation effects. However, the use of antioxidants to reduce the effects of chronic ROS-mediated injury in post-radiation treatments has not been sufficiently studied and therefore the mechanisms by which these effects are mediated are ill-defined. Below, we provide some examples of what is known from studies of post-radiation administration of protein and non-protein based antioxidants on late radiation effects in tissues.

20.4.1 Inflammatory Mediators

Under conditions of chronic oxidative stress, as would be encountered in irradiated tissues, the generation of ROS triggers an inflammatory response through the activation of cytokines and other inflammatory mediators [49]. The administration of antioxidants in animal and human studies has the effect of reducing the inflammatory response through the modulation of cytokine levels in tissues. For example, epicatechin, trans-resveratrol, and theaflavin were shown to reduce the production of interleukin 1-β (IL-1β), tumor necrosis factor-α (TNF-α) and interleukin-8 (IL-8), respectively, after stimulation of an inflammatory response [50–52]. Conversely, catalase and NAC have the ability to upregulate interleukin-10 expression which, in turn, decreased the synthesis of other cytokines [53]. Evidence also exists suggesting the role of antioxidants in reducing ROS and inflammation in late radiation-induced tissue injury. Protection of late radiation-induced lung injury was observed in mice over-expressing a transgene for human MnSOD [54]. In another approach, increasing SOD levels through the treatment of rats 15 min prior to irradiation and 5 days post-radiation with a SOD mimetic resulted in both a reduction of collagen deposition in lung tissue and a 1.2–2.1 fold reduction in transforming growth factor-β (TGF-β) 10 to 14 weeks post-radiation [55]. In a pig model, treatment with Cu/Zn SOD or MnSOD 3 times a week for 3 weeks post-radiation resulted in a softening and shrinkage of fibrotic tissue in a cutaneous radiation field that had received a single 160 Gy dose 6 months prior to treatment with antioxidant [56]. Also in a pig model, the co-administration of pentoxifylline [PTX] and α-tocopherol for 26 weeks, after a 26 week post-radiation development of a subcutaneous fibrosis, resulted in a decrease of TGFβ-1 levels in residual scar tissue (26 weeks post-irradiation) as compared with groups receiving pentoxifylline + irradiation or irradiation alone [57]. In human studies, the combined treatment with PTX and α-tocopherol post-irradiation appeared more effective at reducing radiation-induced fibrotic tissue in skin than when PTX or α-tocopherol were given alone [58]. PTX is expected to reduce reperfusion injury and was shown in clinical studies to lower the levels of circulating bFGF (basic fibroblast growth factor) and TNF-α toward non-irradiated control levels [59]. In cultured fibroblasts harvested from normal or radiation-induced fibrotic human skin, treatment of the fibroblasts harvested from fibrotic tissue with liposomal Cu/Zn SOD resulted in increased expression of MnSOD and decreased levels of TGF-β1, but no significant changes in the levels of these parameters were observed in treated fibroblasts harvested from normal skin [60]. Thus, it can be seen that strategies for increasing SOD levels in post-irradiated tissues result in the protection of the tissues from late radiation-induced effects through an apparent reduction in the ROS-mediated damage and the decreased expression of at least one cytokine, TGF-β. Finally, the best treatment for many chronic radiation-induced soft tissue injuries is hyperbaric oxygen. The mechanisms of action of hyperbaric dives is not certain but includes natural induction of SOD and other antioxidants, and is associated with inhibition of inflammation and improved tissue vascularization.

Interestingly, cytokines can be radioprotective through induction of SOD levels. For example, pretreatment of mice with interleukin-1 twenty hours before receiving a lethal dose (8 Gy) of radiation was found to enhance the radioresistance of bone marrow cells [61]. It has been suggested that one reason for the radioprotective effect of the cytokine is the increased expression of MnSOD in bone marrow cells that resulted from the cytokine pretreatment. Similarly, tumor necrosis factor-α (TNF-α) has been shown to induce MnSOD in hematopoietic stem cells with a concomitant radioprotective effect [62].

At the level of tissue vasculature, irradiation of endothelial cells results in the increased expression of intercellular cell adhesion molecule-1 (ICAM-1). Expression of ICAM-1 contributes to an inflammatory response that mediates the adhesion and movement of leucocytes to and through the vascular endothelium. ROS are assumed to be involved in the increased expression of ICAM-1, presumably through the AP-1 signaling pathway [63,64]. Therefore, it is expected that a reduction in ROS by reactions with antioxidants, for example, should reduce radiation-induced expression of ICAM-1. However, pretreatment of human umbilical vein endothelial cells with the thiols NAC and pyrrolidine dithiocarbamate (PDTC) followed by a 7 Gy dose found that neither thiol reduced the radiation-induced expression of ICAM-1 at a post-radiation time of 48 h. Instead, thiol treatment increased expression by up to 2-fold over cells irradiated alone [64]. In fact, just pretreatment of cells without exposure to radiation resulted in increased expression of ICAM-1 that was 1–2 fold higher than cells irradiated alone within 48 h post-treatment. This latter result suggests that these thiols can be considered, under certain conditions, as pro-inflammatory agents. However, it is not known whether this extends to other antioxidants, especially thiol-based antioxidants including amifostine.

20.5 Conclusion

The radioprotective effects of antioxidants and the mechanisms by which these effects are mediated depend on the properties of both the antioxidant and the compartment (e.g., cellular or tissue targets) where the radioprotective effects are measured. There is a large volume of data on the radioprotective effects of antioxidants at the cellular level, especially at the level of nuclear DNA, where the radical scavenging by the antioxidant protects this and other sensitive cellular targets. Many antioxidants have been shown to also protect the cell by acting to increase cellular antioxidant capacity through their ability to elevate the levels of natural antioxidants (e.g., GSH) and antioxidant enzymes (e.g., GPx, GRd and MnSOD). Interestingly, exposure to chronic, low-dose-rate ionizing radiation can also lead to the induction of antioxidant enzymes. For example, exposure of mice to a 0.5 Gy at a dose rate of 1.2 mGy/h for 23 days increased the gene expression of catalase and MnSOD by a factor of 2.5 [65]. However, at higher doses of 1.0 and 1.3 Gy accumulated at the same dose rate, gene expression either increased by only approximately 1.4 or was not significantly different from unirradiated controls, respectively. Therefore, care is needed in low-dose-rate studies in discerning to what extent various agents, like antioxidants, have on modifying the levels of antioxidant enzymes. Even so, based on what is currently known, specific chemical and/or physical properties of antioxidants can be designed to take advantage of biochemical properties or a specific cellular target. In addition, there are a number of in vitro and in vivo studies that show increased radioresistance in normal tissues when antioxidants are given in combination compared with antioxidants given individually [30,31]. There are a number of hypotheses that have been suggested to explain the enhanced radioprotective effect of combined antioxidant treatments related to the regulation and response to ROS, including the regeneration of vitamin E and other antioxidants by vitamin C, induction of cellular antioxidant systems, and interaction with inflammatory mediators.

The impact of radiation on the mitochondrial DNA and thus long-term reproductive health of the mitochondria, reproduction of the cell, and on cellular redox and energy state has not been studied in detail. The long-term consequences of radiation may be very dependent on this mechanism of radiation toxicity and may be greatly alleviated by properly designed antioxidants.

Regarding what is known about the radioprotective effects of antioxidants on late radiation effects in tissues, especially for non-protein antioxidants, there is only a limited understanding of these effects at a mechanistic level. Therefore, additional studies are needed of current and new antioxidant compounds to look at these and other radioprotective effects in antioxidants in irradiated cells and tissues to support rational approaches in the design of antioxidants as radioprotectors.

Acknowledgment This research was supported by the Center for Medical Countermeasures against Radiation Program, U19-AI067733, National Institute of Allergy and Infectious Diseases.

References

1. J. P. Freyer, K. Jarrett, S. Carpenter, et al., Oxygen enhancement ratio as a function of dose and cell cycle phase for radiation-resistant and sensitive CHO cells, *Radiat. Res.* **127**:297–307 (1991).
2. D. J. Grdina, J. S. Murley, and Y. Kataoka, Radioprotectants: Current status and new directions, *Oncology* **63**(suppl. 2):2–10 (2002).
3. K. D. Held, Models for thiol protection of DNA in cells, *Pharmac. Ther.* **39**:123–131 (1988).
4. G. D. Smoluk, R. C. Fahey, and J. F. Ward, Interaction of glutathione and other low-molecular weight thiols with DNA: evidence for counterion condensation and coion depletion near DNA, *Radiat. Res.* **114**:3–10 (1988).
5. S. Zheng, G. L. Newton, J. F. Ward, et al., Aerobic radioprotection of pBR322 by thiols: effect of thiol net charge upon scavenging of hydroxyl radicals and repair of DNA radicals, *Radiat. Res.* **130**:183–193 (1992).
6. D. Murray, A. Prager, S. C. Vanankeren, et al., Comparative effect of the thiols dithiothreitol, cysteamine and WR-151326 on survival and on the induction of DNA damage in cultured Chinese hamster ovary cells exposed to γ-radiation, *Int. J. Radiat. Biol.* **58**:71–91 (1990).
7. S. S. Kumar, T. P. A. Devasagayam, B. Jayshree, et al., Mechanism of protection against radiation-induced DNA damage in plasmid pBR322 by caffeine, *Int. J. Radiat. Biol.* **77**:617–623 (2001).
8. R. J. Reiter, D. Tan, J. C. Mayo, et al., Melatonin as an antioxidant: biochemical mechanisms and pathophysiological implications in humans, *Acta. Biochem. Pol.* **50**:1129–1146 (2003).
9. D. K. Maurya, V. P. Salvi, and C. K. K. Nair, Radioprotection of normal tissues in tumor-bearing mice by troxerutin, *J. Radiat. Res.* **45**:221–228 (2004).
10. P. Uma Devi, K. S. Bisht, and M. Vinitha, A comparative study of radioprotection by Ocimum flavonoids and synthetic animothiol protectors in the mouse, *Brit. J. Radiol.* **71**:782–784 (1998).
11. J. F. Weiss, and M. R. Landauer, Protection against ionizing radiation by antioxidant nutrients and phytochemicals, *Toxicology* **189**:1–20 (2003).
12. B. Frei, and J. V. Higdon, Antioxidant activity of tea polyphenols in vivo: evidence from animal studies, *J. Nutr.* **133**:3275S–3284S (2003).
13. J. M. Yuhas, M. E. Davis, D. Glover, et al., Circumvention of the tumor membrane barrier to WR-2721 absorption by reduction of the drug hydrophilicity, *Int. J. Radiat. Oncol.* **8**:519–522 (1982).
14. D. Q. Brown, J. M. Yuhas, L. J. MacKensie, et al., Differential radioprotection of normal tissues by hydrophilic chemical protectors, *Int. J. Radiat. Biol.* **10**:1581–1584 (1984).
15. G. L. Newton, J. A. Aguilera, T. Kim, et al., Transport of aminothiol radioprotectors into mammalian cells: passive diffusion versus mediated uptake. *Radiat. Res.* **146**:206–215 (1996).

16. V. Santini, and F. J. Giles, The potential of amifostine: from cytoprotectant to therapeutic agent, *Haematologica* **84**:1035–1042 (1999).
17. J. M. Yuhas, S. M. J. Afzal, and V. Afzal, Variation in normal tissue responsiveness to WR-2721, *Int. J. Radiat. Oncol.* **10**:1537–1539 (1984).
18. H. I. Quiñones, A. F. List, and E. W. Gerner, Selective exclusion by the polyamine transporter as a mechanism for differential radioprotection of amifostine derivatives, *Clin. Cancer Res.* **8**:1295–1300 (2002).
19. P. M. Calabro-Jones, J. A. Aguilera, J. F. Ward, et al., The limits to radioprotection of Chinese hamster V79 cells by WR-1065 under aerobic conditions, *Radiat. Res.* **149**:550–559 (1998).
20. C. R. Cully, and C. M. Spencer, An update on its clinical status as a cytoprotectant in patients with cancer receiving chemotherapy or radiotherapy and its potential therapeutic application in myelodysplasia syndrome, *Drugs* **61**:641–684 (2001).
21. J. S. Murley, Y. Kataoka, D. Cao, et al., Delayed radioprotection by NFκB-mediated induction of SOD2 (MnSOD) in SA-NH tumor cells after exposure to clinically used thiol-containing drugs, *Radiat. Res.* **162**:536–546 (2004).
22. J. S. Murley, Y. Kataoka, C. J. Weydert, et al., Delayed radioprotection by nuclear transcription factor κB-mediated induction of manganese superoxide dismutase in human microvascular endothelial cells after exposure to the free radical scavenger, WR1065, *Free Rad. Biol. Med.* **40**:1004–1016 (2006).
23. D. C. Wallace, The mitochondrial genome in human adaptive radiation and disease: on the road to therapeutics and performance enhancement, *Gene* **354**:169–180 (2005).
24. L. J. Marnett, Oxy radicals, lipid peroxidation and DNA damage, *Toxicology* **181**:219–222 (2002).
25. N. M. Gandhi, and C. K. K. Nair, Radiation protection by diethyldithiocarbamate: protection of membrane and DNA in vitro and in vivo against γ-irradiation, *J. Radiat. Res.* **45**:175–180 (2004).
26. N. M. Gandhi, U. V. Gopalaswamy, and C. K. K. Nair, Radiation protection by disulfiram: protection of membrane and DNA in vitro and in vivo against γ-radiation, *J. Radiat. Res.* **44**:255–259 (2003).
27. K. Shimoi, S. Masuda, B. Shen, et al., Radioprotective effects of antioxidative plant flavonoids in mice, *Mutation Res.* **350**:153–161 (1996).
28. N. Cherdyntseva, A. Shishkina, I. Butorin, et al., Effect of tocopherol-monoglucoside (TMG), a water-soluble glycosylated derivate of vitamin E, on hematopoietic recovery in irradiated mice, *J. Radiat. Res.* **46**:37–41 (2005).
29. J. R. Woods, M. A. Plessinger, and R. K. Miller, Vitamins C and E: missing links in preventing preterm premature rupture of membranes. *Am. J. Obstet. Gynecol.* **185**:5–10 (2001).
30. K. N. Prasad, Rationale for using high-dose multiple dietary antioxidants as an adjunct to radiation therapy and chemotherapy, *J. Nutr.* **134**:3182S–3183S (2004).
31. K. N. Prasad, Rationale for using multiple antioxidants in protecting humans against low doses of ionizing radiation, *Br. J. Radiol.* **78**:485–492 (2005).
32. L. Malakhova, V. G. Bezlepkin, V. Antipova, et al., The increase in mitochondrial DNA copy number in the tissues of γ-irradiated mice, *Cell. Mol. Biol. Lett.* **10**:721–732 (2005).
33. S. P. LeDoux, G. L. Wilson, E. J. Beecham, T. Stevnsner, K. Wassermann, and V. A. Bohr, Repair of mitochondrial DNA after various types of DNA damage in Chinese hamster ovary cells, *Carcinogenesis* **13**:1967–1973 (1992).
34. G. L. Dianov, N. Souza-Pinto, S. G. Nyaga, T. Thybo, T. Stevnsner, V. A. Bohr, Base excision repair in nuclear and mitochondrial DNA, *Prog. Nuc. Acid Res. Mol. Biol.* **68**:285–297 (2001).
35. A. May, and V. A. Bohr, Gene-specific repair of γ-ray-induced DNA strand breaks in colon cancer cells: no coupling to transcription and no removal from the mitochondrial genome, *Biochem. Biophys. Res. Commun.* **269**:433–437 (2000).

36. H. Budworth, and G. L. Dianov, Mode of inhibition of short-patch base excision repair by thymine glycol within clustered DNA lesions, *J. Biol. Chem.* **278**:9378–9381 (2003).
37. H. Budworth, G. Matthewman, P. O'Neill, and G. L. Dianov, Repair of tandem base lesions in DNA by human cell extracts generates persisting single-strand breaks, *J. Mol. Biol.* **351**:1020–1029 (2005).
38. N. Yang, M. A. Chaudhry, and S. S. Wallace, Base excision repair by hNTH1 and hOGG1: a two edge sword in the processing of DNA damage in γ-irradiated human cells, *DNA Repair* **5**:43–51 (2006).
39. L. Tretter, É. Rónai, G. Szabados, et al., The effect of the radioprotector WR-2721 and WR-1065 on mitochondrial lipid peroxidation, *Int. J. Radiat. Biol.* **57**:467–478 (1990).
40. M. W. Epperly, C. A. Sikora, S. J. DeFilippi, et al., Manganese superoxide dismutase (SOD2) inhibits radiation-induced apoptosis by stabilization of the mitochondrial membrane, *Radiat. Res.* **157**:568–577 (2002).
41. R. I. Salganik, The benefits and hazards of antioxidants: controlling apoptosis and other protective mechanisms in cancer patients and the human population, *J. Am. Coll. Nutr.* **20**:464S–472S (2001).
42. D. E. McClain, J. F. Kalinich, and N. Ramakrishnan, Trolox inhibits apoptosis in irradiated MOLT-4 lymphocytes, *FASEB* **9**:1345–1354 (1995).
43. G. Hernández-Flores, P. C. Gómez-Contreras, J. R. Domínguez-Rodríguez, et al., γ-irradiation induced apoptosis in peritoneal macrophages by oxidative stress. Implications of antioxidants in caspase mitochondrial pathway, *Anticancer Res.* **25**:4091–4100 (2005).
44. H. Kondo, S-H. Park, K. Watanabe, et al., Polyphenol (-)-epigallocatechin gallate inhibits apoptosis induced by irradiation in human HaCaT keratinocytes, *Biochem. Biophys. Res. Commun.* **316**:59–64 (2004).
45. N. N. Khodarev, Y. Kataoka, J. S. Murley, et al., Interaction of amifostine and ionizing radiation on transcriptional patterns of apoptotic genes expressed in human microvascular endothelial cells (HMEC), *Int. J. Radiat. Oncol. Biol. Phys.* **60**:553–563 (2004).
46. H. B. Stone, C. N. Coleman, M. S. Anscher, et al., Effects of radiation on normal tissue: consequences and mechanisms, *Lancet Oncol.* **4**:529–36 (2003).
47. M. E. C. Robbins, and W. Zhao, Chronic oxidative stress and radiation-induced late normal tissue injury: a review, *Int. J. Radiat. Biol.* **80**:251–59 (2004).
48. C. Borek, and W. Troll, Modifiers of free radicals inhibit in vitro the oncogenic actions of x-rays, bleomycin, and the tumor promoter 12-*O*-tetradecanoylphorbol 13-acetate, *Proc. Natl. Acad. Sci.* **80**:1304–307 (1983).
49. M. S. Anscher, L. Chen, Z. Rabbani, et al., Recent progress in defining mechanisms and potential targets for prevention of normal tissue injury after radiation therapy, *Int. J. Radiat. Oncol. Biol. Phys.* **62**:255–259 (2005).
50. M. Mitjans, V. Martínez, J. del Campo, et al., Novel epicatechin derivatives with antioxidant activity modulate interleukin-1β release in lipopolysaccharide-stimulated human blood, *Bioorg. Med. Chem. Lett.* **14**:5031–5034 (2004).
51. J-P. Marier, K. Chen, P. Prince, G. Scott, J. R. E. del Castillo, and P. Vachon, Production of ex vivo lipopolysaccharide-induced tumor necrosis factor-α, interleukin-1β, and interleukin-6 is suppressed by trans-resveratrol in a concentration-dependent manner, *Can. J. Vet. Res.* **69**:151–154 (2005).
52. R. Aneja, K. Odoms, A. G. Denenberg, and H. R. Wong, Theaflavin, a black tea extract, is a novel anti-inflammatory compound, *Crit. Care Med.* **32**:2097–2103 (2004).
53. J. J. Haddad, and C. S. Fahlman, Redox- and oxidant-mediated regulation of interleukin-10: an anti-inflammatory, antioxidant cytokine? *Biochem. Biophys. Res. Commun.* **297**:163–176 (2002).
54. M. W. Epperly, J. Bray, S. Kraeger, et al., Prevention of late effects of irradiation lung damage by manganese superoxide dismutase gene therapy, *Gene Ther.* **5**:196–208 (1998).

55. Z. Vujaskovic, I. Batinic-Haberle, Z. N. Rabbani, et al., A small molecule weight catalytic metalloporphyrin antioxidant with superoxide dismutase (SOD) mimetic properties protects lungs from radiation-induced injury, *Free Rad. Biol. Med.* **33**:857–863 (2002).
56. J-L. Lefaix, S. Delanian, J-J. Leplat, et al., Successful treatment of radiation-induced fibrosis using Cu/Zn-SOD and Mn-SOD: an experimental study, *Int. J. Radiat. Oncol. Biol. Phys.* **35**:305–312 (1996).
57. J-L. Lefaix, S. Delanian, M-C. Vozenin, et al., Striking regression of subcutaneous fibrosis induced by high doses of gamma rays using a combination of pentoxifylline and α-tocopherol: an experimental study, *Int. J. Radiat. Oncol. Biol. Phys.* **43**:839–847 (1999).
58. S. Delanian, R. Porcher, S. Balla-Mekias, et al., Randomized, placebo-controlled trial of combined pentoxifylline and tocopherol for regression of superficial radiation-induced fibrosis, *J. Clin. Oncol.* **21**:2545–2550 (2003).
59. P. Okunieff, E. Augustine, J. E. Hicks, et al., Pentoxifylline in the treatment of radiation-induced fibrosis, *J. Clin. Oncol.* **22**, 2207–2213 (2004).
60. S. Delanian, M. Martin, A. Bravard, et al., Cu/Zn superoxide dismutase modulates phenotypic changes in cultured fibroblasts from human skin with chronic radiotherapy damage, *Radiother. Oncol.* **58**:325–331 (2001).
61. J. Eastgate, J. Moreb, H. S. Nick, et al., A role for manganese dismutase in radioprotection of hematopoietic stem cells by interleukin-1, *Blood* **81**:639–646 (1993).
62. J. Moreb, and J. R. Zucali, The therapeutic potential of interleukin-1 and tumor necrosis factor on hematopoietic stem cells, *Leuk. Lymphoma* **8**:267–275 (1992).
63. C. Muňoz, M. C. Castellanos, A. Alfranca, et al., Transcriptional up-regulation of intracellular adhesion molecule-1 in human endothelial cells by the antioxidant pyrrolidine dithiocarbamate involves the activation of activating protein-1, *J. Immunol.* **157**:3587–3597 (1996).
64. M. Walther, W. Kaffenberger, and D. van Beuningen, Influence of clinically used antioxidants on radiation-induced expression of intracellular cell adhesion molecule-1 on HUVEC, *Int. J. Radiat. Biol.* **75**:1317–1325 (1999).
65. K. Otsuka, T. Koana, H. Tauchi, and K. Sakai, Activation of antioxidant enzymes induced by low-dose-rate whole-body γ irradiation: adaptive response in terms of initial DNA damage, *Radiat. Res.* **166**, 474–478 (2006).

Chapter 21
Anti-Cancer Effect of Resveratrol is Associated with Induction of Apoptosis via a Mitochondrial Pathway Alignment

Weimin Sun, Wei Wang, Jung Kim, Peter Keng, Shanmin Yang, Hengshan Zhang, Chaomei Liu, Paul Okunieff, and Lurong Zhang[1]

Abstract Resveratrol, a phytoalexin found in the skin of grapes, is believed to have multiple bioactivities including anti-cancer, anti-carcinogenesis and anti-inflammatory. The mechanisms by which resveratrol might produce these effects are not well understood. In this study, malignant human pancreatic cancer cells were treated without or with resveratrol in combination with ionizing radiation (IR), and then the mitochondrial function of treated cells was evaluated using several standardized assays. They include the Calcein AM method for mitochondria transition pore; the JC-1 staining method for mitochondria membrane potential; the CM-H2DCFDA method for reactive oxygen species; and the Annexin V/propidium iodide (PI) method for apoptosis/cell death. Our results indicated that (1) pore function was partially intact after resveratrol, but resveratrol probably interfered with the accumulation of intracellular Calcein AM; (2) depolarization of the mitochondria membrane was increased in the resveratrol treated cells, consistent with mitochondrial dysfunction; (3) ROS was slightly increased with resveratrol, a phenomenon that was greatly increased when this agent was combined with IR; and (4) in parallel with the above changes in mitochondrial and drug transport, cells treated with resveratrol showed increased apoptosis as measured by Annexin V/PI staining. In summary, the anti-cancer effect of resveratrol is associated with the damage of mitochondrial function that leads to increased ROS, apoptosis, and possibly intracellular drug accumulation via inhibition of proteins involved in multi-drug resistance (MDR).

21.1 Introduction

Resveratrol {3, 4′,5-trihydroxy-*trans*-stilbene 5-[(1E)-2-(4-hydroxyphenyl)-ethenyl] -1,3-benzenediol} is a small compound (molecular weight 228.24) that can be purified from several types of plants, most commonly the skin of

[1]Weimin Sun, Wei Wang, Jung Kim, Peter Keng, Shanmin Yang, Hengshan Zhang, Chaomei Liu, Paul Okunieff, and Lurong Zhang, Department of Radiation Oncology, University of Rochester Medical Center, Rochester, NY 14642 USA.

grapes [1–3]. It serves as a polyphenolic antibiotic to protect the plants (grapes, and nuts) from pathogenic microorganisms, such as bacteria and fungi [4].

A wide range of biological activities have been ascribed to resveratrol including anti-inflammatory, anti-oxidant, anti-platelet aggregation, cell growth-modulation, anticarcinogenesis, anti-atherogenic, estrogen-like effects, immuno-modulation, and chemoprevention [4–6]. Among these, there is mounting evidence indicating that resveratrol is a promising natural compound for cancer prevention and for treatment of a variety of human cancers [7]. The molecular and cellular targets responsible for the anti-neoplastic effects are unknown. Possible mechanisms include interference with intracellular signaling pathways that regulate cell survival or apoptosis [8], cell cycle arrest, and inhibition of several pathways for kinase activities [9]. The latter mechanism has the potential of augmenting the response to radiation and chemotherapy. We therefore postulate that resveratrol affects mitochondrial function and predisposes cells to apoptosis pathways induced by radiation. We performed studies in a pancreatic cancer model, AsPC-1: a tumor model known to be naturally resistant to apoptosis. We found that resveratrol disrupted the mitochondrial function of AsPC-1 cells, and resulted in increased apoptosis and cell death that was substantially augmented by irradiation.

21.2 Materials and Methods

21.2.1 Reagents and Cells

AsPC-1 cells were cultured in DMEM (Dulbecco's Modification of Eagle's Medium) supplemented with 10% fetal bovine serum at 37°C in an incubator with 5% CO_2. The resveratrol (purity 98.31%) was purchased from Xi An Chongxin Natural Additive Company (Xi An, China). CM-H2DCFDA {5-(and-6)-chloromethyl-2',7'Œ-dichlorodihydro-fluorescein diacetate, acetyl ester}, JC-1, Calcein AM, Annexin V, and propidium iodide (PI) were purchased from Molecular Probes (Eugene, OR).

21.2.2 Resveratrol Treatment of Cells

Flow cytometric analyses of different mitochondrial functions were performed using a FACS Calibur flow cytometer (BD Biosciences, San Jose, CA). Briefly, cells (2×10^5) were cultured in 6 well dishes until 80% confluence, and then treated with resveratrol at a concentration 50 μg/ml with or without 5 Gy radiation (Shepherd Irradiator) at a dose rate of 1.85 Gy/min. Twenty-four hours later, cells were harvested and single cell suspensions were placed separately in 6 ml tubes with 1 ml culture media and cultured further for one hour to allow membrane damage occurring during the harvest process to recover. This was followed by staining in accordance with manufacturer's instructions, as described below.

21.2.3 *Analysis for Transition Pore of Mitochondria*

To determine the effects of resveratrol on the activity of transition pore of mitochondria in AsPC-1 cells, 5 μl of 2 μM Calcein AM and 5 μl of 80 mM $CoCl_2$ were added to 100 μl single cell suspension, incubated at 37°C for 15 min, washed once with PBS, and immediately analyzed by flow cytometer for the percentage of green fluorescent cells. $CoCl_2$ was used to quench the cytosolic fluorescence so that the fluorescent intensity of Calcein AM only represented the activities from mitochondria.

21.2.4 *Analysis for Mitochondria Membrane Potential*

Two μg of JC-1 in 30 μl of saline was added to 100 μl of single cell suspension treated without or with 50 μg/ml of resveratrol and radiation. The mixtures were incubated at 37°C for 10 min, washed with PBS once, and subjected to flow cytometer analysis immediately. The percentages of cells in the high red region or low red and high green region were measured and compared among the different treatment groups.

21.2.5 *Analysis for Reactive Oxygen Species (ROS)*

100 μl of single cell suspension from each treatment was incubated with CM-H2DCFDA (final concentration of 5 μM) at 37°C for 0.5 hour. The intensity of fluorescence was determined by flow cytometry and unstained cells were used for subtraction of auto-fluorescence in the green emission range.

21.2.6 *Analysis for Apoptosis and Cell Death*

AsPC-1 cells treated with resveratrol or radiation were harvested and stained with Annexin V for 30 min and then PI was added immediately before the flow cytometric analysis according to the manufacturer's instructions [10]. The percentage of cells that were Annexin V positive and PI negative was compared among different treatment groups.

21.2.7 *Statistical Analysis*

Student's *t* test was used to determine the significance between the different treatment groups. A P value of < 0.05 was regarded as statistically significant.

21.3 Results

21.3.1 *Resveratrol Allows Increased Levels of Calcein AM to Enter the Cell*

The Calcein AM assay can be used to determine both the activity of the mitochondrial transition pore and the function of multiple drug resistance pump (MDR). The opening of transition pore is an initial event which occurs after cells are damaged. Calcein AM is a non-fluorescent dye that is cleaved to a polar fluorescent molecule by cytoplasmic esterase after it passes through both the cell and mitochondria membranes. $CoCl_2$, is added to quench cytoplasmic fluorescence and allows the detection of mitochondrial fluorescence if the transition pore is closed. The result is shown in Fig. 21.1. Control, non-irradiated cells had little fluorescence. This indicates that the Calcein AM did not accumulate in control cells. In contrast, irradiated cells treated with resveratrol had high fluorescence, indicating resveratrol improved the accumulation of Calcein AM in mitochondria and that the transition pore was intact (closed).

21.3.2 *Resveratrol Depolarizes the Mitochondrial Membrane*

While the transition pore was at least partially functional, more subtle dysfunction of the mitochondria was detected by measurement of its membrane

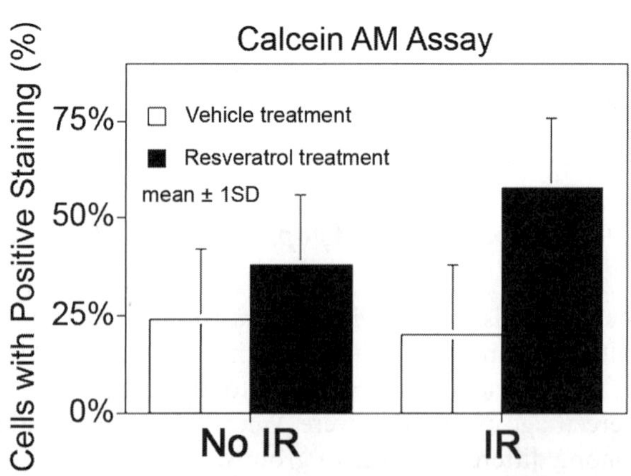

Fig. 21.1 Calcein AM Assay. Control, non-irradiated cells had low fluorescence indicating poor Calcein AM accumulation. Resveratrol aided the accumulation of Calcein AM in the pancreatic carcinoma cells (No IR: P = na; IR: $P<0.05$). Radiation had no effect on Calcein AM accumulation in the cell or on the function of the mitochondrial transition pore. (n = 2).

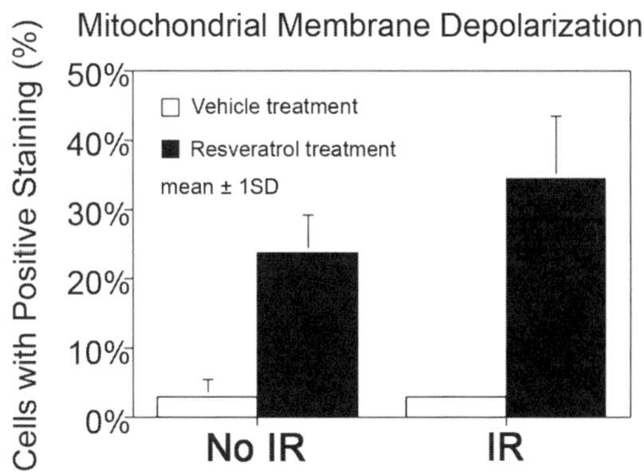

Fig. 21.2 Resveratrol depolarizes the mitochondria membrane. Resveratrol increased the fraction of cells with a green shift indicating depolarization of the mitochondria membrane ($P<0.01$). Radiation had a lesser effect ($P<0.05$) that was not significantly different than resveratrol alone. ($n=2$).

depolarization. To determine if there was membrane dysfunction, JC-1 staining was performed followed by flow cytometry analysis. Figure 21.2 shows that the percentage of cells in the high green and low red region before and after resveratrol. The green shift indicates mitochondrial membrane depolarization in AsPC-1 cells ($P<0.01$). The effect of resveratrol on mitochondrial membrane potential was not diminished by irradiation. The data suggest that resveratrol indeed depolarizes the mitochondria membrane and that the effect produced by a dose of 50 μg/ml resveratrol is more pronounced that that of 5 Gy radiation.

21.3.3 Resveratrol Increases the Production of ROS

To determine if resveratrol at a relatively high dose alters the production of ROS, the CM-H2DCFDA method was used. CM-H2DCFDA passes through plasma membrane and the acetate is cleaved by intracellular esterase and trapped in the cytosol. The polar substrate is then oxidized to the fluorescent form depending on the cellular redox state. The results are summarized in Fig. 21.3. Resveratrol did not change the percentage of AsPC-1 cells in the high fluorescent intensity region. When resveratrol was given in combination with 5 Gy radiation, the percentage of cells in the high fluorescent intensity region increased dramatically, indicating that the combination triggered the production of ROS.

21.3.4 Resveratrol Triggers Apoptosis

The damage of mitochondria can trigger apoptosis [11–13]. Annexin V, a 35–36 kDa, Ca^{2+} dependent, phospholipid binding protein with a high affinity to membrane phospholipid phosphatidylserine, shifts from the inner to the outer leaflet of the plasma membrane during the early stage of apoptosis. The detection of early

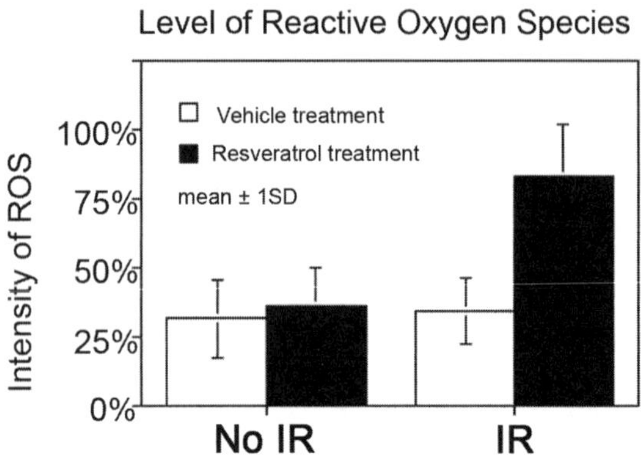

Fig. 21.3 Increased ROS in resveratrol treated cells. The cells were treated with either vehicle alone (as control) or 50 ug/ml of resveratrol and 5 Gy radiation for 24 hours. Resveratrol did not induce ROS, while it did enhance the radiation-induced production of ROS ($P<0.05$). ($n=2$).

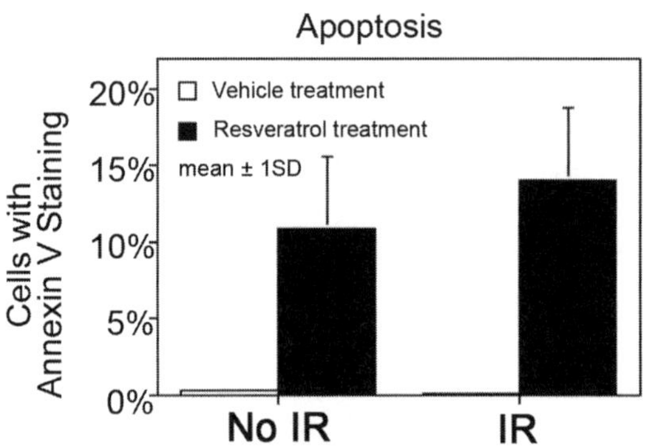

Fig. 21.4 Effect of resveratrol on apoptosis. The AsPC-1 cells treated with resveratrol for 24 hours plus 5 Gy radiation were stained with Annexin V for 30 min and then with PI prior to the flow cytometer analysis. The apoptotic and dead cells increased with the treatment of resveratrol and further increased upon radiation ($P<0.05$). ($n=2$).

apoptosis was supplemented with the PI assay to exclude dead cells. Figure 21.4 shows that the percentage of apoptotic cells significantly increased with resveratrol treatment and further increased when radiation was added. The data are consistent with enhancement of radiation-related cell killing by resveratrol.

21.4 Discussion

In this study, we demonstrated for the first time that the resveratrol alone at high concentration alters mitochondrial function of AsPC-1 malignant pancreatic cancer cells, and when combined with radiation increases ROS and apoptosis. This is also the first documentation of accumulation of Calcein in cancer cells after exposure to resveratrol (Fig. 21.1). Calcein AM is an MDR substrate and did not accumulate in AsPC-1 cells until resveratrol was added. Resveratrol, therefore, might reduce pancreatic carcinoma's resistance to chemotherapy; however, the impact of red wine consumption on chemotherapy remains unclear. In studies of MDR expression in pancreatic carcinoma cell lines, including AsPC-1, mRNA for both MDR1 and MDR3 was present. AsPC-1 cells also had high basal levels of MRP1, MRP3, and MRP5 [14]. Thus, AsPC-1 and other pancreatic tumor cell lines likely owe some of their treatment resistance to these membrane proteins. Inhibition of these proteins by resveratrol deserves further evaluation. Resveratrol is currently being examined in clinical trials as a mitigator of normal tissue toxicity among patients undergoing radiation or chemotherapy for malignancy. Since some normal tissues do have MDR function, these tissues might be sensitized by resveratrol. Consumption of red wine (containing resveratrol) among patients undergoing radiation therapy is not contraindicated; the apoptotic mechanism in normal tissue is seemingly not altered by resveratrol. Further research is necessary to determine if resveratrol produces a differential response between normal tissue and tumor.

Resveratrol alone did not increase the number of cells with high ROS, but in combination with radiation greatly increased ROS (Fig. 21.3). The subsequent increased apoptosis is a likely consequence of this synergistic effect on ROS (Fig. 21.4).

While we were not able to detect dysfunction of the mitochondrial transition pore due to failure of Calcein AM to accumulate in AsPC-1 cell, others have reported that resveratrol modulates mitochondrial transition pore permeability [15, 16]. Since several other aspects of abnormal mitochondrial functions were detected in the current study, mitochondrial still remained to be the major target for resveratrol induced damage. Our hypothesis is further supported by the absence of robust caspase 8 activation but increased cytochrome C with downstream activation of caspases 9 and 3 [7]. Thus the apoptotic signal is likely mitochondrial rather than an extrinsic death pathway. The mechanism of apoptosis after resveratrol, however, is complex since resveratrol-induced apoptosis occurs only in cells expressing wild-type p53 (p53+/+), but not in p53-deficient (p53-/-) cells [17].

In this study, we used a relatively high dose (50 μg/ml) of resveratrol. The rationale for this was: (1) its biological effects are likely dose dependent; (2) high doses are expected to be safe if the agent is to be used clinically; and (3) studies indicate that the resveratrol concentration in red wines can reach as high as 30 μg/ml [18]. It is important to note that many studies have been performed with resveratrol and many biological effects, both cytoprotective and cytotoxic have been claimed. Thus, low doses of the agent may have very different effects than the higher doses we employed.

In conclusion, this study demonstrates the impairment by resveratrol of mitochondrial functions, particularly reduction of the membrane potential, increase of ROS synergistically with irradiation, induction of apoptosis, and increase of radiation induced apoptosis. Resveratrol also improved calcein accumulation in AsPC-1 consistent with inhibition of MDR. While additional studies are indicated, these data suggest resveratrol has a promising future as a modulator of cytotoxic cancer therapies.

Acknowledgment This research was supported by the Center for Medical Countermeasures against Radiation Program, U19-AI067733, National Institute of Allergy and Infectious Diseases. The authors gratefully acknowledge the editing and research assistance of Amy K. Huser.

References

1. P. Waffo-Teguo, M. E. Hawthorne, M. Cuendet, et al., Potential cancer-chemopreventive activities of wine stilbenoids and flavans extracted from grape (Vitis vinifera) cell cultures, *Nutr. Cancer* **40**(2):173–179 (2001).
2. L. Fremont, Biological effects of resveratrol, *Life Sci.* **66**(8):663–673 (2000).
3. G. J. Soleas, and E. P. Diamandis, and D. M. Goldberg, Resveratrol: a molecule whose time has come? And gone? *Clin. Biochem.* **30**(2):91–113 (1997).
4. J. K. Lin, and S. H. Tsai, Chemoprevention of cancer and cardiovascular disease by resveratrol, *Proc. Natl. Sci. Coun. Repub. China B.* **23**(3), 99–106 (1999).
5. J. A. Baur, and D. A. Sinclair, Therapeutic potential of resveratrol: the in vivo evidence, *Natl. Rev. Drug Discov.* **5**(6), 493–506 (2006).
6. S. K. Manna, A. Mukhopadhyay, and B. B. Aggarwal, Resveratrol suppresses TNF-induced activation of nuclear transcription factors NF-kappa B, activator protein-1, and apoptosis: potential role of reactive oxygen intermediates and lipid peroxidation, *J. Immunol.* **164**(12), 6509–6519 (2000).
7. S. Pervaiz, Resveratrol–from the bottle to the bedside? *Leuk. Lymphoma.* **40**(5–6), 491–498 (2001).
8. S. Fulda, and K. M. Debatin, Resveratrol modulation of signal transduction in apoptosis and cell survival: A mini-review, *Cancer Detect. Prev.* **30**(3), 217–223 (2006).
9. D. Delmas, A. Lancon, D. Colin, B. Jannin, and N. Latruffe, Resveratrol as a chemopreventive agent: a promising molecule for fighting cancer, *Curr. Drug Targets* **7**(4), 423–442 (2006).
10. http://probes.invitrogen.com/media/publications/508.pdf
11. N. Dias, and C. Bailly, Drugs targeting mitochondrial functions to control tumor cell growth, *Biochem. Pharmacol.* **70**(1):1–12 (2005).
12. R. Kim, Recent advances in understanding the cell death pathways activated by anticancer therapy, *Cancer* **103**(8):1551–1560 (2005).
13. T. Asakura, and K. Ohkawa, Chemotherapeutic agents that induce mitochondrial apoptosis, *Curr. Cancer Drug Targets* **4**(7):577–590 (2004).
14. S. Eisold, D. Nauheimer, J. Schmidt, T. Giese, E. Klar, and M. Linnebacher, Influence of clinically relevant chemotherapeutics on the expression of multidrug-resistance family members in human pancreatic cell lines, *Society for Surgery of the Alimentary Tract* [abstract]; accessed online 21Dec2006: http://www.ssat.com/cgi-bin/abstracts/06ddw/SSAT_DDW06_38.cgi
15. S. J. Zunino, and D. H. Storms, Resveratrol-induced apoptosis is enhanced in acute lymphoblastic leukemia cells by modulation of the mitochondrial permeability transition pore, *Cancer Lett.* **240**(1), 123–134 (2005).
16. X. M. Tian, and Z. X. Zhang ZX, Resveratrol promote permeability transition pore opening mediated by Ca2+, *Yao. Xue. Xue. Bao.* **38**(2), 81–84 (2003).
17. C. Huang, W. Y. Ma, A. Goranson, and Z. Dong, Resveratrol suppresses cell transformation and induces apoptosis through a p53-dependent pathway, Carcinogenesis **20**(2), 237–242 (1999).
18. J. F. Moreno-Labanda, R. Mallavia, L. Perez-Fons, V. Lizama, D. Saura, and V. Micol, Determination of piceid and resveratrol in Spanish wines deriving from Monastrell (Vitis vinifera L.) grape variety, *J Agric Food Chem.* **52**(17), 5396–5403 (2004).

Part V
Tissue Engineering

Chapter 22
Computationally Determined Shear on Cells Grown in Orbiting Culture Dishes

R. Eric Berson[1], Matthew R. Purcell[1], and M. Keith Sharp[2]

Abstract A new computational model, using computational fluid dynamics (CFD), is presented that describes fluid behavior in cylindrical cell culture dishes resulting from motion imparted by an orbital shaker apparatus. This model allows for the determination of wall shear stresses over the entire area of the bottom surface of a dish (representing the growth surface for cells in culture) which was previously too complex for accurate quantitative analysis. Two preliminary cases are presented that show the complete spatial resolution of the shear on the bottom of the dishes. The maximum shear stress determined from the model is compared to an existing simplified point function that provides only the maximum value. Furthermore, this new model incorporates seven parameters versus the four in the previous technique, providing improved accuracy. Optimization of computational parameters is also discussed.

22.1 Introduction

The effects of hemodynamic forces on cellular responses have been studied for more than thirty years, but the mechanisms linking cause and effect are still not well known. Wall shear stresses are widely accepted as the primary influence affecting characteristics of anchored cells subjected to fluid flow. Endothelial cells, lining the interior walls of arteries and veins, experience shear exerted by the flow of blood and become aligned and elongated with the direction of flow and undergo other physiological and biochemical changes [1–8]. The realization of the relationship between hemodynamic forces on the endothelium and the origins of atherosclerosis and vascular pathology [9–14], in general, has led to considerable attention focusing on the effects of these forces on cellular responses. Detailed, accurate information about the fluid forces acting on

[1]Department of Chemical Engineering, University of Louisville, Louisville, KY.
[2]Department of Mechanical Engineering, University of Louisville, Louisville, KY.

K.A. Kang et al. (eds.), *Oxygen Transport to Tissue XXIX*

cells must be known in order to understand the cause and effect relationship between shear stresses and endothelial responses.

Several studies have been conducted in vitro using experimental apparatus that can provide accurately quantifiable shear stress over a cultured layer of endothelial cells. One type involves inducing a rotating flow field over a stationary layer of cells. In this arrangement, either a flat plate or a cone rotates over a fixed surface where the layer of cells exists. An easily calculated wall shear stress is produced since the rotation rate of the upper plate or cone is known as is the length of the gap between the rotating and stationary surfaces. Another type involves flow in a chamber between two parallel plates, usually where the width and length of the plates are much larger than the gap between the plates. Again, wall shear stress is easily calculated based on the known flow rate and known gap length between the two plates.

These flow devices provide an accurate means of delivering pre-determined shear stresses to a layer of cultured cells but are limited in that only *steady* flow is provided. Also, experiments must be performed consecutively rather than simultaneously unless multiple cone-and-plate or flow chamber devices are available. Another prevalent apparatus for providing fluid motion to cultured cells is the orbital shaker platform. Orbital shakers are ideal for simultaneous cell culture experiments and are widely used throughout the cell culture industry because of their simplicity. More importantly, orbital shakers provide oscillatory flow, somewhat like that experienced by pulsing fluid movement in the human vasculature system.

Despite the prevalence and simplicity of usage, few have attempted to employ the orbiting shaker apparatus as a means for correlating shear stresses to cellular responses due to the complexity involved in accurately calculating wall shear stresses exerted by the fluid. The movement of fluid in a cylinder that derives its motion from an orbiting shaker platform will be oscillatory in nature with a wave whose peak rotates around the cylinder at an angular velocity corresponding to the orbital velocity of the cylinder. Those that have attempted to correlate shear stress in a shaker flask to cellular responses have used simplified means for estimating the magnitude of the shear.

Ley et al [15]. investigated shear-dependent adhesion of human polymorphonuclear neutrophil granulocytes to endothelial cells in cone-and-plate and orbiting shaker experiments. For simplicity, the shear stress they reported in the orbital shaker experiments was an estimate of the maximum wall shear stress at the bottom of the cylinder:

$$\tau_\omega = a \cdot \sqrt{\rho \cdot \mu \cdot (2 \cdot \pi \cdot f)^3} \tag{22.1}$$

where τ_ω is the maximum shear stress on the bottom of the cylinder, a is the radius of orbit, ρ is the fluid density, μ is the fluid viscosity, and f is the frequency of rotation. The authors did not provide a derivation or reference for the origin of Eq. (22.1). Curiously, the equation does not include a term for the fluid height which is ordinarily integral to a shear stress calculation. The use of Eq. (22.1) has become prevalent in the related literature since its introduction and is still widely used [7,8,16,17,18,19].

Computational models that solve the Navier-Stokes equations for flow provide a means to predict fluid dynamic properties that are difficult to or cannot be determined experimentally. In this paper, a computational fluid dynamic (CFD) model is employed to describe fluid behavior in a cylindrical cell culture dish resulting from motion imparted by an orbital shaker apparatus. This allows for the determination of wall shear stresses imposed over the entire area of the bottom surface of the cylinder (representing the growth surface for cells in culture) which was previously too complex for accurate quantitative analysis. Two preliminary cases are presented, and the maximum shear stresses determined from these solutions are compared to Eq. (22.1). Knowledge of the fluid dynamics inside an orbiting cylindrical cell culture dish will significantly enhance the usefulness of simple orbital shaker apparatuses, one of the most common in vitro cell culture apparatuses, in the study of hemodynamic effects on cell cultures.

22.2 Computational Methods

22.2.1 Solver Description

The fluid behavior is modeled using three dimensional Navier-Stokes equations that are solved using Fluent 6.2, a commercial software CFD solver. The solver employs a finite-volume discretization process to numerically solve the governing equations for conservation of mass and momentum. The general form of the mass conservation equation is written as:

$$\frac{\partial \rho}{\partial t} + \nabla \cdot (\rho V) = 0 \tag{22.2}$$

and the momentum conservation equation is written as:

$$\frac{\partial}{\partial t} + (\rho V) + \nabla \cdot (\rho V V) = -\nabla P + \rho g + \nabla(\mu \nabla V) \tag{22.3}$$

Simulating flow in an orbiting cylinder requires creating a dynamic grid that moves through space. Since Fluent can generate rotational and translational motion but not orbiting motion, the motion must be generated by a user-defined function which specifies the orbital frequency, orbital radius, and center of orbit. The user-defined function is an external C++ language subroutine linked to the Fluent solver. A transient solution is required since the location of the fluid domain is changing with time. The sloshing of the fluid as a result of the motion is a free surface flow that requires tracking of the liquid-air interface.

Table 22.1 Optimization of the computational grid. Maximum shear stress as a function of computational cell count

Cell Count	25,300	51,060	98,900	299,230	494,680
Maximum Shear Stress (dyne/cm^2)	3.25	3.44	3.93	4.30	4.39

The surface tracking is accomplished with the Volume of Fluid (VOF) model. In the VOF model, the two fluids across the interface share a single set of momentum equations and the volume fraction of each fluid in each computational cell is tracked throughout the grid. Flow is treated as laminar (Reynolds numbers = ~100) and residual values reached 5e^{-5} or better for each iteration following the earliest time steps.

22.2.2 Optimization of the Computational Grid

Solution accuracy depends on the level of resolution of the computational grid which is a function of the number of computational cells in the computational domain. However, a trade-off occurs in terms of resolution and computational cost. High resolution, transient cases can take on the order of days and weeks to obtain a final converged solution. To determine the optimum cell count, maximum shear stress on the bottom of a dish is compared for a given case over a

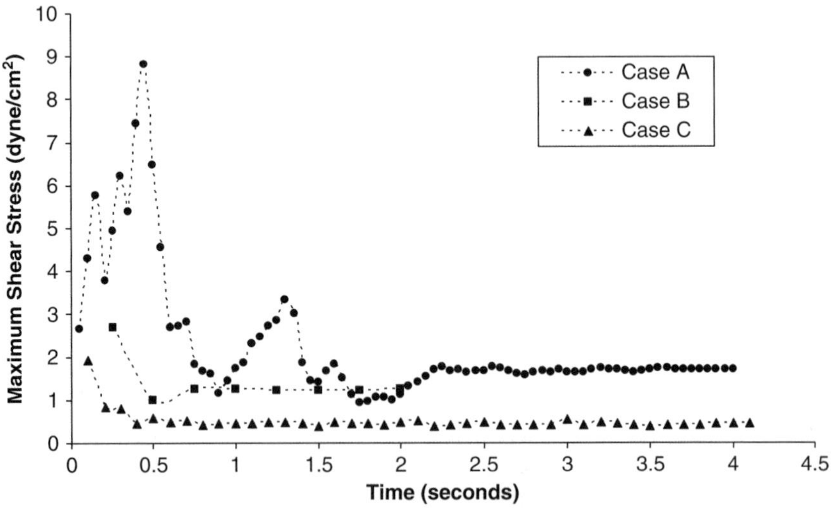

Fig. 22.1 Transient motion analysis – time for fluid motion to reach steady state. (Case A: 60 rpm, 3.50 cm dish diameter;Case B: 137 rpm, 2.88 cm dish diameter; Case C:137 rpm, 0.96 cm dish diameter).

range of cell quantities. The number of cells is increased until the magnitude of shear shows minimal change between increasing cell counts.

Table 22.1 shows how the magnitude of shear stress varies depending on the grid resolution for the case of a dish with a 2.4 cm diameter, a 0.108 cm initial fluid height, a 1.2 cm radius of orbit, and rotating at 82 rpm. Cases were run with cell counts of approximately 25,000, 50,000, 100,000, 300,000, and 500,000. The cell count is adjusted by varying the node spacing on the grid, which eventually determines the final cell count. The trial and error method results in cell counts that deviate slightly from the desired value. The representative maximum shear stress is the value after a minimum of four seconds (and 5.5 rotations) which is longer than the time it was determined for the fluid motion to have achieved steady state (see Fig. 22.1). The ratio of the difference in shear stress to the difference in cell count is smallest between the 300,000 and 500,000 cases, indicating an optimum cell count of 300,000, which is the amount that will be used in all new cases.

22.3 Transient Fluid Motion Analysis

Generating the dish motion requires a transient approach. In order to set the number of time steps in the solver, it is necessary to establish the run time needed for the fluid motion to reach steady state. The time to reach steady state for three cases is shown in Fig. 22.1. The three cases cover a variety of rotation rates and dish geometries; Case A: 60 rpm and 3.50 cm dish diameter, Case B: 137 rpm and 2.88 cm dish diameter, and Case C: 137 rpm and 0.96 cm dish diameter. The maximum shear stress on the bottom surface of the dish is again used as the metric for describing flow characteristics.

Case B and Case C reach steady state in less than one second. Case C reaches steady state in a little more than two seconds. Case B and Case C rotate at a rate of 2.28 times the rate of Case A, which suggests that achieving steady state may be more a function of the number of rotations than of time. Examining cases at higher rotation rates may help determine this. For the cases seen here, it appears that steady state is achieved after about two rotations. To be conservative, each case will encompass a minimum of four rotations prior to analysis.

22.4 Case Studies

Converged solutions provide phase contours representing the fluid motion and spatial resolution of shear stresses over the entire bottom surface of a dish, a significant improvement over the widely used Eq. (22.1) which provides just the maximum value without revealing any information as to the extent of the surface coverage exposed to this shear stress value. Two cases are presented as illustration. The first case has a dish diameter of 2.4 cm, an initial fluid height of 0.11 cm, a radius of orbit equal to 1.2 cm, and rotation rate of 82 rpm. The

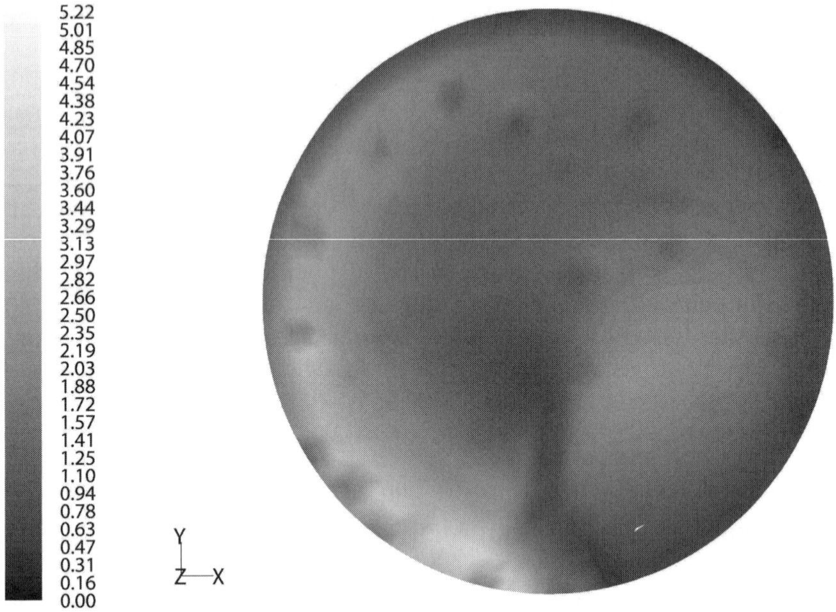

Fig. 22.2 Shear stress contours over bottom surface of dish (dyne/cm^2). [2.4 cm diameter, initial fluid height = 0.11 cm, orbital radius = 1.2 cm, 82 rpm].

second case has a dish diameter of 3.5 cm, an initial fluid height of 0.20 cm, a radius of orbit equal to 1.2 cm, and rotation rate of 60 rpm. The fluid properties (viscosity and density) were treated as water for both cases. Fig. 22.2 and Fig. 22.3 show shear stress contours covering the complete bottom surface of each dish.

While significant detail is lost in the grayscale, some salient features are still visible. The area of maximum shear stress in each contour appears as a relatively small, bright area near the periphery of each dish. Figure 22.4 shows the location of this maximum shear region relative to the leading edge of the traveling wave, which is created by the sloshing of the fluid. As the dish travels in its orbit, the sloshing of the fluid creates a wave with a peak that travels around the dish at a rate relative to the rotation rate of the dish. In Fig. 22.4, the peak of the wave is to the left. The fluid / air interface slopes downward to the right until the fluid depth reaches a minimum at the right side of the dish. The steepest part of this slope occurs just before the fluid reaches a minimum depth. The image is a snapshot in time of a dish that is traveling in the counter-clockwise direction, so this slope is referred to as the leading edge of the wave. Underneath this leading edge, a bright spot appears corresponding to the bright spot in Fig. 22.3. Thus, the region of maximum shear occurs just underneath the leading edge of the traveling wave. It is intuitive that this should be the point of maximum shear since the velocity is maintained while the liquid height is reduced, leading to a high velocity to distance ratio.

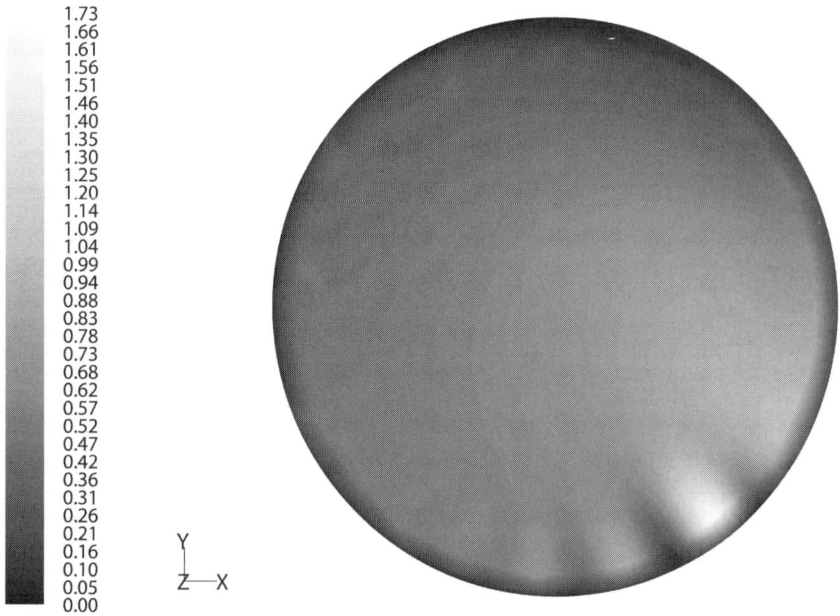

Fig. 22.3 Shear stress contours over bottom surface of dish (dyne/cm^2). [3.5 cm diameter, initial fluid height = 0.20 cm, orbital radius = 1.2 cm, 60 rpm].

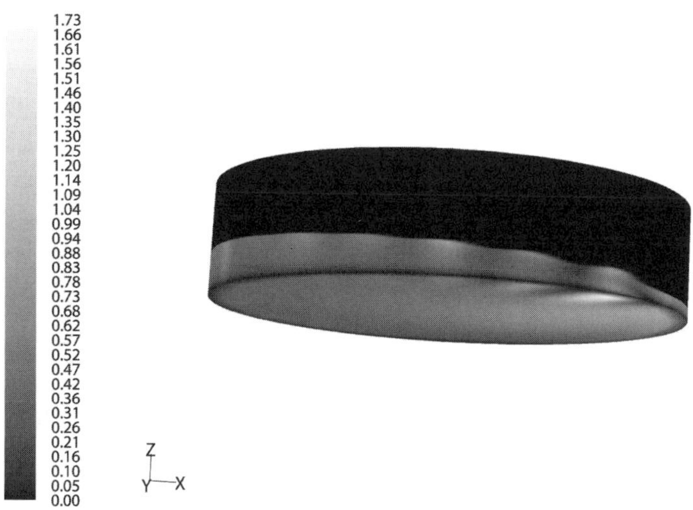

Fig. 22.4 Phase contour showing the fluid / air interface corresponding to the dish in Fig. 22.3.

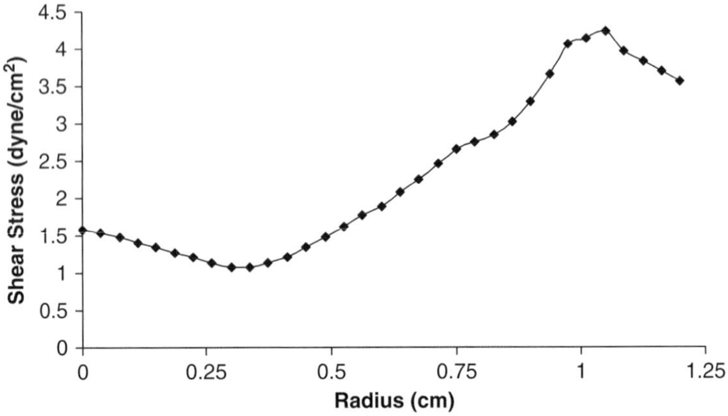

Fig. 22.5 Shear as a function of radius along the bottom surface of the dish corresponding to the contour in Fig. 22.2.

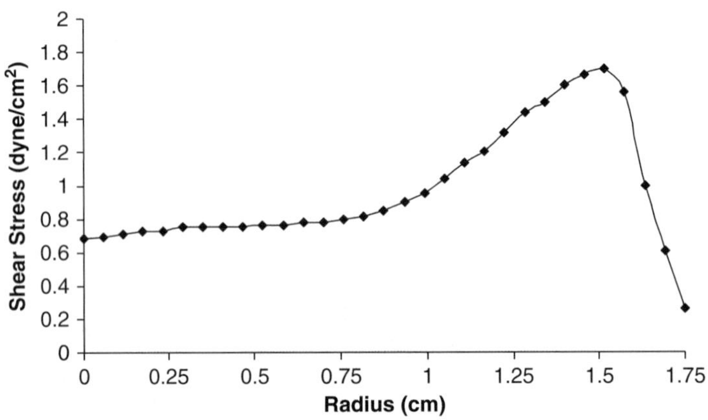

Fig. 22.6 Shear as a function of radius along the bottom surface of the dish corresponding to the contour in Fig. 22.3.

The shear along the bottom surface is plotted as a function of radius in a line extending from the center of the dish towards the maximum shear region (Figs. 22.5 and 22.6). Represented this way, it is easier to see the difference in shear magnitude between the center and the periphery of the dish. In both cases, there is at least a two-fold increase in shear from the center to the periphery, and a steep gradient exists on both sides of the peak. The drop in shear near the side wall of the dish is due to the no slip boundary condition imposed in the solution. Fig. 22.5 and Fig. 22.6 clearly exposes the weakness of Eq. (22.1) when it is used to correlate cellular responses to shear stresses in orbiting dishes: a single point function, such as this equation, is not a valid means to describe the entire cell culture growth area. This is in agreement with Dardik et al [19]. who noticed a difference in shear between a single point near the center and a single point near the periphery using optical velocimetry measurements.

Maximum shear stress as given by Eq. (22.1) is 1.89 dyne/cm^2 and 3.18 dyne/cm^2, respectively, for the two cases presented here compared to 1.70 and 4.30 as determined by our computational model. Eq. (22.1) incorporates fluid properties (viscosity and density), rotational rate, and orbital radius in its solution. In addition to these properties, our computational model incorporates dish geometry, fluid volume, and the effects of gravity. The two methods give a difference in maximum shear of 11% for Case 1 and 26% for Case 2. The additional parameters and stringent convergence criteria associated with the new computational model likely offer a more accurate result.

22.5 Conclusion

A new computational model is presented for determining shear stress on the bottom of a cylindrical cell culture dish that resides on an orbiting shaker apparatus. The model provides significant improvement over an existing method for determining shear stress in that it incorporates seven parameters instead of four, and it provides contours of shear over the entire bottom surface of the dish rather than a maximum value at a single point. This should enhance the usefulness of common orbital shaker apparatuses in the study of cellular responses to hemodynamic forces in culture.

References

1. M.J. Levesque and R.M. Nerem, The elongation and orientation of cultured endothelial cells in response to shear stress, *J Biomech Eng*, **107**(4), 341–347 (1985).
2. R.J. Satcher Jr., S.R. Bussolari, M.A. Gimbrone Jr., and C.F. Dewey Jr., The distribution of fluid forces on model arterial endothelium using computational fluid dynamics, *J Biomech Eng*, **114**(3), 309–316 (1992).
3. P.F. Davies, A. Remuzzi, E.J. Gordon, C.F. Dewey Jr., and M.A. Gimbrone Jr., Turbulent fluid shear stress induces vascular endothelial turnover in vitro, *Proc of the Nat Acad of Sci*, **83**, 2114–2117 (1986).
4. N. DePaola, M.A. Gimbrone Jr., P.F. Davies, and C.F. Dewey Jr., Vascular endothelium responds to fluid shear stress gradients. *Arteriosclerosis and Thrombosis*, **12**(11), 1254–1257 (1992).
5. C.F. Dewey, S.R. Bussolari, M.A. Gimbrone, and P.F. Davies, The dynamic response of vascular endothelial cells to fluid shear stress, *J Biomech Eng*, **103**(3), 177–185 (1981).
6. D.L. Fry, Acute vascular endothelial changes associated with increased blood velocity gradients. *Circ Res*, **22**, 165–197 (1968).
7. L.W. Kraiss, A.S. Weyrich, N.M. Alto, D.A. Dixon, T.M. Ennis, V. Modur, T.M. McIntyre, S.M. Prescott, and G.A. Zimmerman, Fluid flow activates a regulator of translation, p70/p85 S6 kinase, in human endothelial cells, *Am J Physiology*, **278**(5), H1537–1544 (2000).

8. L.W. Kraiss, N.M. Alto, D.A. Dixon, T.M. McIntyre, A.S. Weyrich, and G.A. Zimmerman, Fluid flow regulates E-selectin protein levels in human endothelial cells by inhibiting translation, *J Vasc Surg*, **37**(1), 161–168 (2003).
9. W.E. Stehbens, Hemodynamics and atherosclerosis. *Biorheology*, **19**, 95–101 (1982).
10. R.M. Nerem and M.J. Levesque. Fluid dynamics as a factor in the localization of atherosclerosis. Surface phenomena in Hemorheology: Their theoretical, experimental and clinical aspects, edited by A.L. Copely and G.V.F. Seaman, *Annals of the New York Academy of Science*, **416**, 709–719 (1984).
11. V.S. Repin, V.V. Dolgov, O.E. Zaikina, I.D. Novikov, A.S. Antonov, N.A. Nikolaeva, and V.N. Smirnov, Heterogeneity of endothelium in human aorta. A quantitative analysis by scanning electron microscopy, *Atherosclerosis*, **50**(1), 35–52 (1984).
12. D.P. Giddens, C.K. Zarins, and S. Glagov, The role of fluid mechanics in the localisation and detection of atherosclerosis, *J Biomech Eng*, **115**(4B), 588–594 (1993).
13. S. Glagov, C.K. Zarins, D.P. Giddens, and D.N. Ku, Haemodynamics and atherosclerosis. Insights and perspectives gained from studies of human arteries, *Archives of Pathology and Laboratory Medicine*, 112(10), 1018–1031.
14. A.M. Malek, S.L. Alper, and S. Izumo, Hemodynamic shear stress and its role in atherosclerosis, *J. Amer Med Assoc*, **282**(21), 2035–2042 (1999).
15. K. Ley, E. Lundgren, E. Berger, and K. Arfors, Shear-dependent inhibition of granulocyte adhesion to cultured endothelium by dextran sulfate, *Blood*, **73**(5), 1324–1330 (1989).
16. M. Haga, A. Yamashita, J. Paszkowiak, B.E. Sumpio, and A. Dardik, Oscillatory shear stress increases smooth muscle cell proliferation and Akt phosphorylation, *J Vasc Surg*, **37**(6), 1277–1284 (2003).
17. A.V. Sterpetti, A. Cucina, L.S. D'Angelo, B. Cardillo, and A. Cvallaro, Shear stress modulates the proliferation rate, protein synthesis, and mitogenic activity of arterial smooth muscle cells, *Surgery*, **113**(6), 691–699 (1993).
18. H. Ueba, M. Kawakami, and T. Yaginuma, Shear stress as an inhibitor of vascular smooth muscle cell proliferation: role of transforming growth factor-β1 and tissue-type plasminogen activator, *Arteriosclerosis, Thrombosis & Vascular Biology*, **17**(8), 1512–1516 (1997).
19. A. Dardik, L. Chen, J. Frattini, H. Asada, F. Haziz, F. Kudo, and B. Sumpio, Differential effects of orbital and laminar shear stress on endothelial cells. *J. Vasc Surg* **41**(5), 869–880 (2005).

Chapter 23
Formation of Capillary Tube-like Structures on Micropatterned Biomaterials

Dahai Gao, Girish Kumar, Carlos Co, and Chia-Chi Ho[1]

Abstract The survival of three-dimensional tissue requires a vascular network to provide transport of oxygen and metabolic byproduct. Here, we report a new approach to create capillary blood vessels in vitro on biomaterials suitable for use as scaffolds in engineering tissues. Endothelial cells were cultured on chemical and topographical patterns of micro-sized grooves on gelatin. Selective attachment and spreading of cells within the grooves was ensured by microcontact printing the plateau regions with cell resistant PEG/PLA (polyethyleneglycol-L-polylacticacid). Human microvascular endothelial cells plated on these patterned biomaterials attached and spread exclusively within the grooves. These topographical features promote endothelial cells to form capillary tube-like structures. The results demonstrated that capillary structures formed on biomaterials are useful for engineering vascularized tissues.

23.1 Introduction

Blood vessels play a key role in supplying oxygen and nutrients to tissues. To engineer viable tissues in vitro, it is critical to engineer blood vessels to transfer nutrients and waste to and from the engineered tissues. The differentiation of endothelial cells to form capillary tube-like structures depends on a variety of soluble factors such as VEGF (vascular endothelial growth factor) [1], extracellular matrix molecules, and geometrical cues [2]. Previous studies have demonstrated that microvascular endothelial cells cultured in medium containing endothelial growth factors [3] form capillary tubes spontaneously.

To explore whether capillary tube formation can be promoted by geometric cues, we fabricated 10 μm, 20 μm and 30 μm wide line patterns and grooves on gelatin films by using soft-lithography and microcontact printing techniques [4,5]. We explored the role of geometric confinement on the differentiation

[1]Dahai Gao, Girish Kumar, Carlos Co and Chia-Chi Ho, Department of Chemical Engineering, University of Cincinnati, Cincinnati, OH, USA.

of human microvascular endothelial cells to form capillary structures. We report here the effects of surface topography and spatial distribution of cell adhesive/resistant molecules on guiding endothelial cells to assemble into capillary tube-like structures.

23.2 Materials and Methods

23.2.1 Materials

Gelatin from porcine skin was purchased from Sigma (St. Louis. Mo). PDMS (Sylard 184) was obtained from Dow Corning (Midland, MI). Bovine serum albumin (BSA) fluorescein conjugate was purchased from Molecular Probes (Eugene, OR). (PBS) solution (137mM NaCl, 2. 7 mM KCl, 10mM KH_2PO_4, 10mM $NaHPO_4$, and 10 mM NaOH) was prepared from PBS pellets purchased from Sigma. Microvascular endothelial cell growth medium and fetal bovine serum (FBS) were purchased from Cambrex Bioscience (Walkersville, MD).

PEG-PLA (polyethyleneglycol-L-polylactic acid) was synthesized by the following method. 0.05 mM of Poly(ethylene glycol) monomethyl ether (mPEG-5000, $M_n \sim 5000$ Da) and 69 mM of 3,6-dimethyl-1,4dioxane-2,5-dione (lactide) were each dried by azeotropic distillation from toluene and combined with additional toluene. Following the addition of tin (II) ethyl hexanoate, the mixture was heated at reflux for 4 hours at which time the solvent was removed by distillation. The resulting solid mass was taken up a small amount of methylene chloride, then added drop-wise to ether, causing mPEG-PLA to precipitate as an oily solid (58%) that was isolated by suction filtration. PEG with a M_w less than 50,000 can be excreted from the kidneys [6].

23.2.2 Microfabrication of the Silicon Master Pattern and Transfer of the Topological Patterns onto PDMS

Micropatterns with parallel grooves 60 μm wide and ridges of varying widths (10, 20, and 30 μm) were fabricated on silicon wafers using standard photolithographic techniques. From this silicon master pattern, complementary PDMS replicas were formed by pouring PDMS pre-polymer (mixed in a 10:1 ratio with a crosslinking catalyst) over the Si master and cured at 56 °C for 2 h. The PDMS replicas were used as stamps in subsequent microcontact printing steps or as molds to form topographical patterned gelatin films.

Flat gelatin films were prepared on glass slides by spreading uniformly 0.1 mL gelatin solution in 0.05 M acetic acid over an area of 18.75 $cm.^2$ Topographical patterned gelatin films were prepared by adding gelatin solutions (0.05 M acetic acid) drop wise to the PDMS substrates. After drying, the gelatin film was crosslinked by 2% glutaraldehyde solution for 10 min. The cross-linked gelatin film was then immersed in 0.2% sodium borohydride solution for 30 min to quench the autoflurescence [7]. The gelatin film was soaked in deionized water overnight before seeding cells.

23.2.3 Chemical Patterning of Non-adhesive PEG-PLA Regions

Microcontact printing was used to pattern PEG-PLA over gelatin. The PDMS stamp was first cleaned with ethanol and then air-dried. 20 μL PEG-PLA was dropped onto the stamp. After drying, patterns of PEG-PLA were then transferred onto gelatin films by microcontact printing. Gentle pressure was applied to ensure conformal contact between the stamp and gelatin film. After 15 s, the stamp was removed from the gelatin and the substrate was air-dried.

A flat PDMS stamp was used to pattern topographical gelatin films. Only the plateau regions of the gelatin film were covered by PEG-PLA upon contact printing. Spatial control of protein adsorption onto the PEG-PLA-patterned gelatin films was confirmed by incubating the substrates with BSA fluorescein conjugate and visualized using a Nikon TE-2000 inverted microscope.

23.2.4 Culture of Endothelial Cells

Human microvascular endothelial cells (HMVEC-d, purchased from Cambrex Bioscience,MD) were cultured in endothelial basal medium containing 5% fetal bovine serum, 1 μg/mL of hydrocortisone, 10 μg/mL of epidermal growth factor (EGF), 10 μg/mL of bovine brain extract, 50 μg/mL of gentamycin, and 50 μg/mL of amphotericin-B under 5% CO_2. Prior to incubation with the micropatterned biomaterials, cells were dissociated from the culture dish with trypsin, resuspended in endothelial basal medium containing 10% serum, and allowed to attach onto micropatterned gelatin films(4×10^4 cells/cm^2).

23.2.5 Immunostaining and Image Analysis

After 72 hours of incubation, the attached cells were stained using cell tracker green (1 μM, Molecular Probes, Inc., Eugene, OR) for 30 minutes. For vWF/CD31 immunofluorescence staining, cells were fixed by formaldehyde and permeabilized for 15 min with 0.1% solution of Triton X-100 before incubation with primary antibody vWF and CD31 for 40 min. Secondary antibody Alexa Flor 546 goat anti-rabbit IgG, Alexa Fluor 488 goat anti-mouse IgG and DAPI were then added for 45 min.

Fluorescent microscope images were acquired using a SPOT II CCD camera (SPOT Diagnostic Instruments Inc., version 3.5.1, Sterling Heights, MI) and analyzed with Metamorph (Universal Imaging, version 6. 0r4, Westchester, PA) image analysis software. Confocal images were acquired using a Zeiss LSM 510 laser scanning confocal microscope and analyzed with LSM Image Browser software.

23.3 Results and Discussion

Figure 23.1 shows the procedures used to create micropatterns on the flat (Fig. 23.1A) and topographical patterned (Fig. 23.1D) gelatin films. Patterned PDMS stamps with designed micro-grooves (10 μm, 20 μm and 30 μm) are inked with cell resistant PEG-PLA and stamped directly onto flat gelatin film by microcontact printing. Figure 23.1B shows a phase contrast micrograph of a flat gelatin film patterned with a series of 20 μm wide lines. The 60 μm wide lines

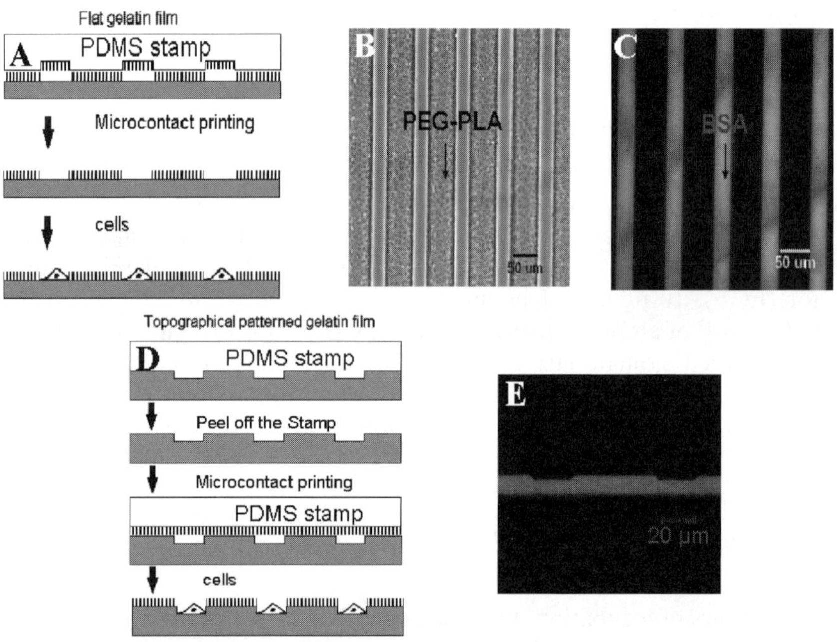

Fig. 23.1 Schematic of the approach used for preparing flat and topographical patterned gelatin films. A). Procedure used to create micropatterns on flat gelatin films. B). Phase contrast images of a patterned flat gelatin film. The 60 μm wide lines were covered by PEG-PLA. C). Selective adsorption of BSA fluorescein conjugate on flat gelatin films. D). Procedure used in fabricating topographically patterned gelatin films. E). Cross-section confocal image of a topographical patterned gelatin film.

separating the 20 μm wide lines were coated with PEG-PLA. BSA selectively adsorbs to the 20 μm lines that were not coated with PEG-PLA (Fig. 23.1C).

Topographical gelatin films were molded from the PDMS mold with reversed features. After gelatin dried in 60°C oven for 4 hours, gelatin film was peeled off from PDMS mold and cross-linked with glutaraldehyde. After drying, the gelatin film was patterned by a flat PDMS stamp inked with PEG-PLA. Only plateau regions of the film were covered by PEG-PLA and form cells resistant area (Fig. 23.1D). The groove depth of the topographical gelatin films was 4.6±0.2 μm (Fig. 23.1E).

To demonstrate the efficacy of patterned PEG-PLA gelatin film to control the spatial distribution of cells, human microvascular endothelial cells were seeded on the patterned films. Figure 23.2 shows the phase contrast images of endothelial cells on PEG-PLA patterned flat (A-C) or topographical patterned (D-F) gelatin with varying line width. Cells selectively attach and spread along the 10, 20 and 30 μm wide lines. The width of the 10 μm lines can accommodate a single cell, while 30 μm wide lines can accommodate two cells side by side.

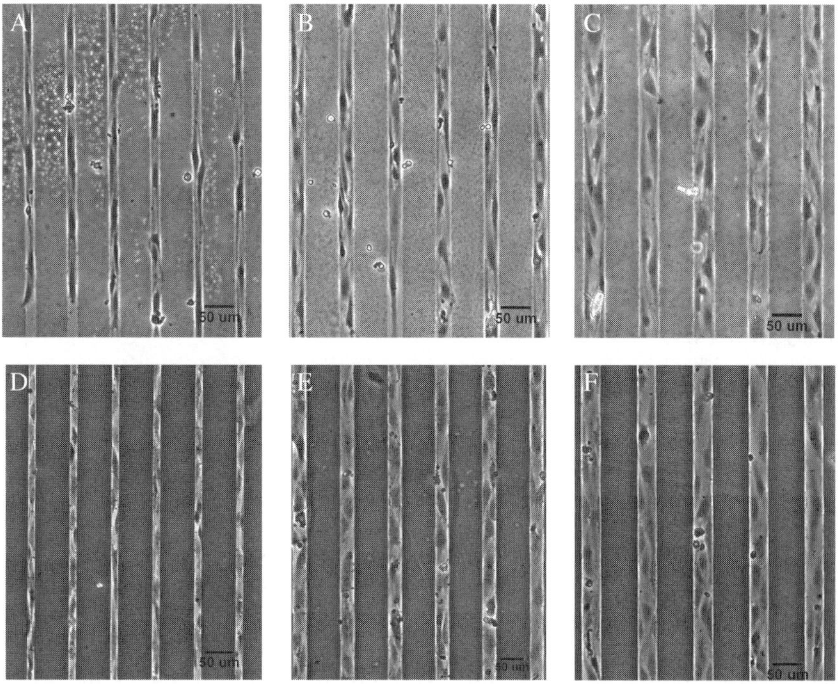

Fig. 23.2 A B C). Endothelial cells on 10 μm, 20 μm, and 30 μm wide lines on flat gelatin film after 3 days. D E F). Endothelial cells on 10 μm, 20 μm, and 30 μm wide lines on topographical patterned gelatin films after 3 days.

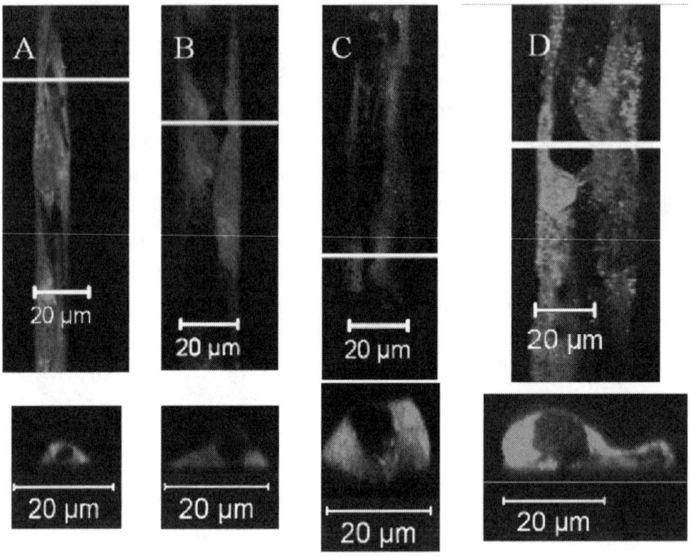

Fig. 23.3 Capillary tube formation by human microvascular endothelial cells. Confocal microscopic images of CMFDA stained cells when viewed in a horizontal (top) or a vertical (bottom) cross section. Endothelial cells form capillary tube-like structures on 10 (A) and 20 μm (B) wide lines on flat gelatin films after 3 days. Endothelial cells form capillary tube-like structures on 20 (C) and 30 μm (D) wide lines on topographical patterned gelatin films after 3 days.

After 3 days, the endothelial cells form capillary tube-like structures in 10 μm and 20 μm wide lines on flat gelatin films (Fig. 23.3A and 23.3B). No capillary structure was observed on 30 μm wide lines.

Endothelial cells grown on topographical patterned gelatin films form capillary structures in 20 μm and 30 μm grooves. No capillary structure was found in 10 μm grooves.

23.4 Conclusions

We have reported here a new approach to control spatially protein adsorption and cell attachment on flat and topographical patterned gelatin films. The patterned gelatin films are stable and have the ability to confine effectively endothelial cells in line patterns. Patterned flat gelatin films formed capillary structures on 10 μm and 20 μm lines while cells on topographical patterned gelatin films form capillary structure on 20 μm and 30 μm grooves. Our results show that under identical medium conditions, cell spreading on patterned substrates can determine whether endothelial cells differentiate to form capillary tube-like structures.

Acknowledgment The authors would like to thank the National Institute of Health (HL-084648) for the financial support of this work.

References

1. K.H. Plate, P. Warnke, Vascular endothelial growth factor. *J. Neurooncol.* **35**(3), 363–370 (1997).
2. L.E. Dike, C.S. Chen, M. Mrksich, J. Tien, M. George, Geometric control of Switching Between Growth, Apooptosis, and Differentiation During Angiogenesis Using Micropatterned Substrates, *In Vitro Cell Dev. Biol. Anim.* **35**, 441–448 (1999).
3. J. Folkman, C. Haudenschild, Angiogenesis in vitro, *Nature* **288**, 551–556 (1980).
4. G. Kumar, Y. Wang, C.C. Co, C.C. Ho, Spatially controlled cell engineering on biomaterials using polyelectrolytes, *Langmuir* **19**(25), 10550–10556 (2003).
5. C.C. Lin, C.C. Co, C.C. Ho, Micropatterning proteins and cells on polylactic acid and poly(lactide-co-glycolide), *Biomaterials* **26**(17), 3655–3662 (2005).
6. T. Yamaoka, Y. Tabata, Y. Ikada, Distribution and tissue uptake of poly(ethylene glycol) with different molecular weights after intravenous administration to mice, *J. Pharm. Sci.* **83**, 601–606 (1994).
7. C. Job, L. Lagnado, Calcium and protein kinase c regulate the actin cytoskeleton in the synaptic terminal of retinal bipolar cells, *J. Cell. Biol.* **143**(6), 1661–1672 (1998).

Part VI
Bio-Instrumentation

Chapter 24
Error Analysis of Finite-Spectral-Linewidth Illumination in Optical Oximetry Systems

Joseph L. Hollmann[1] **and Charles A. DiMarzio**[2]

Abstract Multi-spectral systems consisting of a small number of wavelengths are increasingly using light emitting diodes (LEDs) to reduce the overall costs of the system. However, LEDs typically have broad spectral bandwidths and cannot be modeled as having a single discrete wavelength. This paper puts forth a simple model to analyze the effects of using LEDs to illuminate a single layer of homogenous tissue. Monte Carlo simulations are used to approximate photon propagation through a semi-infinite turbid medium oximetry system using two light emitting diodes with broad spectra for varying oxygen saturations. The results are then compared against diffusion solutions for narrow-band illumination at the same two center wavelengths.

24.1 Introduction

Pulse oximetry was initially developed in the mid-1930s [1]. In its simplest form, tissue is illuminated with two light sources in both the visible and near infrared (NIR) spectrums. Typically the wavelengths are chosen so one is less than, and the other is greater than the isosbestic wavelength (around 800 nm) [2]. The signals' pulsatile components are then filtered and analyzed. Initially, pulse oximeters were designed to monitor signals due to the cardiac cycle which manifests itself primarily in the arteries [1] now some also monitor signal changes due to the respiratory cycle, which primarily affects the veins. Typically, the Modified-Beer Lambert's Law or some variant of it is used to find the oxygen saturation for the pulsatile signal.

Typical pulse oximeter illumination wavelengths are centered at 660 nanometers (nm) and between 880 and 940 nm [3]. For the purposes of this paper, we shall utilize illumination sources centered at 660 nm and 904 nm to analyze

[1]RBC Product Development, Lenexa, Kansas, 66215.
[2]Northeastern University, Department of Electrical and Computer Engineering, Boston, Massachusetts, 02115.

K.A. Kang et al. (eds.), *Oxygen Transport to Tissue XXIX*

the performance of commonly available oximeters; we will also develop the methodology for doing so at other wavelengths.

The basic operating principals of the pulse oximeter have remained largely unchanged even as the hardware used to acquire the measurements has evolved. One of the hardware changes that requires a second look is the evolution from lasers to light emitting diodes (LEDs) for illumination sources. This change is occurring for several reasons: cost, eye safety and the simplicity of the biasing circuit design are examples.

A laser provides narrow illumination spectral width, whereas the full width half max spectrum of an LED ranges from 20 nm in the visible range to 40 nm in the NIR region; the absorption coefficient of hemoglobin can change significantly in this range. However, Beer's Law assumes a single absorption coefficient for each illumination source. Although this is a valid approximation for lasers it may not be accurate for LEDs. This is especially so in portions of the spectrum where the gradient of the absorption spectrum is steep.

24.2 Theory

If oxygenated and deoxygenated hemoglobin are assumed to be homogenously distributed throughout tissue and are the primary absorbers, the absorption of the tissue at a given wavelength is given as

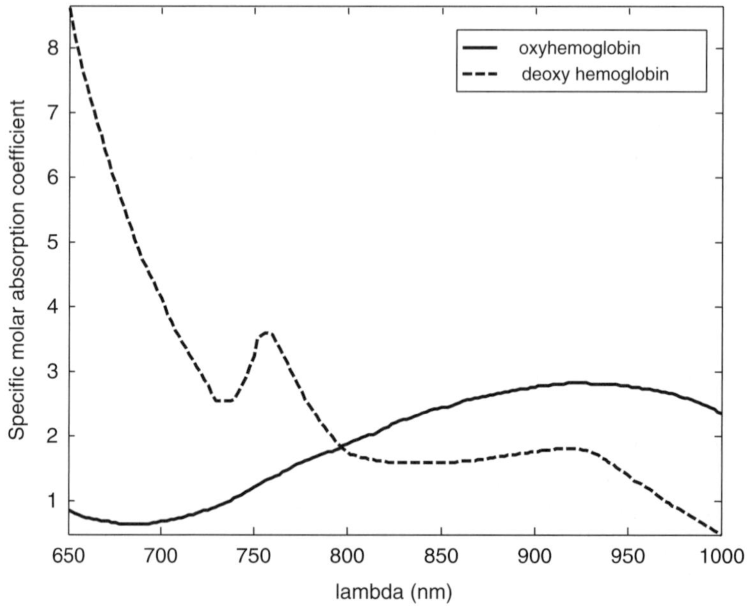

Fig. 24.1 Specific absorption of oxy (solid line) and deoxy hemoglobin (dashed line) versus wavelength.

$$\mu_a(\lambda) = \left(\kappa_{oxy}(\lambda)S_o + \kappa_{deoxy}(\lambda)(1 - S_o)\right)[Hb] \tag{24.1}$$

where $[Hb]$ is the hemoglobin concentration, S_o is the fraction of oxygenated hemoglobin and κ_{oxy} and κ_{deoxy} are the molar absorption coefficients of oxygenated and deoxygenated hemoglobin, respectively. A graph of the molar absorption coefficients over the wavelength range 650 nm–1000 nm can be found in Fig. 24.1.

If each illumination source is monochromatic, Eq. (24.1) would be enough to describe the absorption of light by the tissue. However, this assumption is not met by LEDs. Analyzing the graph in Fig. 24.1 it is easy to see the deoxygenated hemoglobin spectrum has a relatively steep gradient in the wavelength range of 650–730 nm. This may present a problem if the amount of deoxygenated hemoglobin is low.

24.3 Illumination Source

The LED light sources will be modeled as having Gaussian spectral envelopes: one centered at 660 nm with a 20 nm BW and the other at 904 nm with a 40 nm BW. The full width-half max (FWHM) of a Gaussian curve is related to the standard deviation (σ) by

$$\sigma = 2.3548 \cdot FWHM \tag{24.2}$$

The power spectral density at discrete wavelengths is given as

$$P_\lambda(\lambda) = \frac{P_{input}}{\sigma\sqrt{2\pi}} \exp\left(-(\lambda - \lambda_c)^2 \big/ 2\sigma^2\right) \tag{24.3}$$

where λ_c is the center wavelength of the LED's spectral envelope. The total power of the LED, P_{input}, is given by

$$P_{input} = \int_{\lambda\_lo}^{\lambda\_hi} P_\lambda(\lambda) d\lambda, \tag{24.4}$$

where $P_\lambda(\lambda)$is the LED power per wavelength. The limits of integration, $\lambda_{hi}$and λ_{lo}, are the maximum and minimum wavelengths of the LED's spectral bandwidth (BW), respectively. For the purposes of our simulations, the LED's spectral envelope encompassed wavelengths up to one BW away from the center wavelength. The laser beam was modeled as emitting a single discrete, wavelength centered at either 660 or 904 nm.

24.4 Tissue Model

To evaluate the difference between the narrow band and broadband model the pulse oximeter was simulated as being on the ear lobe. The earlobe is simulated as a 5 mm thick infinite slab with a normally incident, infinitesimally narrow, collimated light beam incident on one side and a 1 mm diameter circular detector on the other side of the lobe. Figure 24.2 shows the geometry. The blood vessel in the figure represents the arterial flow.

The microvasculature and cartilage in the ear lobe was modeled as a homogenous medium (as shown in Fig. 24.3). Varying the hemoglobin concentrations in the tissue simulated the diastolic and systolic pressures in the microvasculature of the earlobe.

When the microvasculature was at its minimum volume or diastolic state, the hemoglobin concentration was 0.12 milliMolar (mM) with low oxygen saturation. After the pulse, at the microvasculature's maximum or diastolic volume, the concentration is doubled with the addition of highly oxygenated hemoglobin. A table of values for the diastolic and systolic pressures can be found in Table 24.1.

Thescattering coefficient for the earlobe was chosen to be 400 cm^{-1} at 660 nm and 350 cm^{-1} at 940 nm [4]. The anisotropy factor was 0.8. Since the scattering coefficient changes gradually with wavelength, it was assumed to be unchanging over the bandwidth of the LEDs. It should be noted that if both illumination

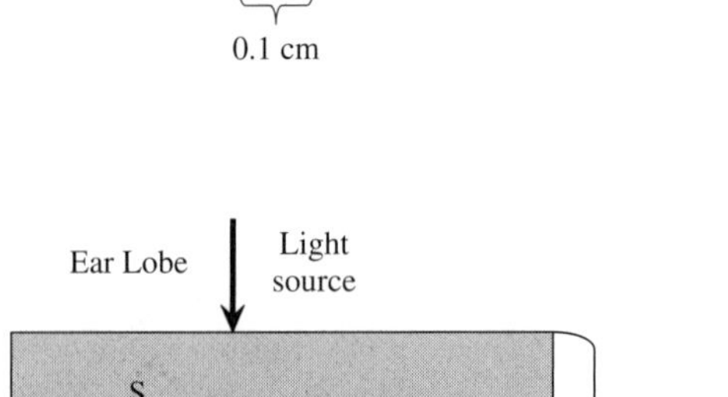

Fig. 24.2 Earlobe model where $[Hb_p]$ and $[Hb_t]$ are the molar concentration of hemoglobin and S_{po} and S_{to} are the pulsatile and tissue oxygen saturation, respectively.

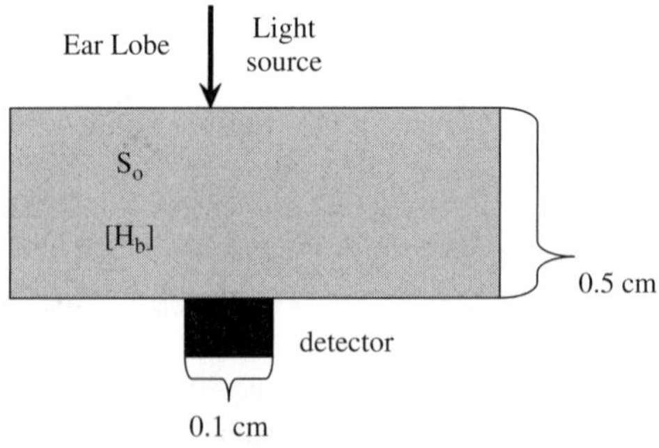

Fig. 24.3 Homogenous model of the earlobe where S_o is the average oxygen saturation of the ear and [Hb] is the total hemoglobin concentration.

Table 24.1 List of oxygen saturations and the resulting measured power for systolic and diastolic pressures

		Measured Power (mW)			
		Diastolic [Hb] = 0.12 mM		Systolic [HB] = 0.24 mM	
		65%	75%	82.5%	87.5%
Broad Band	660	7.8392e-005	1.1337e 004	5.3160c-005	7.4098c-005
	904	1.4857e-004	1.4318e-004	4.2293e-005	4.1030e-005
Narrow Band	660	8.6977e-005	1.2214e-004	5.4018e-005	7.4855e-005
	904	1.5063e-004	1.4915e-004	4.0397e-005	4.1481e-005

wavelengths were chosen closely together, they could be modeled as having a single scattering coefficient.

24.5 Propagation of Light

Beer's law describes the propagation of light through a non-scattering medium. However, it does not accurately describe the effects of scattering. Since most tissues, including the earlobe, are highly scattering Monte Carlo simulations were used. A Monte Carlo simulation propagates light through a turbid medium with tissue simulating optical properties using a weighted random walk. For this work the Multi-Layer Monte Carlo program (MCML) was utilized [5].

The MCML program was run with 1 million photons for each wavelength. The program outputs the transmission in spatially resolved probability (T). If we assume a 1 mW input, the output has units of mW/cm^2. The power incident on the detector is therefore solved for by integrating the output power over the area of the detector or

$$P_{\text{detector}}(\lambda) = \int_{-.5cm}^{.5cm} \int_{-.5cm}^{.5cm} P_{input}(\lambda) T(x, y) dx dy. \tag{24.5}$$

However, MCML assumes radial symmetry and provides an output in cylindrical coordinates, so the power incident on the detector is,

$$P_{\text{detector}}(\lambda) = 2\pi \int_{0}^{0.5cm} P_{input}(\lambda) \rho T(\rho, \lambda) d\rho \tag{24.6}$$

where $\rho = \sqrt{x^2 + y^2}$.

The result provides the spectrally resolved power at the detector.

To calculate the power at a detector due to a LED, the integral of the Gaussian weighted values of $P_{detector}(\lambda)$ was computed.

$$P_{total} = \int_{\lambda\_lo}^{\lambda\_hi} P_{detector}(\lambda) T(\lambda) d\lambda \tag{24.7}$$

The spectrum of$P(\lambda)$was simulated by discrete contribution with a spacing of 2 nm. The calculated P_{total} for a laser illumination source was given by $P_{detector}$ since it is monochromatic. For the integrations in Eqs. 24.6 and 24.7, Simpson's numerical quadrature was employed [6].

The calculated $P_{detector}$ values were utilized for the inverse problem to find the pulsatile oxygen saturation using Modified Beer-Lambert's law

$$S_{po} = \frac{\kappa_{deoxy}^{904} R - \kappa_{deoxy}^{660}}{\kappa_{oxy}^{660} - \kappa_{deoxy}^{660} - R\left(\kappa_{oxy}^{904} - \kappa_{deoxy}^{904}\right)} \tag{24.8}$$

where

$$R = \frac{\ln\left(P_{total}^{diastolic}(660) / P_{total}^{systolic}(660)\right)}{\ln\left(P_{total}^{diastolic}(904) / P_{total}^{systolic}(904)\right)}. \tag{24.9}$$

24.6 Results and Discussion

Table 24.1 shows the absolute reflectance measurements for two hemoglobin concentrations with various oxygen saturations. As mentioned above, the power measured at the detector for laser illumination was calculated using Eq. 24.6 and the power for the LED was calculated using (7. It should be noted that the error between the measured LED and laser power is largest at low saturations. From the discussions above we can see this is expected. As the amount of deoxy-hemoglobin increases so does the effect of the gradient along the illumination spectra.

The oxygen saturations were calculated for a diastolic oxygen saturation of 75% and a pulsatile saturation of 90% and 100%. Table 24.2 displays the resulting error in oxygen saturation calculations as calculated by Eq. 24.8.

It is important to note that the inverse oxygen saturations do not match the actual oxygen saturations. This is due to the scattering properties of the medium; however, these effects are normally calibrated out by taking test data over several subjects. It is important to note that the narrow band and broadband calculations are within a few percent of each other.

Table 24.2 Computed oxygen saturations as calculated by Eq. 8

	Calculated Saturations	
Oxygen Saturation	Narrow Band	Broad Band
90%	86.5%	86.4%
100%	97.0%	95.4%

24.7 Conclusion

As discussed, most pulse oximeter algorithms assume a monochromatic illumination source when solving for the oxygen saturation in tissue. A narrow bandwidth source such as a laser can be approximated as a monochromatic source; however a broadband LED cannot. This paper addressed the issue of utilizing LEDs in pulse oximetry applications.

This analysis was done for a pulse oximeter on a simulated earlobe. The spectrally resolved power was then propagated through the earlobe using Monte Carlo simulations. The LEDs' spectral envelopes were modeled as Gaussians and used to solve for the power arriving at the detector with Eq. 24.6. It should be noted that any arbitrary spectral envelope can be utilized.

The simulated absolute power was then solved for each LED and laser illumination source and the results were displayed in Table 24.1. The absolute power measurements were then utilized to calculate the oxygen saturations shown in Table 24.2. The results show the error associated with a broadband LED is minimal within physiological conditions.

However, there are other possible error sources associated with utilizing LEDs, which have not been analyzed here; such as uncollimated light. The simulations utilized assumed the light illuminating the tissue was collimated, an assumption more closely met by lasers. Future work will merge the analysis conducted in this paper with an examination of the error associated with utilizing uncolimated light sources.

References

1. Y. Mendelson, Pulse oximetry: theory and applications for noninvasive monitoring, *Clin. Chem.*, **38**(9), 1601–1607 (1992).
2. S. Fantini and M. A. Franceshini, in: *Handbook of Optical Biomedical Diagnostics*, edited by V. V. Tuchin (SPIE Press, Bellingham, 2002), pp. 427–431.
3. S. M. L. Silva, M. L. D. Castilla, and J. P. S. Martin, Near-infrared transmittance pulse oximetry with laser diodes, *J. Bio. Opt.*, 8(3), 525–533 (2003).
4. M. J. C. Van Gemert, , S. L. Jacques, H. J. C. M. Sterenborg, and W. M. Star, Skin Optics, *IEEE Trans. Bio. Eng.*, 36 (12), 1146–1154 (1989).
5. L.-H. Wang, S. L. Jacques, and L.-Q. Zheng, MCML – Monte Carlo Modeling of Photon Transport in Multi-layered Tissues, Computer Methods and Programs in Biomedicine 47, 131–146 (1995).
6. M. N. O. Sadiku, *Numerical Techniques in Electromagnetics, 2nd edition* (CRC Press, Boca Raton, 2001) pp. 197–199.

Chapter 25
Changes in the Attenuation of Near Infrared Spectra by the Healthy Adult Brain During Hypoxaemia Cannot be Accounted for Solely by Changes in the Concentrations of Oxy- and Deoxy-Haemoglobin

Martin M. Tisdall[1], Ilias Tachtsidis[2], Terence S. Leung[2], Clare E. Elwell[2], and Martin Smith[1]

Abstract It has been suggested that changes in oxidised cytochrome c oxidase concentration ([oxCCO]) measured using cerebral near infrared spectroscopy (NIRS) may be algorithm artefacts. We examine the change in near infrared (NIR) attenuation by the healthy adult brain ($n = 10$) during hypoxaemia. Broadband spectroscopic data were collected during normoxia, and hypoxaemia. The UCL*n* algorithm was used to fit (a) oxy- (HbO_2) and deoxy-haemoglobin (HHb) spectra (2 component fit), and (b) HbO_2, HHb and oxidised-reduced cytochrome c oxidase difference spectra (3 component fit) to the mean change in NIR attenuation between baseline and hypoxaemia. The sum of squares of the residuals was 100×10^{-7} OD^2 for the 2 component fit and 8×10^{-7} OD^2 for the 3 component fit, and the two sets of residuals differed from each other ($p = 0.0003$). We compare experimental and simulated data and suggest that the 2 component residuals indicate a change in [oxCCO]. Changes in near infrared attenuation by the healthy adult brain during hypoxaemia cannot be accounted for solely by changes in oxy- and deoxy-haemoglobin concentrations. Including [oxCCO] in the algorithm improves its fit quality. These data suggest that changes in cerebral cytochrome c oxidase redox occur during hypoxaemia and that they can be detected using NIRS.

25.1 Introduction

The use of near infrared spectroscopy (NIRS) to measure changes in the optical characteristics of living tissue was first described by Jöbsis in 1977 [1]. NIRS makes use of the fact that biological tissue is relatively transparent to light between 700–900 nm, thus allowing interrogation of structures beneath the

[1]Department of Neuroanaesthesia and Neurocritical Care, The National Hospital for Neurology and Neurosurgery, Queen Square, London, UK.
[2]Department of Medical Physics and Bioengineering, University College London, London, UK.

K.A. Kang et al. (eds.), *Oxygen Transport to Tissue XXIX*

tissue surface [1]. Light passing through biological tissue is subject to multiple scattering interactions, and this complicates the interpretation of the intensity of detected light. However, if light lost due to scattering is assumed to remain constant, and the average pathlength of light through tissue is known, the modified Beer-Lambert law can be used to convert change in light attenuation to absolute change in chromophore concentrations [2].

In both animals and humans, NIRS has been used to measure change in concentrations of oxy- ($\Delta[HbO_2]$), and deoxy-haemoglobin (Δ[HHb]) and oxidised cytochrome c oxidase (Δ[oxCCO]) [3–5], however controversy still remains as to the validity of the measured Δ[oxCCO] [5].

Cytochrome c oxidase (CCO) is the terminal electron acceptor of the mitochondrial electron transfer chain and catalyses over 95% of oxygen metabolism. The reduction of dioxygen provides the proton motive force to drive aerobic adenosine triphosphate synthesis [6]. The difference spectrum between the oxidised and reduced forms of CCO has a distinct band in the near infrared (NIR) region, with a broad peak located around 830 nm [1]. Assuming the total concentration of CCO remains constant, changes in the CCO signal represent changes in the CCO redox state. The CCO signal is an attractive target for clinical monitoring, as it offers the potential to provide a non-invasive marker of the adequacy of mitochondrial oxygen delivery.

However, detection of changes in the CCO signal is complicated by the fact that the concentration of CCO in the brain is approximately one order of magnitude less than that of either oxy- or deoxy-haemoglobin [7]. This raises the possibility that measured Δ[oxCCO] might simply be an artefact produced by the algorithms, used to convert measured attenuation changes into chromophore concentration changes, being unable to adequately separate the CCO and haemoglobin signals [5]. Furthermore, controversy remains as to the degree of hypoxaemia required to produce changes in CCO redox state.

Despite these issues, Δ[oxCCO] has been shown to correlate with nuclear magnetic resonance 31P spectroscopy measured reduction in phosphocreatine and nucleoside triphosphate in an animal model of cerebral ischaemia [8], and in cardiac surgery in humans it has been shown to predict adverse neurological outcome [4].

In this study we measure the change in NIR light attenuation by the healthy human brain during hypoxaemia using a broadband NIR spectrometer. We analyse the residual errors produced by the fitting procedure for the conversion of light attenuation into chromophore concentrations in order to determine if the change in NIR attenuation can be accounted for solely by Δ[HHb] and $\Delta[HbO_2]$, or whether Δ[oxCCO] must also be considered.

25.2 Materials and Methods

This study was approved by the Joint Research Ethics Committee of the National Hospital for Neurology and Neurosurgery and the Institute of Neurology. We studied 10 healthy subjects (7 male, 3 female, median age

32 years, range 30–39). Broadband spectrometer (BBS) optodes were placed 3.5 cm apart in a black plastic holder, and fixed to the right side of the forehead in the midpupilary line. Light from a stabilised tungsten halogen light source was passed through 610nm long-pass and heat absorbing filters, and transmitted to the head via a 3.3 mm diameter glass optic fibre bundle. Light incident on the detector optode was focused via an identical fibre bundle onto the 400 μm entrance slit of a 0.27 m spectrograph (270M, Instruments SA, France) with a 300g/mm grating. NIR spectra between 650 and 980 nm were collected at 1Hz on a cooled charge coupled device detector (Wright Instruments, UK) giving a spectral resolution of ~5 nm. An oximeter probe (Novametrix Medical Systems Inc., USA) measured arterial oxygen saturation (SaO_2). A modified anesthetic machine delivered gas to the subject via a mouthpiece. Inspired oxygen concentration (FiO_2) was measured using an inline gas analyser (Hewlett Packard, UK). The study commenced with five minutes monitoring at normoxia. We then added nitrogen to the inspired gases to induce a gradual fall in SaO_2 to 80% and, immediately after this was achieved, the FiO_2 was returned to normoxia for five minutes. This cycle was repeated three times.

The points just prior to the start of each hypoxaemia (baseline), and at the end of each hypoxaemia (hypoxaemia), were identified manually using the SaO_2 data. At each of the two points the mean of the preceding ten seconds of data was taken. Data from the three experimental cycles were averaged to give mean values for SaO_2 and NIR light intensity at baseline and hypoxaemia for each subject. Optical pathlength was calculated using second differential analysis of the 740nm water feature [9]. Change in NIR attenuation was then calculated from:

$$\Delta A = \log_{10}\left(I_{base}/I_{hypox}\right) \tag{25.1}$$

where ΔA = change in attenuation from baseline to hypoxaemia, I_{base} = light intensity at baseline and I_{hypox} = light intensity at hypoxaemia measured in units of optical density (OD). The UCLn algorithm [10], a multiple regression analysis utilising the Beer-Lambert law was then used to fit chromophore extinction coefficients, corrected for the wavelength dependence of the optical pathlength [11], to the group mean change in attenuation, using 120 wavelengths between 780 and 900 nm. Chromophore specific extinction coefficients were downloaded from the medical physics UCL website [12]. First, only oxy- (HbO_2) and deoxy-haemoglobin (HHb) spectra (2 component fit – Eq. (25.2) and then HbO_2, HHb and the oxidised-reduced CCO difference spectra (oxCCO) (3 component fit – Eq. (25.3)) were fitted to the group mean change in attenuation. After interpolation of the residuals to the spectral resolution of the BBS (5 nm), the sums of the squares of the residuals from these two analyses were calculated, and the distributions of the two sets of residuals were compared.

$$\begin{pmatrix} \Delta[HbO_2] \\ \Delta[HHb] \end{pmatrix} = \frac{1}{PL} \begin{pmatrix} \varepsilon_{HbO_2}(\lambda_i) & \varepsilon_{HHb}(\lambda_i) \\ \vdots & \vdots \\ \varepsilon_{HbO_2}(\lambda_j) & \varepsilon_{HHb}(\lambda_j) \end{pmatrix}^{-1} \begin{pmatrix} \Delta A(\lambda_i) \\ \vdots \\ \Delta A(\lambda_j) \end{pmatrix}$$

$$\begin{pmatrix} \Delta[HbO_2] \\ \Delta[HHb] \\ \Delta[oxCCO] \end{pmatrix} = \frac{1}{PL} \begin{pmatrix} \varepsilon_{HbO_2}(\lambda_i) & \varepsilon_{HHb}(\lambda_i) & \varepsilon_{oxCCO}(\lambda_i) \\ \vdots & \vdots & \vdots \\ \varepsilon_{HbO_2}(\lambda_j) & \varepsilon_{HHb}(\lambda_j) & \varepsilon_{oxCCO}(\lambda_j) \end{pmatrix}^{-1} \begin{pmatrix} \Delta A(\lambda_i) \\ \vdots \\ \Delta A(\lambda_j) \end{pmatrix} \quad (25.3)$$

where Δ[HHb], $\Delta[HbO_2]$ and Δ[oxCCO] are changes in the concentrations of oxy-, and deoxy-haemoglobin and oxidised cytochrome c oxidase in μM, PL = pathlength in cm, ϵ is the specific extinction coefficient of the subsequent chromophore in OD/μM/cm and ΔA is the change in attenuation, at wavelengths λ_i to λ_j.

We then produced a simulated attenuation spectrum calculated using assumed Δ[HHb], $\Delta[HbO_2]$ and Δ[oxCCO] and their respective specific extinction coefficients, and ignoring change in attenuation due to other chromophores, using Eq. (4). We fitted a 2 component model to this spectrum and compared the resultant residuals with those from a 2 component fit to the group mean experimental spectrum. 2 component fits to the experimental and simulated data for each individual were then compared.

$$\Delta A(\lambda_j) = PL\{\Delta[HbO_2] \times \varepsilon_{HbO_2}(\lambda_j) + \Delta[HHb] \times \varepsilon_{HHb}(\lambda_j) + \Delta[oxCCO] \times \varepsilon_{oxCCO}(\lambda_j)\} \quad (25.4)$$

Statistical analysis was carried out using SAS software (v8.2, SAS Institute, USA) and p values <0.05 were considered significant. Group changes between baseline and hypoxaemia were compared using Wilcoxon signed rank test and the distributions of the residuals from the various fitting procedures were compared using a 2 sample Siegel-Tukey test [13].

25.3 Results

Results are presented as median (interquartile range). The median time of hypoxia required to reach an SaO_2 of 80% was 4.48 mins (3.92 to 5.04). During the study SaO2 fell from a baseline value of 99.0% (98.2 to 99.2) to 82.4% (80.1 to 84.7) at the end of hypoxaemia (p = 0.002) (Fig. 25.1). Note that due to the ten second averaging window this median SaO_2 is higher than 80%. There was no change in optical pathlength between baseline and hypoxaemia (p = 0.23). Group mean change in attenuation from baseline to hypoxaemia is shown in Fig. 25.2

The residuals from the 2 and 3 component fits to the group mean experimental spectrum differed from each other (p = 0.0003) (Fig. 25.3). The sum of

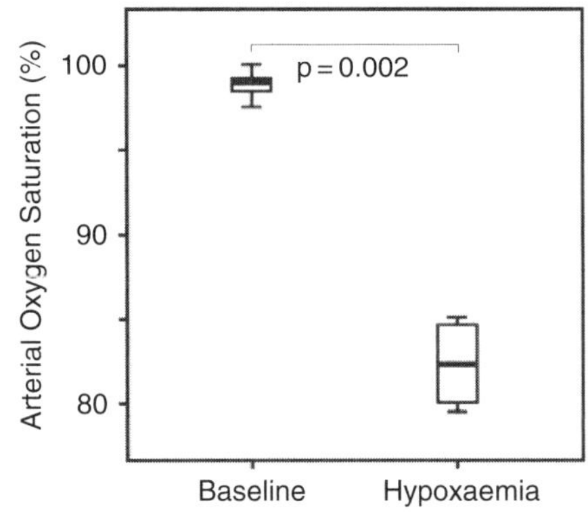

Fig. 25.1 Boxplot showing arterial oxygen saturation data at baseline and hypoxaemia.

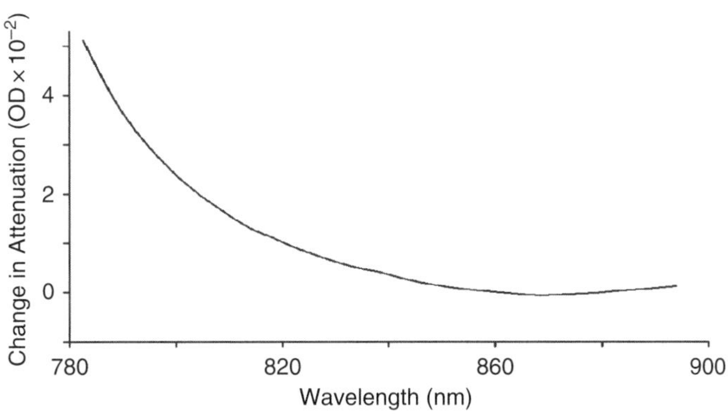

Fig. 25.2 Group mean change in near infrared attenuation between baseline and hypoxaemia.

the squares of the residuals was 100×10^{-7} OD^2 for the 2 component fit to the experimental spectrum and 8×10^{-7} OD^2 for the 3 component fit to the experimental spectrum. There was no difference between the residuals from the 2 component fits to the experimental and simulated spectra ($p=0.61$) (Fig. 25.3). The 2 component fits to the experimental and simulated data for each subject are shown in Fig. 25.4. In eight out of the ten subjects, there were no differences between the two sets of residuals ($p>0.05$).

25.4 Conclusions

The quality of a multiple regression fit can be determined by assessing the residuals of the fitting procedure. The better the fit the smaller will be the sum of the square of the residuals, with the perfect theoretical fit having residuals all

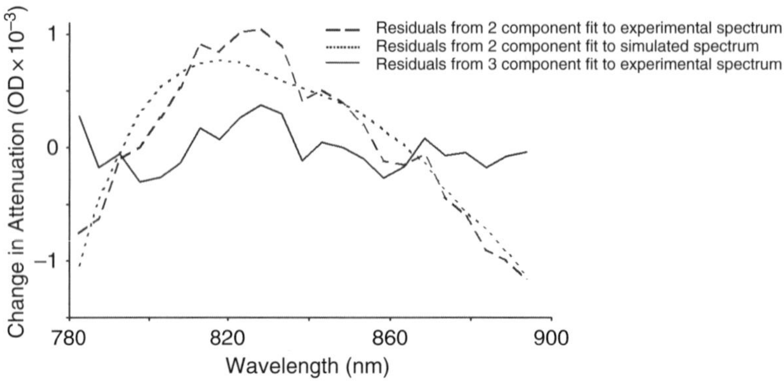

Fig. 25.3 Residuals from 2 and 3 component fits to group mean change in near infrared attenuation between baseline and hypoxaemia and 2 component fit to simulated spectrum.

equal to zero. Furthermore, any residuals which are present should be randomly distributed around zero. The presence of residuals which are not randomly distributed suggests that there is a component missing from the fitting analysis.

The residuals from the 2 component fit to the group mean experimental spectrum do not appear independent and show a broad peak located around 830 nm which is similar to the oxCCO difference spectrum. This suggests that the attenuation of NIR spectra by the healthy human brain during hypoxaemia cannot be accounted for solely by Δ[HHb] and Δ[HbO_2]. When the 3 component model is used, to also fit for Δ[oxCCO], the sum of the squares of the residuals is reduced and the residuals appear random, thus improving the fit. The simulated spectrum assumes that Δ[HHb], Δ[HbO_2] and Δ[oxCCO] are occurring and that no other chromophores are causing a change in optical attenuation. The residuals from the 2 component fit to this spectrum are those that would be expected from the 2 component fit to the experimental data if Δ[HHb], Δ[HbO_2] and Δ[oxCCO] were occurring in the brain during this hypoxaemic challenge. The residuals from the 2 component fit to the experimental spectrum do not differ from those resulting from the 2 component fit to the simulated spectrum. This strongly suggests that Δ[oxCCO] is occurring and this accounts for the residuals from the 2 component to the experimental data.

There is no statistical difference between the individual 2 component fits to the experimental and simulated data in eight out of ten of the individual subjects. This demonstrates the optical effect of Δ[oxCCO] at the individual as well as the group level. In the two subjects who exhibited differences between the residuals to the experimental and simulated data, the simulated residuals are very close to zero. This results in a very low dispersion in the residuals to the simulated data. It is possible that in these two individuals the physiological challenge was insufficient to produce a significant change in Δ[oxCCO].

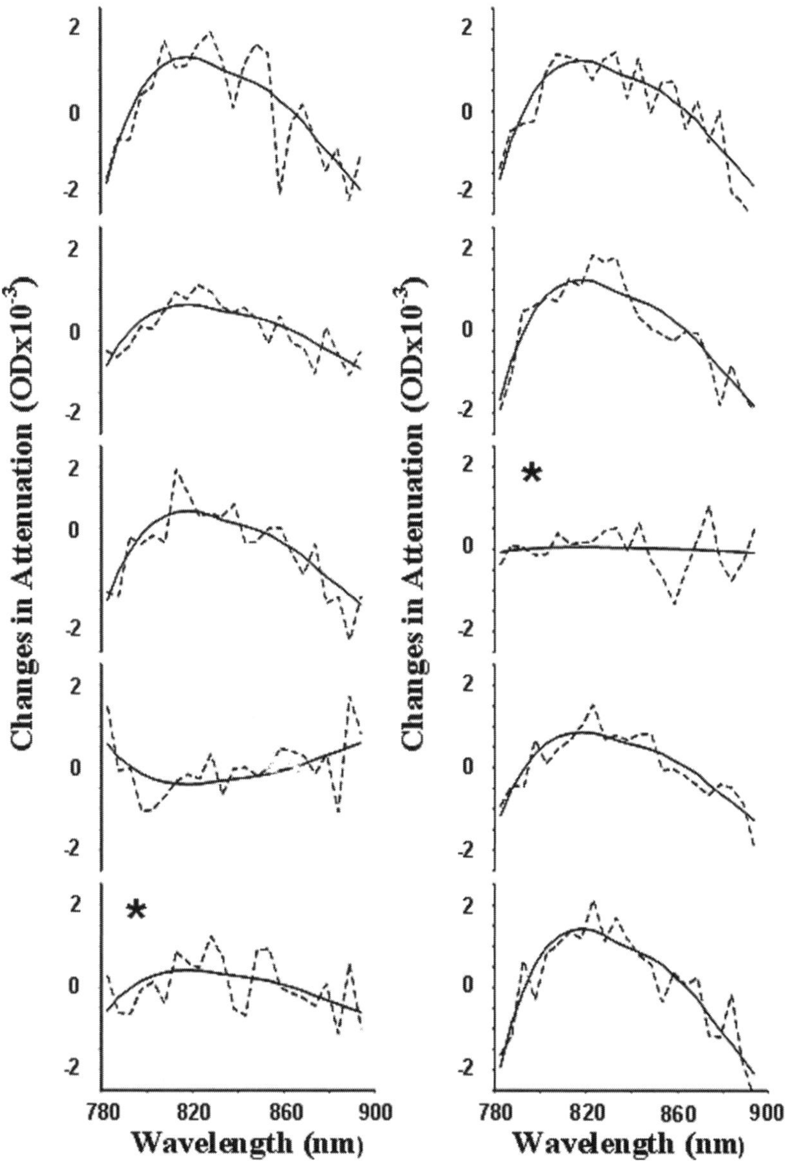

Fig. 25.4 Residuals from 2 component fits to change in near infrared attenuation between baseline and hypoxaemia (- -) and 2 component fit to simulated spectrum (—) for each individual subject. Two subjects had significant differences between the two sets of residuals (marked with *).

We postulate, therefore, that changes in cerebral CCO redox state occur during moderate hypoxaemia, and that we can detect these changes using non-invasive BBS. We suggest that one should fit the oxidised-reduced CCO difference spectra when using NIRS to monitor the brain during hypoxaemia. It has been suggested that NIRS algorithms using a small number of discrete wavelengths are less capable of separating the HbO_2, HHb and oxCCO signals.[10] This broadband spectroscopy dataset collected using multiple wavelengths will allow us to test various sets of wavelengths in order to determine which subsets perform best.

We are currently using BBS to study changes in human cerebral CCO redox state occurring after traumatic brain injury. This measurement may be able to provide clinically relevant information with which to guide neuroprotective treatment of acute brain injury on the neurocritical care unit.

Acknowledgment MMT is a Welcome Research Fellow, Grant No 075608. IT is supported by UCL/UCLH Trustees.

References

1. F. F. Jöbsis, Noninvasive, infrared monitoring of cerebral and myocardial oxygen sufficiency and circulatory parameters, *Science* **198**(4323), 1264–1267 (1977).
2. D. T. Delpy, M. Cope, P. van der Zee, S. Arridge, S. Wray, and J. Wyatt, Estimation of optical pathlength through tissue from direct time of flight measurement, *Phys. Med. Biol.* **33**(12), 1433–1442 (1988).
3. C. E. Cooper, D. T. Delpy, and E. M. Nemoto, The relationship of oxygen delivery to absolute haemoglobin oxygenation and mitochondrial cytochrome oxidase redox state in the adult brain: a near-infrared spectroscopy study, *Biochem. J.* **332**(3), 627–632 (1998).
4. Y. Kakihana, A. Matsunaga, K. Tobo, S. Isowaki, M. Kawakami, I. Tsuneyoshi, Y. Kanmura, and M. Tamura, Redox behavior of cytochrome oxidase and neurological prognosis in 66 patients who underwent thoracic aortic surgery, *Eur. J. Cardiothorac. Surg.* **21**(3), 434–439 (2002).
5. T. Sakamoto, R. A. Jonas, U. A. Stock, S. Hatsuoka, M. Cope, R. J. Springett, and G. Nollert, Utility and limitations of near-infrared spectroscopy during cardiopulmonary bypass in a piglet model, *Pediatr. Res.* **49**(6), 770–776 (2001).
6. O. M. Richter, and B. Ludwig, Cytochrome c oxidase–structure, function, and physiology of a redox-driven molecular machine, *Rev. Physiol. Biochem. Pharmacol.* **147**, 47–74 (2003).
7. G.C. Brown, M. Crompton, and S. Wray, Cytochrome oxidase content of rat brain during development, *Biochem. Biophys. Acta* **1057**(2), 273–275 (1991).
8. R. J. Springett, M. Wylezinska, E. B. Cady, V. Hollis, M. Cope, and D. T. Delpy, The oxygen dependency of cerebral oxidative metabolism in the newborn piglet studied with 31P NMRS and NIRS, *Adv. Exp. Med. Biol.* **530**, 555–563 (2003).
9. S. J. Matcher, M. Cope, and D. T. Delpy, Use of the water absorption spectrum to quantify tissue chromophore concentration changes in near-infrared spectroscopy, *Phys. Med. Biol.* **39**(1), 177–196 (1994).
10. S. J. Matcher, C. E. Elwell, C. E. Cooper, M. Cope, and D. T. Delpy, Performance comparison of several published tissue near-infrared spectroscopy algorithms, *Anal. Biochem.* **227**(1), 54–68 (1995).

11. M. Essenpreis, M. Cope, C.E. Elwell, S.R. Arridge, P. van der Zee P, and D.T. Delpy, Wavelength dependence of the differential pathlength factor and the log slope in time-resolved tissue spectroscopy, *Adv. Exp. Med. Biol.* **333**, 9–20 (1993).
12. http://www.medphys.ucl.ac.uk/research/borl/research/NIR_topics/spectra/spectra.htm.
13. S. Siegel, and N.J. Castellan Jr, *Nonparametric Statistics for the Behavioural Sciences* 2nd ed. (McGraw-Hill, Singapore, 1988).

Chapter 26
Assessment of Oxygenation and Perfusion in the Tongue and Oral Mucosa by Visible Spectrophotometry and Laser Doppler Flowmetry in Healthy Subjects

D.B. Singh[1], **G. Stansby**[2], **and D.K. Harrison**[1]

Abstract Use of Visible Light Spectrophotometry (VLS) and Laser Doppler Flowmetry (LDF) is currently being studied by the authors to assess the viability of tissue margins in colon resection and to assess mucosal oxygenation in the colon. Thus, as a preliminary study it was necessary to evaluate whether there is any systematic inter-probe variability of the measurements by VLS and LDF. The oral mucosa was used as a model.

Methods SO_2 with VLS (Whitland Research RM200) and blood flow with LDF (Moor Instruments DRT4) were measured at 10 sites each on the tongue and oral mucosa of 10 healthy volunteers at 0, 6 and 24 hours using 3 different probes for VLS and 2 probes for LDF.

Results The results showed that the SO_2 measurements by VLS using the different probes on the tongue and mucosa were significantly correlated ($P<0.05$). SO_2 values at 6 hours were significantly higher than at 0 and 24 hours ($P<0.05$) in all but one case. SO_2 measurements were not correlated with LDF. LDF measurements by the 2 probes were correlated significantly ($P<0.05$) but the standard deviations were very large.

Conclusions SO_2 measurements on the oral mucosa are reproducible. Due to the large variations in LDF, VLS is likely to be the more clinically useful tool for identifying mucosal ischaemia

26.1 Introduction

A clinical study is currently being undertaken by the authors on the use of Visible Light Spectrophotometry (VLS) and Laser Doppler Flowmetry (LDF) to assess the viability of tissue margins in colon resection. Adequate tissue blood flow and oxygenation are known to be essential for anastomotic healing

[1]University Hospital of North Durham, Durham, UK.
[2]Northern Vascular Unit, Freeman Hospital, Newcastle upon Tyne, UK.

following colorectal resections in both animal and human clinical studies [1]. In a comparison of colonic pouch-anal and conventional straight end-to-end anastomosis, the microcirculation at the site of the anastomosis in the pouch was shown to be better preserved, as measured by higher tissue blood flow, than at the end of the conventional straight anastomosis [2]. This may account for the reported lower rates of leakage with a colonic J pouch reconstruction following low anterior resection [3].

A further study is also assessing mucosal oxygenation in the colon by VLS, during colonoscopy, for diagnosing bowel ischaemia.

The technique of lightguide spectrophotometry to measure tissue oxygen saturation has been validated for use in assessing microvascular blood flow in skin in peripheral vascular disease and has been shown to be of clinical use in defining the levels of amputation in patients with peripheral arterial disease [4]. Use of VLS to assess tissue oxygen saturation (SO_2) in the colon is a relatively new procedure [5]. In the current clinical studies different configurations of probes are being used for the measurements of SO_2 and LDF in the luminal surface and serosal surface of the bowel. Thus, as a preliminary study it was necessary to evaluate whether there is any systematic inter-probe variability of the measurements by VLS and LDF. Similarly, it was necessary to establish the range and reproducibility of normal values for mucosal SO_2 and to investigate whether there were any diurnal variations. The oral mucosa was used as a model.

26.2 Aim

The main aim of this study was to evaluate whether SO_2 measurements by different probes as measured by spectrometry are reproducible and whether there is any inter-probe variability. The reproducibility of Laser Doppler Flux measurements was also evaluated.

26.3 Materials

The visible light spectrophotometer used was a Whitland Research RM200 SO_2 monitor. A Moor Instruments DRT4 was used for Laser Doppler flux measurements.

Three types of probes (2 endoscopic catheter probes and 1 surface probe) were used for SO_2 measurements by VLS. Endoscopic catheter probes, 4 metres in length, had outer diameters of 2.1mm and 1.35mm (Moor Instruments, UK, DP6sd). The endoscopic probes used in this study were side delivery ones which meant that the light was emitted and received from the side within 5mm of the tip rather than the end. These probes are specifically designed for laser Doppler flux measurements, but with the help of an adaptor (manufactured with the help of Moor Instruments, UK) it was possible to use the same probes for SO_2

measurements. The surface probe used is a standard probe by Whitland Research used for measuring SO_2. An external light source was used with the catheter probes in the form of a halogen lamp (400watts) as the LED light source in the spectrophotometer was not sufficiently powerful for the catheter probe. For the surface probe we used the spectrophotometer's in-built light source. In order to eliminate systemic spectral effects such as lightguide transmission characteristics, variations in photodiode array sensitivity and the different light sources used, a dark and white balance was carried out with each probe prior to measurement.

Two types of probes were used for LDF measurements – one endoscopic catheter probe with outer diameter 2.1 mm (Moor Instruments DP6asd) and a surface probe (Moor Instruments, DP1T-V2).

26.4 Methods

Ten healthy volunteers were recruited for the study after securing ethical approval from the Local Research Ethics Committee. Participants with known medical conditions and smokers were excluded from the study. SO_2 with VLS (Whitland Research RM200) and laser Doppler flux (Moor Instruments DRT4) were measured on the tongue and oral mucosa at 0, 6 and 24 hours. The three types of probes described above were used for VLS. The endoscopy catheter probe with outer diameter 2.1 mm and surface probe (DP1T-V2) were used for LD flux measurements. Measurements were taken with each probe at 10 points both on the tongue and oral mucosa at 0, 6 and 24 hours and the mean values recorded. Participants were requested not to have any hot drinks for at least an hour before the measurements were taken. An ambient temperature of 22–23°C was maintained throughout the measurements.

26.5 Results

The median age of the participants was 43 (range 37–63) years with 7 females. The results showed that the SO_2 measurements by VLS using the different probes on the tongue and mucosa (Table 26.1) were significantly correlated ($P<0.05$).

Bland and Altman [6] analysis of the original data showed that 95% of the difference between the small catheter and surface probes were within 2 standard deviations (SD). For the large catheter and surface probes, 90% of the difference were within 2SD. However, this raw data may contain bias due to differences between the probes because of physical factors such as lightguide diameter. In order to investigate whether this systematic bias could be corrected, regression analysis was carried out whereby the surface probe was considered as the

Table 26.1 Mean SO_2 % in tongue and mucosa by different probes at 0, 6 and 24 hours

	TongueMean SO_2 (SD) %			MucosaMean SO_2 (SD) %		
	0 hour	6 hours	24 hours	0 hour	6 hours	24 hours
Large catheter probe	81.8 (5.6)	85.3 (4.3)	83.2 (3.6)	85.8 (4.0)	87.8 (3.3)	87.3 (3.1)
Small catheter probe	75.3 (9.2)	79.8 (8.3)	75.3 (7.0)	80.3 (6.6)	84.6 (5.6)	81.4 (3.6)
Surface probe	81.2 (5.6)	81.2 (5.8)	80.7 (2.6)	82.2 (5.1)	83.5 (3.9)	83.8 (2.9)

standard. The correlations are shown in Figs. 26.1 and 26.2. The appropriate regression equations shown in the figures were used to "calibrate" the SO_2 values measured using the catheter probes against the surface probe values. After the corrections, further Bland and Altman [6] plots showed that the measurements using all 3 probes were in agreement (Figs. 26.3 and 26.4). Ninety five percent of the differences were within 2 standard deviations.

The measurements from all probes were combined and produced mean (SD) SO_2 values for the tongue at 0, 6 and 24 hrs of 78.5 (10.0)%, 81.7 (7.7)% and 78.9 (6.3)% respectively. Corresponding figures for oral mucosa were 82.6 (6.6)%, 85.8 (5.4)% and 84.4 (4.4)% (Table 26.3).

SO_2 measurements were normally distributed, so the Student t-test was used to analyse any differences between 0, 6 and 24 hours values in the tongue and mucosa. The analysis showed significant differences between tongue and mucosal SO_2 across all times (except at 0 hours) with mucosal SO_2 being higher. SO_2 values at 6 hours were significantly higher than at 0 and 24 hours ($P<0.05$) in all but one case.

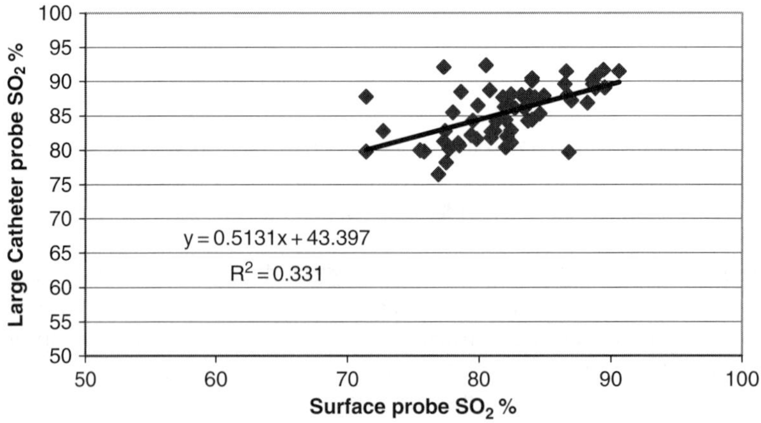

Fig. 26.1 Correlation between surface and large catheter SO_2 probes.

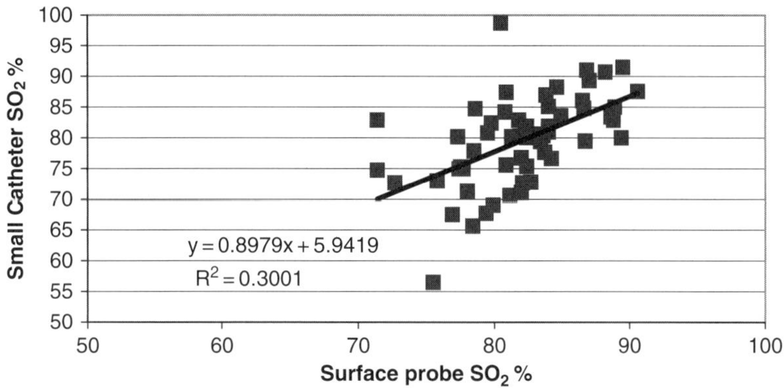

Fig. 26.2 Correlation between surface and small catheter SO_2 probes.

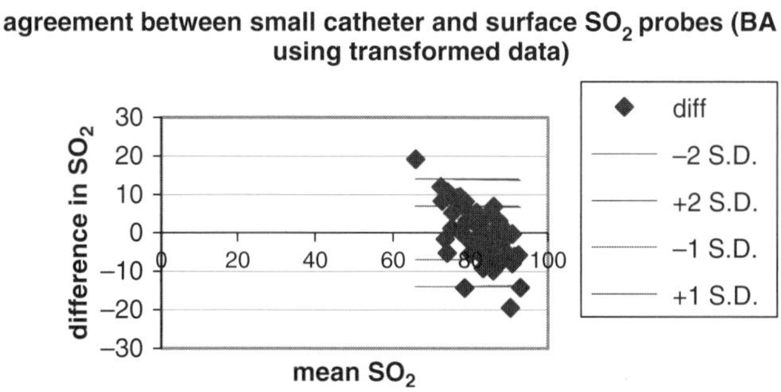

Fig. 26.3 Bland Altman plot for surface and small catheter SO_2 probes.

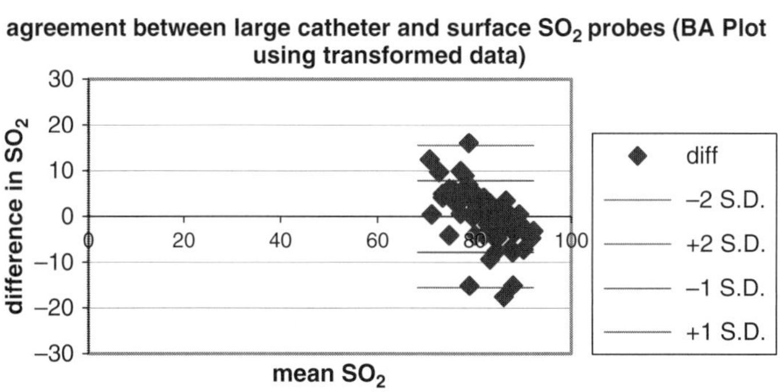

Fig. 26.4 Bland Altman plot for surface and large catheter SO_2 probes.

Table 26.2 Mean Laser Doppler Flux in tongue and mucosa measured by different probes at 0, 6 and 24 hours

	Tongue Mean flux (SD)			Mucosa Mean flux (SD)		
	0 hour	6 hours	24 hours	0 hour	6 hours	24 hours
Surface probe	339.1 (74.7)	312.0 (76.4)	312.0 (61.6)	362.9 (54.9)	364.5 (90.4)	397.6 (109.4)
Large catheter probe	218.0 (42.3)	263.5 (63.4)	255.3 (26.3)	271.3 (40.2)	292.9 (54.1)	269.5 (19.8)

Table 26.3 Mean SO_2 and Flux values on the tongue and mucosa at 0, 6 and 24 hours

Time	Tongue Mean SO_2 (Std Dev)%	Mucosa Mean SO_2 (Std Dev)%	Tongue Mean Flux (Std Dev)	Mucosa Mean Flux (Std Dev)
0 hour	78.5 (10.0)	82.6 (6.6)	282.77 (100.98)	371.84 (106.57)
6 hours	81.7 (7.7)	85.8 (5.4)	332.31 (170.41)	412.81 (158.75)
24 hours	78.9 (6.3)	84.4 (4.4)	317.29 (79.3)	386.2 (91.9))

SO_2 measurements were not correlated with LDF. LDF measurements using the 2 probes (Table 26.2) were correlated significantly ($P<0.05$) but the standard deviations were very large. For the purpose of data analysis we considered the surface Doppler probe as the standard and applying the correlation we corrected the values measured with the catheter probe. The corrected flux also had a high standard deviation (163.3).

26.6 Discussion

This study confirms that after calibration of the catheter probes against the surface probe, the physiological SO_2 measurements by three different probes are in agreement as 95% of the differences between the probes were within 2 SD and are within clinically acceptable limits. For the catheter probe a 400W external light source was used to achieve an adequate signal. This loss is partly be due to the adaptor used to connect the Laser Doppler probe to the spectrophotometer. Further significant losses were experienced when using the small diameter catheter probe and the quality of the spectra obtained was poor. For this reason, the small catheter probe has been discontinued from use in our further clinical work. The difference in mean SO_2 for oral mucosa at 0 and 6 hours was statistically significant but the actual difference was only 3%, which would not be clinically significant. Friedland et al [5]. showed in their study that the colon SO_2 decrease by 40% on induction of ischaemia. Fournell et al [8]. hypothesized that an SO_2 value below 60% probably mark the threshold for mesenteric ischaemia.

Laser Doppler flux measurements on the tongue and oral mucosa were highly variable. There was no correlation between SO_2 and LDF. This poor correlation can be explained by the fact that SO_2 does not change significantly until perfusion becomes much lower than any values measured in this study (Caddick et al) [7]. The measurements using the surface probe were higher than with the endoscopic probe at all times and sites. Even after correction of the difference by applying the correlation, the differences between these two probes were high with a SD of 144.5. There is a possibility that this difference may be due to the side viewing configuration of the catheter probe. Even with only slight rotation, the port would look away from the tissue, thus giving a low reading.

26.7 Conclusions

The results indicate that SO_2 measurements on the oral mucosa are reproducible, and values recorded using different probes are comparable. Due to the large variations in LDF, VLS is likely to be the more clinically useful tool for identifying mucosal ischaemia.

References

1. A. Vignali, L. Gianotti, M. Braga, G. Radaelli, L. Malvezzi and V. Di Carlo. Altered microperfusion at the rectal stump is predictive for rectal anastomotic leak. *Dis Colon Rectum* **43**, 76–82 (2000).
2. M. Sailer, E. S. Debus, K. H. Fuchs, J. Beyerlein and A. Thiede. Comparison of anastomotic microcirculation in coloanal J-pouches versus straight and side-to-end coloanal reconstruction: an experimental study in the pig. *Int. J. Colorectal Dis.* **15**, 114–117 (2000).
3. O. Hallbook, K. Johansson, and R. Sjodahl, Laser Doppler blood flow measurement in rectal resection for carcinoma–comparison between the straight and colonic J pouch reconstruction. *Br. J. Surg.* **83** 389–392 (1996).
4. D. K. Harrison, P. T. McCollum, D. J. Newton, P. Hickman and A. S. Jain, Amputation level assessment using lightguide spectrophotometry. *Prosthet. Orthot. Int.* **19** 139–147 (1995).
5. S. Friedland, D. Benaron, I. Parachikov and R. Soetikno, Measurement of mucosal capillary hemoglobin oxygen saturation in the colon by reflectance spectrophotometry. *Gastrointest. Endosc.* **57** 492–497 (2003).
6. J. M. Bland and D.G. Altman, Statistical methods for assessing agreement between two methods of clinical measurement. *Lancet* **1** 307–310 (1986).
7. J. Caddick C. Raine, D. Harrison and M. Erdmann, Lightguide spectrophotometry to monitor free TRAM flaps. *Adv. Exp. Med. Biol.* **578** In Press (2006).
8. A. Fournell, S. Pourhassan, K. Franke, L. A. Schwarte, T. W. Scheeren and W. Sandmann, Reflectance spectrophotometry: A novel diagnostic approach to assess intestinal oxygenation in patients with mesenteric ischaemia. Presented at 32nd Conference of the International Society on Oxygen to Tissue, Bari, Italy 21st–26th August 2004.

Chapter 27
Cerebral Tissue Oxygen Saturation Calculated Using Low Frequency Haemoglobin Oscillations Measured by Near Infrared Spectroscopy in Adult Ventilated Patients

Terence S. Leung[1], Martin M. Tisdall[2], Ilias Tachtsidis[1], Martin Smith[2], David T. Delpy[1], and Clare E. Elwell[1]

Abstract Oxy- (HbO_2) and deoxy- (HHb) haemoglobin signals measured by near infrared (NIR) spectroscopy over the human frontal lobes frequently contain respiratory and low frequency oscillations (LFOs). It has been suggested previously that venous oxygen saturation (S_vO_2) can be calculated from these respiratory oscillations. In this paper, we investigated the use of a Fourier transform based algorithm to calculate an oxygen saturation measure known as $S_{osc}O_2$ which may be a close estimate of the underlying S_vO_2. $S_{osc}O_2$ was calculated using three different frequency ranges, (1) respiratory oscillations only, (2) LFOs only, and (3) both respiratory oscillations and LFOs. At each frequency range $S_{osc}O_2$ was calculated using either (1) the modified Beer-Lambert law (MBL) or (2) spatially resolved spectroscopy (SRS). In total six different measurements of $S_{osc}O_2$ were investigated here. Experiments were performed in six adult ventilated patients with traumatic brain injury. The patients' inspired oxygen fraction (F_iO_2) was raised in two hyperoxic phases. The calculated $S_{osc}O_2$ values were compared with other cerebral oxygenation measures including an intraparenchymal catheter based brain tissue oxygen tension (PbrO2) and the NIR based tissue oxygenation index (TOI). It was found that the $S_{osc}O_2$ calculated using the combined respiratory and LFO frequency range and the SRS method resulted in the highest detection rates of hyperoxic changes. This measure of $S_{osc}O_2$ may provide a viable, continuous, non invasive, bedside measure of cerebral venous oxygen saturation.

27.1 Introduction

Measurements of cerebral S_vO_2 using non-invasive NIRS have previously been discussed in the literature. The central idea is to identify changes in blood volume which can be attributed to the venous compartment and then to

[1]Department of Medical Physics and Bioengineering, University College London, London, UK.
[2]Department of Neuroanaesthesia and Neurocritical Care, The National Hospital for Neurology and Neurosurgery, Queen Square, London, UK.

calculate S_vO_2 based on the relative proportion of $\Delta[HbO_2]$ and $\Delta[HHb]$ within those changes. In adults and neonates, changes in venous blood volume can be initiated through head tilting [1] or jugular vein occlusion [2,3]. They can also be associated with respiration [4–6] which is one of the main focuses in this work. Respiration influences cardiovascular activities such as heart rate, stroke volume, arterial pressure, venous pressure and blood flow [7]. Strictly speaking, respiratory oscillations are associated with both arterial and venous blood volume changes. However, since the veins are much more compliant than the arteries [8], it is expected that these oscillations occur predominantly within the venous compartment [6]. In this study, all patients were being ventilated using intermittent positive pressure, in which, inspiration generally causes an increase in intrathoracic pressure leading to a reduction of venous return and an increase in peripheral blood volume [5]. The patients' ventilation rates were between 10 and 14 breaths/min (0.17 and 0.23 Hz).

Other types of oscillation often found in the cerebral haemodynamic signals ($\Delta[HbO_2]$ and $\Delta[HHb]$) in healthy humans occur at a frequency of around 0.1 Hz and are termed either vasomotion-waves or spontaneous low frequency oscillations [9–12]. In this paper, we have adopted the term low frequency oscillations (LFOs). Similar oscillations can also be found in arterial blood pressure and heart rate [13,14] and are known as Mayer-waves which are generally thought to be generated by baroreflex activity [15].

In this paper we describe a cerebral oxygen saturation measure which utilises these respiratory and low frequency oscillations. We have termed this saturation, $S_{osc}O_2$ with the subscript "osc" indicating the oscillatory basis of the signals. The aim of this paper is to investigate whether $S_{osc}O_2$ can be used to measure cerebral S_vO_2. We compare our $S_{osc}O_2$ value with other cerebral oxygenation measures such TOI and $PbrO_2$ during a hyperoxic study in six adult ventilated patients with traumatic brain injury. TOI is a mixed arterial and venous oxygen saturation measurement that is also dependent on the arterial to venous volume ratio (which in the brain is assumed to be 1:3). Another cerebral oxygenation measurement, which is invasive, and often used in the care of traumatic brain injury patients on the intensive care unit, is $PbrO_2$ which provides a local partial pressure of oxygen in the extra-cellular fluid of the brain tissue and reflects the availability of oxygen for aerobic metabolism. As such $PbrO_2$ can be thought of as reflecting the balance between oxygen delivery and consumption [16].

27.2 Methods

27.2.1 Experiments

The study was approved by the Joint Research Ethics Committee of the National Hospital for Neurology and Neurosurgery and the Institute of

Neurology. We studied six adult ventilated patients with traumatic brain injury. An NIR monitor (NIRO300, Hamamatsu Photonics KK.) was used to measure $\Delta[HbO_2]$, $\Delta[HHb]$ and TOI in the less injured frontal lobe. $PbrO_2$ was measured using a Licox PMO catheter inserted in the peri-contusional brain tissue. Arterial oxygen saturation (S_aO_2) was measured using a pulse oximeter (Novametrix) placed on the finger. During the study, F_iO_2 was increased so that comparisons could be made between the baseline and two hyperoxic levels. We investigated which of the cerebral oxygenation measures, namely $S_{osc}O_2$, TOI and $PbrO_2$, could detect an increase (1) from baseline to hyperoxic phase 1, and (2) from hyperoxic phase 1 to 2. Baseline F_iO_2 was determined by the minimum level required to produce an arterial partial pressure of oxygen (p_aO_2) larger than or equal to 13 kPa. After 30 minutes of baseline, F_iO_2 was increased to 60% for 60 minutes (phase 1) and then 100% for 60 minutes (phase 2), before being returned to baseline for a further 30 minutes. If baseline F_iO_2 was larger than 60% then phase 1 was omitted.

27.2.2 Theory

An algorithm based on the Fourier transform of the data was used to estimate $S_{osc}O_2$:

$$S_{osc}O_2 = \frac{\sum\limits_i \sqrt{P_{HbO_2}[i]}}{\sum\limits_i \sqrt{P_{HbO_2}[i]} + \sum\limits_i \sqrt{P_{HHb}[i]}} \times 100\% \qquad (27.1)$$

where $P_{HbO2}[i]$ and $P_{HHb}[i]$ are the power spectral densities (PSD) of the $\Delta[HbO_2]$ and $\Delta[HHb]$ signals, and the index i corresponds to different frequency ranges. Three frequency ranges have been used here, (1) the LFO range: from 0.018 to 0.1 Hz, (2) the ventilation/respiration range: a bandwidth of 0.02Hz around the ventilation/respiration frequency (different in each patient) and (3) the combined LFO and ventilation/respiration range: from 0.018 to 0.3 Hz.

The NIR spectrometer used in this work (NIRO300) is able to make measurements based on both the modified Beer Lambert law (MBL) and spatially resolved spectroscopy (SRS) [17]. The SRS measurements (i.e. $k[HbO_2]$ and $k[HHb]$ where k is a constant accounting for scattering) have previously been shown to be more sensitive to intracerebral changes than those based on MBL [18]. Two versions of $S_{osc}O_2$ can thus be calculated, using either the MBL ($S_{osc}O_2^{MBL}$) or the SRS ($S_{osc}O_2^{SRS}$) in the three frequency ranges previously mentioned. The SRS version of $S_{osc}O_2$ was calculated simply by using $k[HbO_2]$ and $k[HHb]$ in calculating $P_{HbO2}[i]$ and $P_{HHb}[i]$ in equation (27.1). In total six versions of $S_{osc}O_2$ were calculated for each set of data.

27.2.3 Data Analysis

To implement equation (1), the $\Delta[HbO_2]/\Delta[HHb]$ (for MBL) or $k[HbO_2]/k[HHb]$ (for SRS) signals were first linearly detrended over 10 minutes. Their power spectral densities $P_{HbO2}[i]$ and $P_{HHb}[i]$ were then estimated using the Welch spectral estimation method with a 1024 point Fast Fourier Transform, 50% overlap and a 1024 point Hanning windowing function. Subsequently, $S_{osc}O_2$ was calculated using Eq. (27.1). Each 10 minute block of data resulted in one value of $S_{osc}O_2$. Each calculation was then repeated on a block of data with the same length but shifted along by 1 minute. Altogether ten $S_{osc}O_2$ measurements were calculated for each phase (baseline, hyperoxic phase 1 and 2). For hyperoxic phases 1 and 2, the initial 20 minutes of data after the increase of F_iO_2 were excluded to allow for stabilisation. The ten measurements from each patient were used to calculate the individual mean value in each phase. One averaged value of $PbrO_2$ and TOI was obtained per 10-minute block for each parameter and 10 values (separated by 1 minute) were calculated for each phase.

27.3 Results

In the baseline phase, the mean value of F_iO_2 was measured at the mouth to be 32±5% (range: 24 – 39%). In the first hyperoxic phase, F_iO_2 was raised and the mean value was measured as 58±1% (range: 56 – 59%). In the second hyperoxic phase, F_iO_2 was raised further and the mean value was measured as 96±3% (range: 90 – 98%).

In this study, we found consistently strong LFOs at around 0.02 Hz in the $\Delta[HbO_2]$ and $\Delta[HHb]$ signals in all our patients with brain injury. The frequency was lower than those reported previously for healthy human subjects [9].

Using spectral analysis, we often found high peaks in the LFO frequency range in the $\Delta[HbO_2]$ and $\Delta[HHb]$ amplitude spectra. The existence of a strong $\Delta[HHb]$ spectral peak is most interesting. It is expected that the LFOs in $\Delta[HbO_2]$ and $\Delta[HHb]$ are both due to blood volume and possibly flow changes in the arterial and venous sites. Arterial blood is highly oxygenated at around 98% which means that the amplitude of the $\Delta[HHb]$ LFOs should be very low in the arterial site. The strong LFOs found in $\Delta[HHb]$ are therefore most likely to arise from venous changes. Examples of the $\Delta[HbO_2]$ and $\Delta[HHb]$ signals and their amplitude spectra are shown in Fig. 27.1.

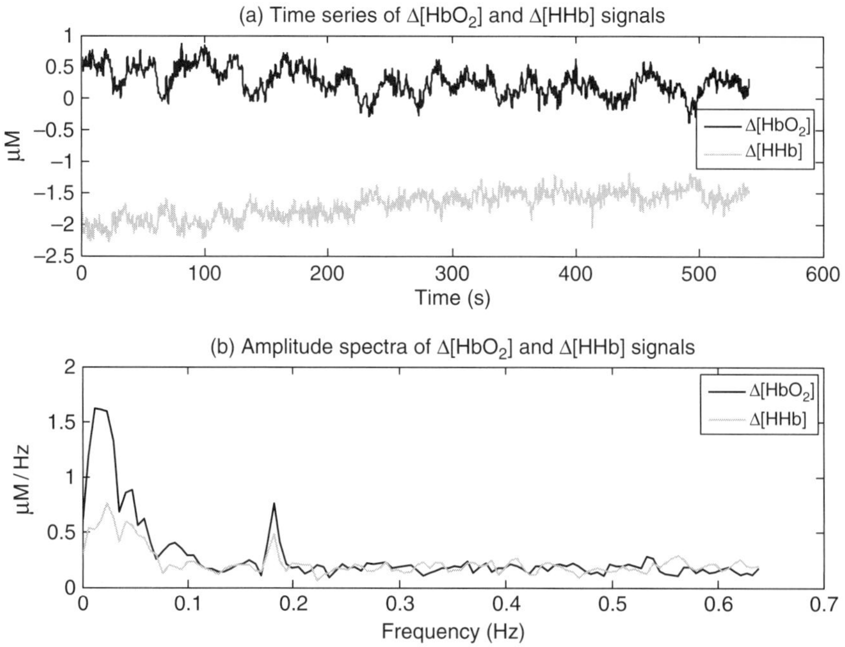

Fig. 27.1 Time series and amplitude spectra of the $\Delta[HbO_2]$ and $\Delta[HHb]$ signals.

The group mean and standard deviation of $S_{osc}O_2$, TOI and $PbrO_2$ for all the patients were calculated from the individual means (Table 27.1). Table 27.2 shows the number of patients whose $S_{osc}O_2$, TOI and $PbrO_2$ show statistically significant increases (1) from baseline to hyperoxic phase 1, and (2) from hyperoxic phases 1 to 2, based on the ten measurements in each phase. As mentioned earlier, there are six versions of $S_{osc}O_2$ in total each being calculated using the same method as described above.

27.4 Discussion

The mean values of the six versions of $S_{osc}O_2$ were between 55 and 71% in the baseline phase. These values are comparable to S_vO_2 measured in a study in which the mean of the jugular venous saturation of normoxic subjects (n = 6) was measured as 69% [19]. All patients were in a stable condition when the studies were carried out and it is assumed that the cerebral metabolic rate of oxygen ($CMRO_2$) was constant during the experiment. In the first hyperoxic phase, S_aO_2 was increased from 98 to 100% and the underlying S_vO_2 was also expected to increase because of a stable metabolic rate. A small increase in dissolved oxygen in the plasma should also increase the underlying S_vO_2. In the

Table 27.1 Group means and standard deviations of S_aO_2, TOI and $S_{osc}O_2$ (6 versions) in the three phases of the hyperoxic experiments (n = 6)

	S_aO_2(%)	TOI(%)	$S_{osc}O_2{}^{MBL}$(%)			$S_{osc}O_2{}^{SRS}$(%)		
			LFO range	Resp. range	LFO & Resp.Range	LFO range	Resp.range	LFO & Resp.Range
Baseline	98±1	67±11	64±10	63±14	59±7	59±5%	58±7%	55±3%
HyperoxicPhase 1	100±1	69±13	67±10	64±11	61±8	63±6%	59±7%	58±4%
Hyperoxic Phase 2	100±1	71±15	67±7	65±11	61±5	63±5%	62±8%	60±4%

LFO range: 0.018–0.1 Hz
Resp. range: Respiratory frequency with a bandwidth of 0.02 Hz
LFO & Resp. range: 0.018–0.3 Hz

Table 27.2 Number of patients (detection rate %) showing statistically significant increases (t-test, $p<0.05$) over the previous phase ($n = 6$)

	$PbrO_2$	TOI	$S_{osc}O_2^{MBL}$			$S_{osc}O_2^{SRS}$		
			LFO range	Resp range	LFO & Resp.range	LFO range	Resp range	LFO & Resp.range
From Baseline to Hyperoxic Phase 1	6 (100%)	4 (67%)	6 (100%)	2 (33%)	5 (83%)	4 (67%)	2 (33%)	6 (100%)
From Hyperoxic Phase 1 to 2	6 (100%)	5 (83%)	2 (33%)	1 (17%)	3 (50%)	2 (33%)	3 (50%)	5 (83%)

LFO range: 0.018–0.1 Hz
Resp. range: Respiratory frequency with a bandwidth of 0.02 Hz
LFO & Resp. range: 0.018–0.3 Hz

second hyperoxic phase, F_iO_2 was raised further. Although S_aO_2 was fully saturated at 100%, there should be a small increase in dissolved oxygen in the plasma and hence an increase in the underlying S_vO_2. Therefore, any increase in $S_{osc}O_2$ during hyperoxic phase 1 to 2 is likely to represent this increase in S_vO_2.

Table 27.2 shows that PbrO2 increased in all patients during both hyperoxic phases, indicating an increased oxygenation at the tissue level. It is interesting to note however that other cerebral oxygenation measures such as TOI and $S_{osc}O_2$ do not always increase. As shown in Table 27.2, TOI increased from the baseline to hyperoxic phase 1 in only 4 out of 6 patients. This could be due to the fact that vasoconstriction occurred during hyperoxia which in turn lowered the arterial to venous volume ratio. While the underlying S_aO_2 and S_vO_2 may both slightly increase, the overall effect could be a lowered TOI. The same mechanism could also explain the fact that TOI increases from hyperoxic phase 1 to 2 in only 5 out of 6 patients.

The six versions of $S_{osc}O_2$ performed differently in the hyperoxic tests. Despite previously being used in both adults and neonates [3,5], $S_{osc}O_2$ based on the respiratory frequency range (both MBL and SRS versions) had low detection rates in the two hyperoxic phases as shown in Table 27.2. In fact, not all patients exhibit strong respiratory oscillations. In two patients, the spectral peaks at the respiratory frequency were very weak (just above the noise floor) in the $\Delta[HbO_2]/\Delta[HHb]$ (and $k\Delta[HbO_2]/k[HHb]$) PSDs.

By comparison, $S_{osc}O_2$ based on the LFO frequency range has higher detection rates (both MBL and SRS versions) in the two hyperoxic phases. This could be due to the fact that there are consistently higher spectral peaks in the LFO range in the $\Delta[HbO_2]/\Delta[HHb]$ and $k\Delta[HbO_2]/k[HHb]$ PSDs for all the patients and all phases.

Relatively high detection rates in the two hyperoxic phases were achieved by $S_{osc}O_2$ based on the combined respiratory and LFO range. In particular, those using SRS have the highest detection rates (100% for hyperoxic phase 1 and 83% for phase 2) compared with TOI and other versions of $S_{osc}O_2$. This may be explained by the fact that SRS measurements have been shown to have a higher sensitivity to intracerebral changes [18]. It is possible that $S_{osc}O_2$ is dominated by the venous blood and is thus less susceptible to changes in the arterial to venous volume ratio.

In this preliminary analysis, we have taken the empirical approach that $S_{osc}O_2$ can be calculated using Eq. (27.1). The use of this equation however has not been fully justified in this paper, especially for the LFOs. Previous studies [5–6] have shown that this equation is only valid when the $\Delta[HbO_2]$ and $\Delta[HHb]$ oscillations (both respiratory and LFO) are caused by blood volume change alone. As shown in Fig. 27.1, the $\Delta[HbO_2]$ and $\Delta[HHb]$ LFOs are sometimes out of phase, suggesting that a blood flow change may also have occurred, violating this assumption. However we found that the $S_{osc}O_2$ values obtained in our studies consistently fall within the expected range of venous saturation values. We are currently working on a theoretical model to explain the behaviour of the

$\Delta[HbO_2]$ and $\Delta[HHb]$ LFO signals which may in turn improve the calculation of $S_{osc}O_2$ as an estimate of the underlying S_vO_2.

Acknowledgment The authors would like to thank Hamamatsu Photonics KK, the Wellcome Trust (Grant no. 075608), the UCL/UCLH trustees, Association of Anaesthetists of Great Britain and Ireland and the EPSRC/MRC (Grant no. GR/N14248/01).

References

1. L. Skov, O. Pryds, G. Greisen, H. Lou, Estimation of cerebral venous saturation in newborn infants by near infrared spectroscopy, *Pediatr Res* **33**(1), 52–55 (1993).
2. C. W. Yoxall, A. M. Weindling, The measurement of peripheral venous oxyhemoglobin saturation in newborn infants by near infrared spectroscopy with venous occlusion, *Pediatr Res* **39**, 1103–1106 (1996).
3. C. E. Elwell, S. J. Matcher, L. Tyszczuk, J. H. Meek, D. T. Delpy, Measurement of cerebral venous saturation in adults using near infrared spectroscopy, *Adv Exp Med Biol* **411**, 453–460 (1997).
4. C. E. Elwell, H. Owen-Reece, J. S. Wyatt, M. Cope, E. O. R. Reynolds, D. T. Delpy, Influence of respiration and changes in expiratory pressure on cerebral hemoglobin concentration measured by near-infrared spectroscopy, *J Cereb Blood Flow Metab* **16**(2), 353–357 (1996).
5. M. Wolf, G. Duc, M. Keel, P. Niederer, K. von Siebenthal, H-U. Bucher, Continuous noninvasive measurement of cerebral arterial and venous oxygen saturation at the bed-side in mechanically ventilated neonates, *Crit Care Med* **25**(9), 1579–1582 (1997).
6. M. A. Franceschini, D. A. Boas, A. Zourabian, S. G. Diamond, S. Nadgir, D. W. Lin, J. B. Moore, S. Fantini, Near-infrared spiroximetry: noninvasive measurements of venous saturation in piglets and human subjects, *J Appl Phyiol* **92**, 372–384 (2002).
7. L. Nilsson, A. Johansson, S. Kalman, Macrocirculation is not the sole determinant of respiratory induced variations in the reflection mode photoplethysmographic signal, *Physiol Meas* **24**, 925–937 (2003).
8. R. M. Berne, M. N. Levy, *Cardiovascular Physiology (7th ed.)*, St. Louis, MO: Mosby Year Book (1997).
9. C. E. Elwell, R. Springett, E. Hillman, D. T. Delpy, Oscillations in cerebral haemodynamics – implications for functional activation studies, *Adv Exp Med Bio* **471**, 57–65 (1999).
10. H. Obrig, M. Neufang, R. Wenzel, M. Kohl, J. Steinbrink, K. Einhaupl, A. Villringer, Spontaneous low frequency oscillations of cerebral hemodynamics and metabolism in human adults, *Neuroimage* **12**, 623–639 (2000).
11. I. Tachtsidis, C. E. Elwell, T. S. Leung, C. W. Lee, M. Smith, D. T. Delpy, Investigation of cerebral haemodynamics by near infrared spectroscopy in young healthy volunteers reveals posture dependent spontaneous oscillations, *Physiol Meas* **25**(2), 437–445 (2004).
12. T. Katura, N. Tanaka, A. Obata, H. Sato, A. Maki, Quantitative evaluation of interrelations between spontaneous low-frequency oscillations in cerebral hemodynamics and systemic cardiovascular dynamics, *Neuroimage,* **31**, 1592–1600 (2006).
13. A. C. Guyton, J. E. Hall, *The textbook of medical physiology*, 10th ed. W.B.Sunders Company, Philadelphia (2000).
14. K. Siebenthal, J. Beran, M. Wolf, M. Keel, V. Dietz, S. Kundu, H. U. Bucher, Cyclical fluctuations in blood pressure, heart rate and cerebral blood volume in preterm infants, *Brain Dev* **21**(8), 529–534 (1999).
15. H. Nilsson, C. Aalkjaer, Vasomotion: mechanisms and physiological importance, *Mol Interv* **3**(2), 79–89 (2003).

16. J. Nortje, A. K. Gupta, The role of tissue oxygen monitoring in patients with acute brain injury, *Brit J Anaesthesia* **97**(1), 95–106 (2006).
17. S. Suzuki, S. Takasaki, T. Ozaki, Y. Kobayashi, A Tissue Oxygenation Monitor using NIR Spatially Resolved Spectroscopy, *Proc SPIE* **3597** 582–592 (1999).
18. P. G. Al-Rawi, P. Smielewski, P. J. Kirkpatrick, Evaluation of a near-infrared spectrometer (NIRO 300) for the detection of intracranial oxygenation changes in the adult head. *Stroke* **32**(11), 2492–2500 (2001).
19. H. M. Watzman, C. D. Kurth, L. M. Montenegro, J. Rome, J. M. Steven, S. C. Nicolson, Arterial and venous contributions to near-infrared cerebral oximetry, *Anesthesiology* **93**, 947–953 (2000).

Chapter 28
Biosensor for Diagnosing Factor V Leiden, A Single Amino Acid Mutated Abnormality of Factor V

Yongjie Ren, Samin Rezania, and Kyung A. Kang[1]

Abstract Factor V Leiden (FVL) is an abnormality with a single amino acid mutation of Factor V (FV) and is the most common, hereditary blood coagulation disorder. FVL is currently diagnosed by DNA analysis, which takes a long assay time, high cost, and a specially trained person. We are developing a rapid, accurate, and cost-effective biosensing system to quantify both FV and FVL in blood plasma, to diagnose FVL and also to evaluate the seriousness of the disease status. This system is based on a sandwich immuno-reaction on an optical fiber. To produce the monoclonal antibody against only FV or only FVL without cross-reacting with the other molecule and with a higher probability, a 20 amino acid sequence (20-mer) of FV or FVL around the mutation region was injected into mice and then hybridoma cell lines specific to each 20-mer were selected. When these antibodies were tested with native FV or FVL molecules, they were found to be cross-reacting with the other molecules, but some with higher affinity to FV (FV preferred) and some to FVL (FVL preferred). Using these antibodies, two different sensors were developed: FV preferred and FVL preferred sensors. These two sensors allowed us to quantify FV and FVL in plasma with a maximum error of 4%. The plasma levels of both molecules provide us not only FVL diagnosis but also the level of the seriousness. The same principles may be used for developing diagnostic tools for other diseases with a single point mutation.

28.1 Introduction

Factor V (FV) is an essential factor of the blood coagulation cascade [1]. Its molecular weight is 330 kDa and it is composed of a heavy chain (MW = 105 kDa) and a light chain (MW = 71∼74 kDa). When the gene for the arginine at the position 506 (R506Q) on the heavy chain of FV is mutated and replaced with glutamine, FV loses the cleavage site for activated protein C (APC;

[1]Yongjie Ren, Samin Rezania and Kyung A. Kang, Department of Chemical Engineering, University of Louisville, Louisville, KY 40292, USA.

anticoagulant) [2]. This mutation (i.e., factor V Leiden; FVL) prevents APC from the efficient inactivation of FV and facilitates overproduction of thrombin, leading to excess fibrin generation and blood clotting [3].

FVL is the most common hereditary blood coagulation disorder in the United States. It is present in 5% of the Caucasian population and 1.2% of the African American population [7]. FVL increases the risk of venous thrombosis 3–8 folds for the heterozygous and 30–140 folds for the homozygous individuals [7]. Currently, the FVL diagnosis is made by DNA analysis or by the clotting test that measures the degree of prolongation of plasma clotting time after the addition of APC [3]. DNA analysis is, however, expensive and time-consuming, and the clotting test is not specific only for FVL.

The fiber optic biosensing system is a rapid, accurate, and cost-efficient method to detect the level of the specific protein in plasma [4, 9, 10, 11]. This method performs a sandwich assay on the surface of an optical fiber.

Here, a fiber optic biosensing system for FVL diagnosis is presented.

28.2 Materials, Instruments and Methods

28.2.1 Materials and Instruments

For biosensing: The fluorometer (Analyte 2000) and the quartz fibers used for sensors were from Research International (Monroe, WA). Factor V and a monoclonal antibody against FV light chain were purchased from Haematologic Tech. (Essex Junction, VT) and another monoclonal antibody against FV light chain was from Fitzgerald (Concord, MA). The homozygous patient plasma was obtained from a FVL homozygous patient, following approval by the University of Louisville Institutional Review Board (IRB). Alexa Fluor® 647 (AF647) was purchased from Invitrogen (Carlsbad, CA). ImmunoProbe™ Biotinylation Kits, avidin, hydrofluoric acid, phosphate buffered saline (PBS), triethylamin, γ-maleimideobutyric acid N-hydroxysuccinimide ester, (3-mercaptopropyl) – trimethoxysilane were from Sigma-Aldrich (St. Louis, MO). FV free plasma was from American Diagnosica Inc. (Stamford, CT).

For enzyme linked immunoassay (ELISA): Bovine Serum Albumin (BSA) and *o*-phenylenediamine dihydrochloride (OPD) tablet were from Sigma-Aldrich. Fc specific, horseradish peroxidase-conjugated rabbit anti-mouse IgG was from Jackson ImmunoResearch Laboratories, Inc. (West Grove, PA). The ELISA plate reader was from Bio-Rad (Hercules, CA).

28.2.2 Methods

For biosensing by fiber optic sensing system: All sensors were prepared following the protocol established by previous researchers [4, 9, 10, 11]. Briefly, the antibody against FV or FVL (1°MAb) that we have developed using 20-mers was immobilized on the fiber surface by the avidin-biotin linkage and then the fiber is enclosed in a sensing chamber. When a sample is injected into the chamber, the FV/FVL molecule in the sample is captured by the 1°MAb. After washing the fiber surface to remove unbound bio-molecules, the antibody against FV light chain (2°MAb) conjugated with AF647 is applied to the sensing chamber. After the sandwich complex is formed, the excitation light is applied to the sensor and the emitted fluorescence is measured by a fluorometer. The fluorescence intensity is correlated with the amount of FV/FVL in the sample. Regarding the sample and 2°MAb incubation time, for testing affinity of the generated antibodies, 10 and 10 minutes (10/10 min.) were used. For sensing analytes, 3 and 2 minutes (3/2 min.) were used.

For ELISA: To test the affinity of the antibodies generated, ELISA was performed as follows: 96 wells of an ELISA plate (Dow corning, NY) were coated with 100 μl of FV in plasma (2 μg-FV/ml-FV free plasma) or 100 μl of homozygous FVL plasma (2 μg/ml). First, the well surface was blocked with 250 μl, 1% BSA each well for 90 minutes at room temperature, then 100 μl of anti-FV antibodies (1 μg/ml) was applied on the first column wells and the ½ serial dilution was performed. After incubation at 37°C for 90 minutes, 100 μl of 1:1000 Fc specific, HRP conjugated rabbit anti-mouse IgG was applied for 20 minutes at 37°C. After washing the plate and adding 100 μl of OPD solution to each well, the plate was incubated at room temperature for 30 minutes, and then the optical density was measured at 450 nm by the ELISA reader.

28.3 Results and Discussion

28.3.1 Production of 1°MAb Against FV and FVL

Currently, neither pure FVL molecule nor the antibody against FVL without cross-reacting with FV is available. Generating antibodies specifically against a single point mutation site of a molecule is extremely difficult. To increase the probability for generating antibodies specifically against the mutation site, the 20 amino acid sequences (20-mer) of FV or FVL around the mutation site were generated (Fig. 28.1) [8]. The 20-mers were then conjugated with a carrier protein, the conjugated molecules were injected into mice, and the hybridoma cell lines were generated [9, 12]. The resulting antibodies were first screened with the 20-mers and those with high affinity to FVL molecules without cross-reacting with FV were selected, and vice versa.

FV 20-mer:	H-I-C-K-S-R-S-L-D-R-**R**-G-I-Q-R-A-A-D-I-E-Q-NH_2
FVL 20-mer:	H-I-C-K-S-R-S-L-D-R-**Q**-G-I-Q-R-A-A-D-I-E-Q-NH_2

Fig. 28.1 The amino acid sequence of 20-mers for FV and FVL (the mutation sites are in bold).

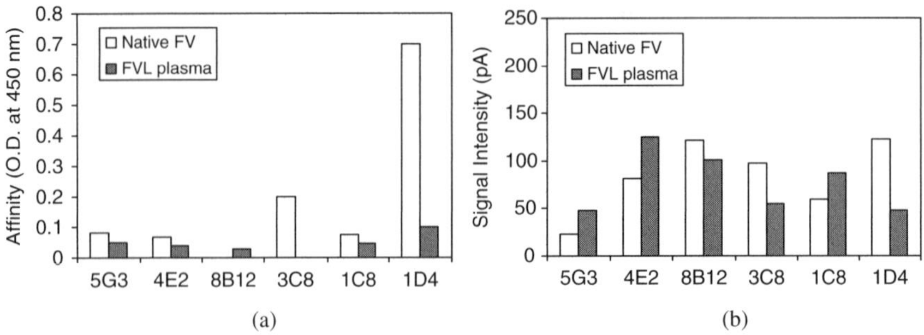

Fig. 28.2 Relative affinity values (in O.D.) of six selected antibodies against FV (□) and FVL (■) (a) by ELISA and (b) a biosensor. [Experiment conditions: 10 cm sensors, 10/10 min. sample/2°MAb incubation times, 1.2 cm/s circulation flow velocity during incubation].

The screened monoclonal antibodies were then tested with native FV molecules and homozygous FVL plasma by ELISA [Fig. 28.2(a)]. They were found to have some cross-reactivity with the other molecule. However, some had higher affinity with FV and some with FVL. The affinities of these antibodies were also tested as the 1°MAb for the sensing system, and a commercial antibody against the FV light chain was used as the 2°MAb. For the biosensor, the experiments was performed for the sample 5 μg FV/FVL in 1 ml plasma and the sensor also showed the cross-reactivity [Fig. 28.2(b)]. Interestingly, the antibody 5G3 had a higher affinity for FV in ELISA but it shows the highest relative affinity for FVL (FVL: FV = 2.17). The antibody 1D4 shows the highest relative affinity for FV (FV: FVL = 2.55), consistent with the ELISA result. Therefore, the antibody 5G3 and 1D4 are selected to be 1°MAbs for the FVL and the FV preferred sensors, respectively.

28.3.2 Sensing Performance of the FV and FVL Preferred Sensors

Heterozygous patients have both FV and FVL molecules in their blood. Quantifying both FV and FVL in their plasma provides information on the degree of the abnormality. First, the FV and the FVL preferred sensors were tested for their behaviors for FV or FVL, separately in the sample, with the physiological range of 0~15 μg/ml-plasma (Fig. 28.3). For both FV preferred and FVL preferred sensors, the relationships between the analyte concentration and the signal intensity were linear. For the FV preferred sensor, the slope of the standard curve for FV was 9.3, which is higher than the slope of FVL 6.8, as expected [Fig. 28.3(a)]. For the FVL preferred sensor, the slopes were 8.6 for FVL and 5.4 for FV [Fig. 28.3(b)].

Next, the response of these sensors was studied for a mixture with both FV and FVL molecules. The FV preferred sensor was tested for a sample with both FV and FVL but at a constant FVL concentration (8 μg/ml) [Fig. 28.4(a)]. The

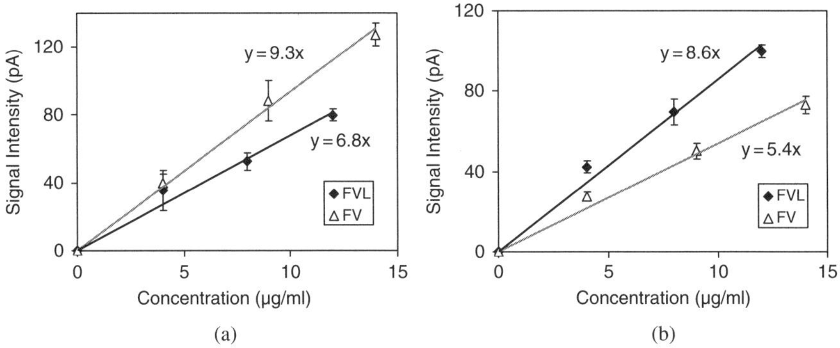

Fig. 28.3 Standard curves for FV and FVL by (a) FV preferred and (b) FVL preferred sensors. [Experiment conditions: 3 cm sensors, 3/2 min. sample/2°MAb incubation times, 1.2 cm/s circulation flow velocity during incubation].

signal intensity of the mixture was linear with the change in the FV concentration, with the slope of 8.9, which is 96% of that for FV only (9.3) [Fig. 28.3(a)]. The signal intensity of the y intercept was 54.4, the same for 8 μg/ml of FVL only. In other words, the signal intensity of the mixture was found to be the addition of the signal intensities by FV and by FVL, showing that FVL molecules in a sample do not affect the affinity of FV for the FV preferred sensor. Figure 28.4(b) confirms that FV molecules in the sample do not affect the affinity of FVL in FVL preferred sensor, either. In summary, the FV and the FVL molecules in a sample contribute to the sensor signal intensity independently, without interfering with each other.

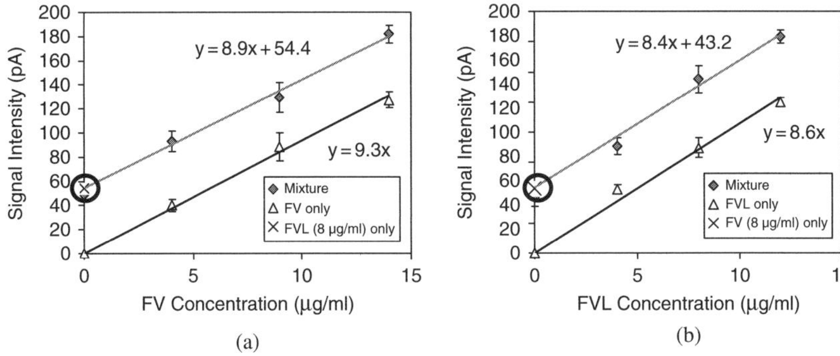

Fig. 28.4 Signal intensities of (a) the FV preferred sensor with the change in FV concentration in the sample when 8 μg/ml of FVL is added in the sample and (b) the FVL preferred sensor with the change in FVL concentration when 8 μg/ml of FV is added in the sample. [Experiment conditions: 3 cm sensors, 3/2 min. sample/2°MAb incubation times, 1.2 cm/s circulation flow velocity during incubation].

28.3.3 Quantification of FV and FVL in Blood Plasma Sample

Since our sensors may not detect FV and FVL molecules exclusively, a mathematical manipulation is needed for the quantification of these two molecules in a sample. The relationship between the two sensor signal intensities and the concentrations of FV and FVL molecules in a sample can be expressed as follows:

$$SI_1 = A_1 C_{FV} + B_1 C_{FVL} \tag{28.1}$$

$$SI_2 = A_2 C_{FV} + B_2 C_{FVL} \tag{28.2}$$

where, *SI* is the signal intensity (pA) generated by a sensor; *A* is the slope of the standard curve for FV, in pA/(μg/ml); *B* is the slope of the standard curve for FVL, in pA/(μg/ml); *C* is the concentrations of FV or FVL in the sample, in μg/ml; Subscripts 1 and 2 represent FV preferred and FVL preferred sensor, respectively. The concentrations of FV (C_{CFV}) and FVL (C_{CFVL}) can be, therefore, expressed as Eq. 28.3 and Eq. 28.4, respectively.

$$C_{FV} = \frac{\frac{SI_2}{B_2} - \frac{SI_1}{B_1}}{\frac{A_2}{B_2} - \frac{A_1}{B_1}} \tag{28.3}$$

$$C_{FVL} = \frac{\frac{SI_2}{A_2} - \frac{SI_1}{A_1}}{\frac{B_2}{A_2} - \frac{B_1}{A_1}} \tag{28.4}$$

The sensing system is composed of one FV preferred sensor and one FVL preferred sensor. A summary of the assay for FV and FVL is shown in Fig. 28.5:

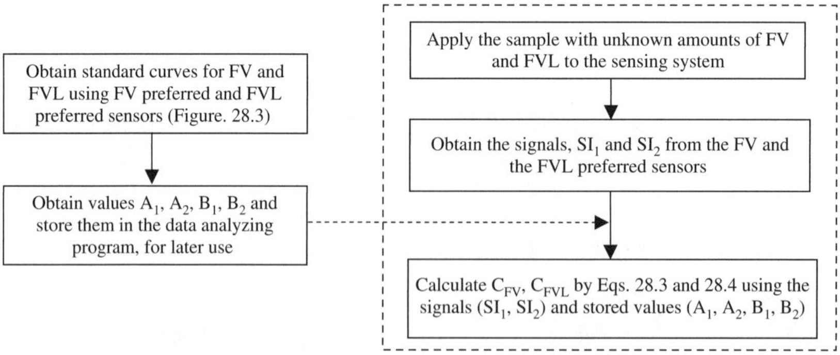

Fig. 28.5 A schematic diagram describing the procedure for obtaining the amount of FV and FVL in a sample, using our sensing system. The part within a dashed line is the actual protocol for an assay.

Table 28.1 An example of the quantification of FV and FVL in a sample. (a) the parameters of the FV/FVL sensing system and (b) the sensing result of an example

(a)

	A [pA/(μg/ml)]	B [pA/(μg/ml)]	SI (pA) from measurement #1	SI (pA) from measurement #2
FV preferred sensor	9.3	8.6	89.5	79.5
FVL preferred sensor	6.8	5.4	83.0	69.5

(b)

	Actual concentration (μg/ml)	Calculated concentration (μg/ml)	Relative error (%)
FV	5	4.8±0.1	–4
FVL	6	5.8±1.1	–3

First, the parameters need to be obtained from the standard curves of the samples of FV or FVL, and stored in the system; then the actual assay for an unknown sample will be performed as described in the block framed by a dashed line. The total assay time is about 8 minutes.

As an example, a mixture of 5 μg of FV and 6 μg of FVL in 1 ml plasma was tested by our sensing system (Table 28.1). As can be seen from the table, the FV and FVL in the sample were quantified with a relative error less than 5%.

28.4 Conclusions

The amino acid sequences of FV and FVL are different by only one amino acid. In order to increase probability for generating antibodies specific only to FV or only to FVL, 20-mers around the mutation site for each molecule were used in the hybridoma cell generation. Two antibodies with a higher affinity to FV and to FVL were selected as 1°MAb for the FV and the FVL preferred sensors, respectively. For a sample containing both FV and FVL, the total signal intensity was found to be the addition of the signals by FV and by FVL. The signals from these two sensors were used to quantify FV and FVL in a plasma sample accurately with an error of only 4%. Also, an entire assay can be completed within 10 minutes. This system is a rapid and cost-effective tool for FVL diagnosis.

The system can also be used for the diagnosis of factor V deficiency. The same principles may be applied for developing diagnostic tools for other diseases with a single point mutation.

Acknowldgment The authors thank the National Institutes of Health (5R21EB003485-02) for the financial support, and Dr. Sharma at the Hematology and Oncology Department of the University of Louisville for obtaining the plasma from a FVL homozygous patient.

References

1. R. W. Colman, et al., *Hemostasis and Thrombosis: Basic Principles and Clinical Practice*, J.B. Lippincott Company, Philadelphia, 3rd edition, 113–120 (1993).
2. R. M. Bertina, B. P. Koeleman, T. Koster, F. R. Rosendaal, R. J. Dirven, H. de Ronde, P. A. van der Velden, and P. H. Reitsma, Mutation in blood coagulation factor V associated with resistance to activated protein C. *Nature*, 369, 64–67 (1994).
3. A. Tripodi, B. Negri, R. M. Bertina, and P. M. Mannucci, Screening for the FV: Q506 mutation–evaluation of thirteen plasma-based methods for their diagnostic efficacy in comparison with DNA analysis, *Thromb Haemost*, 77(3), 436–439 (1997).
4. J. O. Spiker, K. A. Kang, W. N. Drohan, and D. F. Bruley, Preliminary study of biosensor optimization for the detection of protein C, *Adv. Exp. Med. Biol.*, 454, 681–688 (1998).
5. H. J. Kwon, H. I. Balcer, and K. A. Kang, Protein C biosensor sensitivity for biological samples and sensor reusability, *Comp. Biochem. Physiol.*, Part A, 132, 231–238 (2002).
6. L. Tang, and K. A. Kang, Preliminary study of simultaneous multi-anticoagulant deficiency diagnosis by a fiber optic multi-analyte biosensor, *Adv. Exp. Med Biol.*, 566, 303–309 (2005).
7. J. C. Mattson, D. Crisan, Inherited thrombophilia due to factor V Leiden mutation. *Mol. Design*, 3, 55–61 (1998).
8. R. J. Jenny, et al., Complete cDNA and derived amino acid sequence of human factor V, *Proc. Natl. Acad. Sci. Biochemistry*, 84, 4846–4850 (1987).
9. H. J. Kwon, *Theoretical and experimental investigation on sensing performance of protein C immuno-optical sensor for physiological samples*. Dissertation. Chemical Engineering, University of Louisville, Louisville, KY (2002).
10. L. Tang. *Multi-analyte, fiber-optic immuno-biosensing system for rapid disease diagnosis: model systems for anticoagulants and cardiac markers*. Dissertation. Chemical Engineering, University of Louisville, Louisville, KY (2005).
11. B. Hong, K. A. Kang, Biocompatible, Nano-gold-particle fluorescence enhancer for fluorophore mediated, optical immunosensor, *Biosensor and Bioelectronics*, 21(7), 1333–1338 (2006).
12. H. J. Kwon, S. C. Peiper, and K. A. Kang, Fiber optic immunosensors for cardiovascular disease diagnosis: quantification of protein C, Factor V Leiden, and cardiac Troponin T in plasma, *Adv. Exp. Med Biol.*, 510, 115–119 (2003).

Chapter 29
Scanning Laser Ophthalmoscope-particle Tracking Method to Assess Blood Velocity During Hypoxia and Hyperoxia

Kristen Lorentz, Astrid Zayas-Santiago, Shanti Tummala, and Jennifer J. Kang Derwent[1]

Abstract The main objective was to evaluate a Scanning Laser Ophthalmoscope (SLO) based particle tracking method as a means of quantitative assessment of retinal blood velocity and vessel diameter changes in response to hypoxia and hyperoxia. Retinal blood velocities were measured by tracking fluorescent microspheres (1.0 μm diameter) in anesthetized adult pigmented rats. Velocities were calculated based on microsphere position changes and the recording frame rate. Hypoxia was induced by inspiring a mixture of nitrogen and air and hyperoxia was induced by inspiring 100% oxygen. Average blood velocities during hypoxia obtained for arteries, veins, and small vessels (diameter < 40 μm) were 39.9 ± 9.9, 34.9 ± 2.7, and 8.8 ± 1.8 mm/sec, respectively, whereas during hyperoxia, the average blood velocities obtained were 23.7 ± 6.2, 28.2 ± 2.7, and 7.6 ± 0.7 mm/sec. Hypoxia was found to increase the diameters of arteries by 25% but did not change the diameters of veins; whereas, hyperoxia was found to decrease their diameters by 25% and 18%. Changes detected in vessel diameter and blood velocity suggest that the level of oxygen tension alters retinal hemodynamics. Dynamics of retinal hemodynamics in response to hypoxia and hyperoxia can be assessed using the SLO imaging method.

29.1 Introduction

Many ailments of the eye, including macular degeneration and retinopathy, are related to aberrant blood flow within the ocular vasculature [1–4]. The development of an effective non-invasive technique to measure circulation in vivo in the eye would have a profound impact on assessments of pathophysiology of the ocular circulation, of pathologic vulnerability and of pharmacological treatments of the eye. A direct visualization of retinal vasculature and blood

[1]Kristen Lorentz, Astrid Zayas-Santiago, Shanti Tummala, and Jennifer J. Kang Derwent, Department of Biomedical Engineering, Illinois Institute of Technology, Chicago, IL 60616.

flow makes the eye an ideal system, but measurement through thin retina (~250 μm) and coupled with limited noninvasive tools has made assessment of retinal hemodynamics difficult.

Use of a scanning laser ophthalmoscope (SLO) as a noninvasive tool to track particulate flow in the eye is a promising potential technique to assess retinal hemodynamics [5–9]. Direct observation of particulate flow in vivo is possible with an SLO because it allows visibility of fluorescently-labeled cells and other fluorescent particles introduced into the system. Particle dynamics can be tracked and velocities can be calculated. An advantage of this method is that it is non-invasive and it allows to track particles from various different size vessels and retinal regions. The main objective of this study was to test the sensitivity of the SLO imaging technique in detecting changes in blood velocity and vessel diameter resulting from induced hypoxia and hyperoxia.

29.2 Materials and Methods

29.2.1 Experimental Procedure

Five male pigmented Long Evans rats weighing approximately 300–400 g were used in this study. The animals were treated in accordance with the ARVO Statement on the Use of Animals in Ophthalmic and Vision Research. Initial anesthesia was induced with isoflurane and long-term anesthesia was maintained with urethane (800 mg/kg loading dose; 75 mg/kg h maintenance dose). The femoral vein was cannulated with polyethylene tubing (PE-50) for delivery of fluorescent microspheres and maintenance doses of urethane. The femoral artery was also cannulated with PE-50 tubing to allow periodic measurement of arterial blood gas parameters. The animal was paralyzed with pancuronium bromide (Pavulon; 0.3mg/kg loading dose; 0.2 mg/kg h maintenance dose) and artificially ventilated. Pupils were dilated with a drop of 2.5% phenylephrine hydrochloride and 1% atropine sulfate and a topical anesthetic 0.5% proparacaine hydrochloride was applied to the eye. Artificial tears were periodically applied to maintain moisture of the eye. A zero-diopter clear silicon contact was placed on the eye prior to image acquisition.

The commercially available Heidelberg Retina Angiogram (HRA, Heidelberg Engineering, Vista, CA) was used to acquire images of retinal vessels of the eye. Depth discrimination was achieved by scanning the inside of the eye with either an argon ion (488 nm wavelength) or infrared diode (795 nm wavelength) laser in combination with highly selective filters. Heidelberg Eye Explorer (HEE) software automatically digitized live recordings into either 256 × 256-pixel or 512 × 512-pixel images as well as video. A video frame rate of 20.5 per second was used to record particle-tracking movies, typically in a field view of 30°.

Yellow-green 1-μm-diameter polystyrene micropsheres (505 nm maximum absorption and 515 nm maximum emission, Molecular Probes, Eugene, OR) were injected intravenously (0.02–0.06 ml/kg) as tracer particles to measure velocity. Retinal blood velocities were obtained in arteries, veins, and small vessels. "Small vessel" is representative of any vessel with a diameter less than 40 μm. Directional flow of fluorescent microspheres and vessel characterization were determined based on examination of infrared reflectance (IR) and fluorescein angiogram (FA) images of the microsphere paths. IR and FA images were taken before and after each alteration of O_2 gas level in the inspiration mixture and a microsphere FA video was acquired for 2–5 minutes during each altered state. At the end of the experiment, a fluorescein injection (0.1 ml/kg) was intravenously administered to obtain the retinal vascular map.

The inspired gas mixture was altered by adding N_2 to air to give 10% oxygen (hypoxia) and by administering 100% O_2 (hyperoxia). Approximately five minutes after the onset of an episode, an arterial blood gas sample was taken to confirm either hypoxia or hyperoxia. Microsphere FA movies were acquired approximately ten minutes after the onset of each episode. The animal was given sufficient time to recover from each episode before altering the oxygen level to induce a different oxidative state.

29.2.2 Data Analysis

Vessel diameter measurements were obtained from IR images using the HEE software. Blood velocity measurements were obtained by exporting FA movie frames with traceable microspheres into a custom written MATLAB® program. Velocities were calculated by multiplying the distance (millimeters) traversed by the microspheres and the acquisition frame rate (20.5 per second) to obtain instantaneous velocity (millimeters per second). All flows were assumed to be planar, a reasonable assumption given that the SLO only acquires images from a thin region at any given time. Accordingly, as particles exited the plane of focus, they became blurry and disappeared from view [10]. A distance scale was determined by using the width of the image acquisition frame (in millimeters, measured via the HEE software) and the number of pixels. The distance traveled by a particle was converted from a pixel length to a length in millimeters using this scale. Overlaying several consecutive FA microsphere images on an angiogram image allowed assignment of velocity values to particular vessel types. Because distinguishing between small arterioles and post-arteriolar capillaries was somewhat subjective, these vessels were both characterized as "small vessels". Likewise, small venules and pre-venule capillaries were also characterized as "small vessels". Artery and vein measurements were taken from the central branches, which radiate symmetrically outwards from the optic disk. Generally 6–9 arteries and 5–8 veins comprised the outward-extending vessel "spokes" of the rat retina.

29.3 Results

29.3.1 Normoxia

The purpose of this experiment was to determine the feasibility of SLO particle-tracking as an investigative tool to detect small changes in ocular blood flow. Different oxidative states were used to measure the degree of changes in retinal blood velocity with this proposed method. The area of interest was identified with IR images, and then control microsphere FA movies were obtained. Immediately after the injection of microspheres, circulation of particles was extremely pronounced. However, after several minutes of recirculation the population notably decreased, making it easier to track individual particles. Groups of multiple spheres circulating together were not uncommon, especially within the larger vessels. Once the travel sequence images were identified, each frame was individually exported to MATLAB® and distance traveled was measured. Approximately ten measurements were taken in each vessel type during normoxia to represent an average blood velocity, and measurements were restricted to a $7.7 \times 7.7\ mm^2$ area superior to the optic disk in order to limit regional velocity variation.

Diameters of blood vessels were measured from IR images with the HEE software. Mean vessel diameter measurements are displayed in Table 29.1. Three diameter measurements were taken along the vessel length and averaged to represent the mean diameter of that particular vessel. Diameters of small vessels were too small to be measured accurately using the HEE software.

29.3.2 Hypoxia

The level of hypoxia was confirmed by an arterial blood sample taken 5 minutes after the onset of 10% hypoxia and the average P_aO_2 was 32.7 ± 5.2 mm Hg. The effects of 10% hypoxia on blood vessel diameters are readily observed in IR images. As shown in Table 29.1, hypoxia resulted in a statistically significant

Table 29.1 Mean blood vessel diameters measured during normoxia, hyoxia and hyperoxia

	Blood Vessel Diameter (mm) (mean ± SD)		
Vessel Type	Normoxia (n = 5 rats)	Hypoxia (n = 3 rats)	Hyperoxia (n = 5 rats)
Artery	0.08 ± 0.01	0.10 ± 0.01	0.06 ± 0.01
Vein	0.11 ± 0.01	0.11 ± 0.01	0.09 ± 0.01

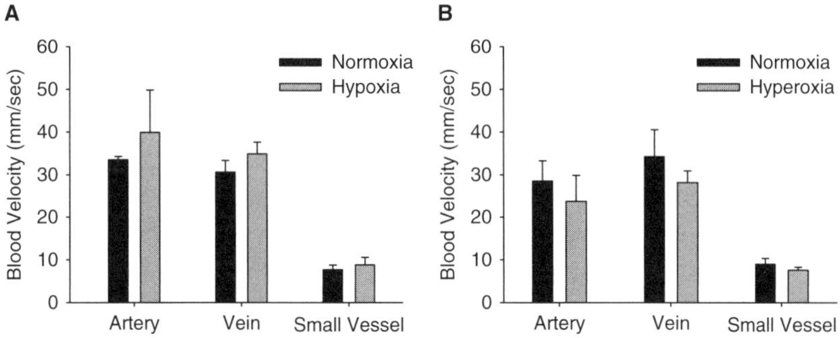

Fig. 29.1 A: Average blood velocity changes obtained from artery, vein, and small vessels during normoxia (black bars) and hypoxia (gray bars). B: Average blood velocity changes obtained from artery, vein, and small vessels during normoxia (black bars) and hyperoxia (gray bars).

increase of approximately 25% in arterial diameters ($P < 0.05$, $n = 3$; paired *t* test), whereas the changes in venous diameter were small or almost none.

After the onset of hypoxia, blood velocities in each vessel also exhibited an increasing trend. Figure 29.1A shows average velocities (approximately 10 measurements for each animal) obtained ten minutes after the onset of hypoxia from arteries, veins, and small vessels of three rats. On average, arterial velocities during hypoxia increased by 19% ($P = 0.038$; paired *t* test), while venous and small vessel velocities increased by 14% and 15%, respectively ($P = 0.033$ and $P = 0.037$; paired *t* test).

29.3.3 Hyperoxia

The average P_aO_2 measured during hyperoxia was 349 ± 146 mm Hg ($n = 5$ rats). Due to an inadequate arterial blood sample, a pulseoximeter was used to monitor the blood oxygen in one of the animals. In contrast to hypoxic effects, after 10 minutes into the onset of hyperoixa, pronounced constriction of the retinal artery is visible. Overall, hyperoxia resulted in statistically significant vessel constriction of approximately 25% in arteries and 18% in veins ($P < 0.05$ and $P < 0.05$, $n = 5$ respectively; paired *t* test) (Table 29.1).

Similar to the normoxia and hypoxia procedures, approximately ten velocity measurements were made in each animal during hyperoxia. Overall, hyperoxia decreased blood velocity (Fig. 29.1B). On average, the arteries decreased by 17% ($P = 0.046$; paired *t* test) and the veins and small vessels decreased by 18% and 16%, respectively ($P = 0.001$ and $P = 0.015$). During extreme cases of hyperoxia (P_aO_2 greater than 430 mm Hg), visibility of microspheres in arteries and small vessels severely decreased, almost to the point of total disappearance.

29.4 Discussion

29.4.1 Normoxia

The major objective of this study was to evaluate the SLO particle-tracking method by measuring the response of retinal blood velocity to changes in arterial PO_2 in the retinal circulation. Microspheres were chosen as the tracer instead of fluorescently labeled cells for several reasons in this study. Microspheres are readily available, bright, uniform tracer particles, whereas fluorescent blood cells involve time-consuming labeling procedures and exhibit less bright and less-uniform fluorescence. Also, while systemically circulating microspheres are biologically inert and pose minimal toxicity hazards [10], excessive fluorescent cell-labeling can lead to adverse effects on biological functions and behaviors of some cells [11]. One of the main concerns about using microspheres is that they are different from red blood cells with regard to their rigidity, size, and shape [9].

In order to address the validity of microspheres as tracers, we compared our results to previously reported blood velocity measurements obtained from either laser Doppler velocimetry or SLO systems (Table 29.2). With our method, we obtained an average velocity of 30.5 ± 4.4 mm/sec in arteries with a mean diameter of 80 ± 10 μm. This average arterial velocity is similar to values obtained with the laser Doppler velocimetry in humans [2,12] as well as in cats [13]. For veins with a mean diameter of 110 ± 10 μm, we obtained an average velocity of 34.2 ± 6.3 mm/sec. This value is somewhat higher than Grunwald et al [2]. but similar to that of Williamson and Baxter[12]. Our central artery and vein velocities are higher than Wajer et al [9]. who used FITC-labeled red blood cells and an SLO to measure the velocities. This difference may be due to the smaller blood vessel size that velocities were

Table 29.2 Comparison of blood velocity measurements

		Arteries		Veins		Capillaries
Authors	Method	Velocity (mm/sec)	Diameter (μm)	Velocity (mm/sec)	Diameter (μm)	Velocity (mm/sec)
Grunwald et al. [2]	Bidirectional Laser Doppler	29 ± 9	108 ± 12	17 ± 4	152 ± 14	N/A
Williamson & Baxter [12]	Color Laser Doppler	31–102	Central	38–58	Central	N/A
Wajer et al. [9]	RBCs/SLO	15.5 ± 0.5	46.7	14.5 ± 3.9	55.6	4.8 ± 1.7
Nagaoka et al. [13]	Bidirectional Laser Doppler	33.7 ± 2.5	84 ± 3	N/A	N/A	N/A
This study	Microspheres/ SLO	30.5 ± 4.4	80 ± 10	34.2 ± 6.3	110 ± 10	9.0 ± 1.3

measured from in the Wajer et al [9]. study (artery diameter of 46.7 μm and vein diameter of 55.6 μm). When we compared our "small vessel" (diameter < 40 μm) velocity to Wajer et al. (2000), our mean velocity of 9.0 ± 1.3 mm/sec fell between the arteriole/venule and capillary velocities reported by Wajer et al [9]. A further study is planned to investigate the similarity and difference of tracers.

One major difference between our work and previous studies is that a different SLO unit was used to make the measurements. Previous works have used Rodenstock SLOs and microscopes with video capability attachments [7–10,14,15]. The Rodenstock SLO system has a different field of view (40°) and uses NTSC video output. One advantage of Rodenstock SLO over our HRA SLO is a higher frame rate, thus allowing for more frames to be used in the analysis. However, agreement of our data with the previous work suggests that it is reasonable to use the HRA SLO to record measurements of blood velocity and vessel diameter. No modification of HRA SLO was done to image rat retina vasculature and to obtain microsphere movies, making it a potential clinical tool. The digital imaging of our HRA is an advantage in that it will remove an additional step of converting video output to digital format in the data analysis. Our data also suggest that microspheres are a reasonable tracer to use for assessment of retinal hemodynamics. Given the positive aspects of SLO particle-tracking method, we believe that this technique can be adopted for a clinical use.

Based on mean vessel diameter and blood velocity, it is also possible to calculate volumetric blood flow rate. For an artery with a diameter of 80 μm and blood velocity of 30.5 mm/sec, the flow rate is 9.2 μL/min along the vessel. For a vein diameter of 110 μm and velocity of 34.2 mm/sec, the flow rate is 19.5 μL/min in this study.

29.4.2 Hypoxia and Hyperoxia

Altering O_2 inspiration had measurable effects on retinal blood vessel diameter and blood velocity. Hypoxia increased both vessel diameter and blood velocity in our experiment. This is consistent with previous work in hypoxia [16,17]. A decrease in blood oxygen leads to vasodilation and increased blood flow to maintain oxygen supply to the retinal cells. Various factors such as nitric oxide, adenosine, and prostanoids have been implicated in the control of blood flow [18–20]. Our measurement method is sensitive enough to detect the changes under hypoxia, and further studies are planned to investigate whether intercellular molecules, such as nitric oxide, play a role in controlling retinal blood flow. In response to hyperoxia, both vessel diameter and blood velocity decreased in our experiments. In contrast to hypoxia, as PO_2 increases during hyperoxia, retinal vessels constrict thus reducing retinal blood flow. Our data are in agreement with previous studies of hyperoxia [21–23].

To our knowledge, this is the first paper to demonstrate the changes in blood velocity and vessel diameter using the HRA unit in response to hypoxia and hyperoxia. Our measurement technique is sensitive enough to detect small changes in blood velocity and would be a valuable clinical tool for assessing retinal hemodynamics. This technique can be further developed to non-invasively measure retinal blood flow changes due to retinal vascular diseases such as retinal occlusions or diabetic retinopathy.

Acknowledgment We would like to thank the Whitaker Foundation for the generous support.

References

1. A. Bill, In: *Handbook of Physiology, Section 2, The Cardiovascular System, Circulation of the eye*, edited by C.C.M.E. Renkin and S.R. Geiger (Am. Physiol. Soc., Bethesda, MD, 1975), pp. 1001–1033.
2. J.E. Grunwald, C.E. Riva, S.H. Sinclair, A.J. Brucker, B.L. Petrig, Laser Doppler velocimetry study of retinal circulation in diabetes mellitus, *Arch Ophthalmol*,**104**, 991–996 (1986).
3. V. Patel, S. Rassam, R. Newsom, J. Wiek, E. Kohner, Retinal blood flow in diabetic retinopathy, *BMJ,* **305**, 678–683 (1992).
4. V. Patel, S.M. Rassam, H.C. Chen, E.M. Kohner, Oxygen reactivity in diabetes mellitus: effect of hypertension and hyperglycaemia, *Clin Sci (Lond)*, **86**, 689–695 (1994).
5. F. Fillacier, G.A. Peyman, Q. Luo, B. Khoobehi, Study of lymphocyte dynamics in the ocular circulation: technique of labeling cells, *Curr Eye Res*, **14**, 579–584 (1995).
6. J. Ben-nun, Comparative flow velocity of erythrocytes and leukocytes in feline retinal capillaries, *Invest Ophthalmol Vis Sci*, **37**, 1854–1859 (1996).
7. R.D. Braun, M.W. Dewhirst, D.L. Hatchell, Quantification of erythrocyte flow in the choroid of the albino rat, *Am J Physiol Heart Circ Physiol*, **272**, 1444–1453 (1997).
8. B. Khoobehi, G.A. Peyman, Fluorescent labeling of blood cells for evaluation of retinal and choroidal circulation, *Ophthalmic Surg Lasers*, **30**, 140–145 (1999).
9. S.D. Wajer, M. Taomoto, D.S. McLeod, R.L. McCally, H. Nishiwaki, M.E. Fabry, R.L. Nagel, G.A. Lutty, Velocity measurements of normal and sickle red blood cells in the rat retinal and choroidal vasculatures, *Microvasc Res*, **60**, 281–293 (2000).
10. B. Khoobehi, B. Shoelson, Y.Z. Zhang, G.A. Peyman, Fluorescent microsphere imaging: a particle-tracking approach to the hemodynamic assessment of the retina and choroids, *Ophthalmic Surg Lasers*, **28**, 937–947 (1997).
11. E.C. Butcher, I.L. Weissman, Direct fluorescent labeling of cells with fluorescein or rhodamine isothiocyanate. I. Technical aspects, *J Immunol Methods*, **37**, 97–108 (1980).
12. T.H. Williamson, G.M. Baxter, Central retinal vein occlusion, an investigation by color Doppler imaging. Blood velocity characteristics and prediction of iris neovascularization, *Ophthalmology*, **101**, 1362–1372 (1994).
13. T. Nagaoka, T. Sakamoto, F. Mori, E. Sato, A. Yoshida, The effect of nitric oxide on retinal blood flow during hypoxia in cats, *Invest Ophthalmol Vis Sci*, **43**, 3037–3044 (2002).
14. H.F. Duijm, A.H. Rulo, M. Astin, O. Maepea, T.J. van den Berg, E.L. Greve, Study of choroidal blood flow by comparison of SLO fluorescein angiography and microspheres, *Exp Eye Res*, **63**, 693–704 (1996).
15. N. Masaoka, K. Nakaya, Y. Koura, M. Ohsaki, Hemodynamic changes in two patients with retinal circulatory disturbances shown by fluorescein angiography using a scanning laser ophthalmoscope, *Retina*, **21**, 155–160 (2001).
16. G. Eperon, M. Johhson, N.J. David, The effect of arterial PO_2 on relative retinal blood flow in monkeys, *Invest Ophtahlmol*, **14**, 342–352 (1975).

17. J. Ahmed, M.K. Pulfer, R.A. Linsenmeier, Measurement of blood flow through the retinal circulation of the cat during normoxia and hypoxemia using fluorescent microspheres, *Microvasc res*, **62**, 143–153 (2001).
18. A. Deussen, M. Sonntag, R.Vogel, L-arginine-derived nitric oxide: A major determinant of uveal blood flow, *Exp Eye Res*, **57**, 129–134 (1993).
19. N. Toda, Y. Kitamura, T. Okamura, Role of nitroxidergic nerve in dog retinal arterioles in vivo and arteries in vitro, *Am J Physiol*, **266**, H1985–H1992 (1994).
20. S. Harino, K. Nishimura, K. Kitanishi, M. Suzuki, P. Reinach, Role of nitric oxide in mediating retinal blood flow regulation in cats, *J Ocur Pharmacol Ther*, **5**, 295–303 (1999).
21. J.E. Grunwald, C.E. Riva, B.L. Petrig, S.H. Sinclair, A.J. Brucker, Effect of pure O_2-breathing on retinal blood flow in normals and in patients with background diabetic retinopathy, *Curr Eye Res*, **3**, 239–241 (1984).
22. B. Kiss, E. Polska, G. Dorner, K. Polak, O. Findl, G.F. Mayrl, H.G. Eichler, M. Wolzt, L. Schmetterer, Retinal blood flow during hyperoxia in humans revisited: concerted results using different measurement techniques, *Microvasc Res*, **64**, 75–85 (2002).
23. C.E. Riva, J.E. Grunwald, S.H. Sinclair, Laser Doppler Velocimetry study of the effect of pure oxygen breathing on retinal blood flow, *Invest Ophthalmol Vis Sci*, **34**, 47–51 (1983).

Part VII
Nano-Bio Technology

Chapter 30
Highly Sensitive Rapid, Reliable, and Automatic Cardiovascular Disease Diagnosis with Nanoparticle Fluorescence Enhancer and Mems

Bin Hong[1], **Junhai Kai**[2], **Yongjie Ren**[1], **Jungyoup Han**[2], **Zhiwei Zou**[2], **Chong H. Ahn**[2], **and Kyung A. Kang**[1]

Abstract Cardiovascular diseases (CVDs) have been the leading threat to human life. An effective way for sensitive and accurate CVD diagnosis is to measure the biochemical markers released from the damaged myocardial cells in the bloodstream. Here, a multi-analyte, fluorophore mediated, fiber-optic immuno-biosensing system is being developed to simultaneously and rapidly quantify four clinically important cardiac markers, myoglobin, C-reactive protein, cardiac troponin I, and B-type natriuretic peptide. To quantify these markers at a pico-molar level, novel nanoparticle reagents enhancing fluorescence were used and signal enhancement was obtained as high as ~230%. Micro-electro-mechanical system (MEMS) was integrated to this system to ensure a reliable and fully-automated sensing performance. A point-of-care, automatic microfluidic sensing system for four cardiac marker quantification was developed with the properties of 3 cm sensor size, 300 μL sample volume, 9-minute assay time, and an average signal-to-noise ratio of 35.

30.1 Introduction

Cardiovascular diseases (CVDs), especially the acute myocardial infarction (AMI; commonly known as heart attack), have been the top killers for human beings [1]. Rapid and accurate diagnosis of CVDs is, therefore, critically important to save lives. This can be realized by rapid, sensitive, and accurate quantification of cardiac markers released from injured cardiac muscles. Creatine Kinase-MB (CK-MB), myoglobin (MG), and cardiac troponin I (cTnI) are important markers for early diagnosis of a heart attack [2]. B-type natriuretic peptide (BNP) and C-reactive protein (CRP) are crucial markers for the diagnosis of congestive heart failure (CHF) and acute coronary syndromes (ACS) and also for the accurate prognosis after an AMI insult [3–4]. Our effort is

[1]Department of Chemical Engineering, University of Louisville, Louisville, KY 40292.
[2]Department of Electrical and Computer Engineering and Computer Science, University of Cincinnati, Cincinnati, OH 45221.

K.A. Kang et al. (eds.), *Oxygen Transport to Tissue XXIX*

focused on developing a highly sensitive, reliable, and user-friendly, point-of-care sensing device, utilizing nanoparticle reagents and micro-electro-mechanical-system (MEMS) technique.

The main challenge in developing a biosensor is the low concentrations of biomarkers in biofluid (very often only a few tens of pico-moles and less) at the early stage of disease [5]. Since our sensing is interrogated by fluorescence, fluorescence enhancement can improve the sensitivity. Nanogold particles (NGPs), possessing strong plasmon polariton fields on the surface, can reroute lone-pair electrons (normally contributing self-quenching) of a fluorophore to NGPs, resulting in fluorescence enhancement [6–7]. Some biocompatible solvents were also found to enhance fluorescence, by shifting the fluorophore excitation/emission wavelengths and/or increasing the number of *trans* carbon double bonds [6–7]. To maximize the enhancement effect, NGPs and solvents were combined, forming nanogold particle reagents (NGPRs). According to our previous results [7], the mixture of 5 nm sized NGPs coated with 2-nm thick self-assembled monolayer (5nmNGP-SAM2nm) in 1-butanol has shown to be an excellent enhancer.

MEMS technique improves the performance of biosensors by providing micro-fabrication tools, the consistency in operation, and compactness, as well as mass production capability. For a reliable and fully automated sensing performance with a minimal system size, MEMs was integrated to our sensing system.

In this paper, a sensitive and accurate cardiac marker sensing system with the application of NGPR and MEMS is reported. With this system, simultaneous four-cardiac marker quantification was completed in 9 minutes at an average signal-to-noise (S/N) of 35.

30.2 Materials, Instruments, and Methods

30.2.1 NGP, Solvent and NGPR-Related Study

The 5 nm nanogold particles coated with tannic acid (Ted Pella, Redding, CA) and 16-mercaptohexadecanoic acid (MHA; Sigma/Aldrich, St. Louis, MO) were used to synthesize 5nmNGP-SAM2nm by self-assembling MHA on the NGP surface [6]. For the butanol based NGPR, 5nmNGP-SAM2nm was then dispersed in pure 1-butanol (Sigma/Aldrich).

30.2.2 Cardiac Marker Sensors and Assay Protocol

Human BNP was purchased from Bachem (Torrance, CA). Monoclonal IgG against human BNP, was from Strategic Biosolutions (Newark, DE). Human cTnI, MG, and CRP, and their respective monoclonal antibodies were obtained from Fitzgerald Industries (Concord, MA). Plasma samples with cardiac markers were prepared by adding a known amount of cardiac markers to the emulated human plasma. The emulated plasma is 103 mg/ml human serum albumin (HSA; Sigma/Aldrich) in the PBS buffered solution [8]. The fluorophore, Alexa Fluor® 647 (AF647; max. excitation/emission wavelengths, 649/666 nm), was from Invitrogen (Carlsbad, CA). Four cardiac marker biosensors were constructed, following the protocol established by Tang et al [5]. The fluorometer with four sensing channels (Analyte 2000™) was from Research International (Monroe, WA). Briefly, the monoclonal antibody (1°Mab) against the respective marker is immobilized on the optical fiber surface *via* streptavidin-biotin bond and the sensor is encased in a chamber. During the assay, the sample is injected to the sensing chamber. The target marker binds specifically to the 1°Mab on the sensor surface. After the sample incubation, unbound molecules are washed away from the sensing chamber. Next, the fluorophore tagged, second monoclonal antibody (fluorophore-2°Mab) is applied to the sensor. When the surface immobilized fluorophores are excited by the laser light, the emitted fluorescence is detected by the fluorometer. For the sensing with NGPR, NGPR is applied before the sample incubation for the baseline [6]. NGPR is also applied after the incubation of fluorophore-2°Mab and sensor washing. The fluorescence signal difference between the baseline and after the sandwich complex formation is correlated to the analyte concentration in the sample. Here, the enhancement is defined as the increase in the fluorescence signals by using NGPR divided by the fluorescence from same sample without using NGPR (control).

30.2.3 Microfluidic Sensing System Utilizing MEMs

To generate micro-turbulence inside the sensing chamber, bumps (or baffles) were added on the upper and bottom sides of the microchamber (i.e., serpentine microchannel). The sensing module with the serpentine microchannels as well as the microchannel network were microfabricated as described by Sohn, et al [9]. The computer software LabVIEW™ (version 7.1) and a data acquisition card DAQ (USB-6008, 8 inputs, 12 bits, 10 ks/s, multifunctional I/O, National Instruments; Austin, TX) were used to control all electronic parts in the flow control unit. Electronically controllable micro-solenoid pump (12 v, 50 μL per stroke, 2 W) and 7 micro-solenoid valves (12 v, 280 mW, Lee Co.; Westbrook, CT) were for the automatic flow control. A drive circuit with a power plug, a power switch, and a power LED were customized by our research group.

30.3 Results and Discussion

Our fluorophore mediated, fiber-optic immuno-sensor is a highly sensitive detection tool and, therefore, it can be used for various human disease diagnosis/prognosis [5,8,10]. In our study for the quantification of BNP in plasma (without using any enhancers), the sensitivity of our system was found to be two orders of magnitude higher than that of enzyme-linked immunosorbent assay (ELISA) (data not shown). However, for rapid cardiac marker quantification, especially for BNP and cTnI, due to their extremely low concentrations in plasma at an early disease stage, additional sensitivity improvement was needed.

30.3.1 Cardiac Marker Sensing Using NGPR

As previously stated, 5nmNGP-SAM2nm in 1-butanol was found to be an excellent fluorescence enhancer. Its enhancement effect was, therefore, tested for a 3-cm BNP sensor. Figure 30.1a shows the sensing performance of BNP sensor with and without the NGPR. With the NGPR, the signal intensity was found to be 410% greater than that without NGPR. This NGPR was also tested with four cardiac marker sensors encased in a four-microchannel sensing module (Fig. 30.1b). The sample was the mixture of four cardiac markers in the emulated human plasma. The concentration of each marker was selected to be at its lower limit in the sensing range, because this is the condition requiring the enhancement the most. Results showed that NGPR is able to increase the signal intensities of BNP, cTnI, MG, and CRP sensors by 60, 50, 180, and 230%, respectively. In general, the signals from the sensors for the markers with higher concentration ranges (MG, 4–40 nM; CRP, 5.6–56 nM) were enhanced more than those for the sensors with lower concentration markers (BNP, 26–260 pM;

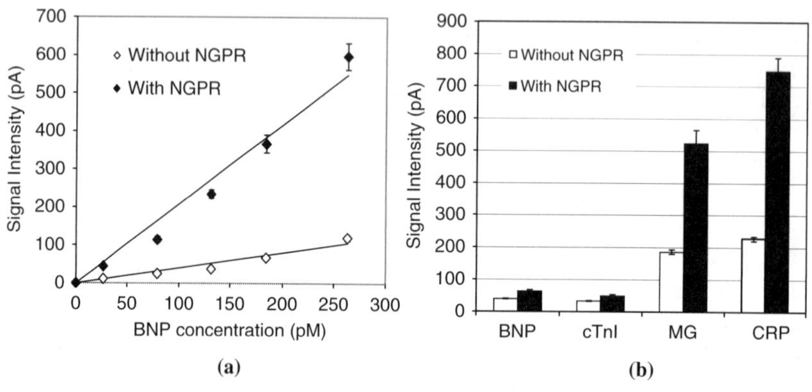

Fig. 30.1 Sensing performances with and without NGPR: (a) BNP sensor in the BNP sensing range and (b) four cardiac marker sensors for their lower sensing limit in the microfluidic sensing system. [Experimental conditions: For (a), 3-cm sensor; 3/4 minutes for the sample and AF647-2°Mab incubation; flow velocity at 1.2 cm/sec, NGPR, 5nmNGP-SAM2nm in 1-butanol, capillary microchannel, automatic sensing. For (b), cardiac markers at their lower limits; mixture of AF647-2°Mab; serpentine sensing module, other operation conditions were the same as (a).]

cTnI, 31–310 pM). The reason for different enhancement levels should be studied further.

30.3.2 Cardiac Marker Sensing Chamber with Microfabricated Serpentine Structure

For biosensors utilizing surface reaction, effective analyte mass transport from bulk media to sensor surface is important for a rapid assay. Convective application of liquid samples/reagents to the sensor surface was proven to improve the sensitivity [10]. However, reasonable flow rates without damaging the microchannels of the sensing system are in a laminar flow range and limit the analyte transport, especially for the sample with a very low analyte concentration [5]. Well-designed microchannels that can create local turbulence facilitate the analyte transport to the surface better [8]. For this purpose, a series of bumps/baffles were microfabricated on the inner surface of the microchannel (serpentine microchannel, Fig. 30.2a and b). Out of various bump configurations that we have tested, the half-circular bump series, with the dimensions of 1200 μm diameter, 400 μm height, and 1200 μm spacing between two adjacent bumps, were found to be very effective [8]. The sensing performance of 3-cm BNP and MG sensors was studied for the effectiveness of this serpentine microchannel and the results were compared with those from the channels without bumps (capillary microchannel; I.D. = 1400 μm). These two molecules were selected for the test because BNP has a low analyte sensing range (26∼260 pM) and MG, a high range (4,000∼40,000 pM). For BNP sensing (Fig. 30.2c), serpentine microchannel presented approximately 30∼90% higher signals than the capillary microchannel. For MG sensor, only a slight signal increase (0∼6%) was exhibited (Fig. 30.2d), probably because, due to its high sensing range, MG is not mass transport limited.

30.3.3 Sensing Operation Utilizing MEMS

Automation of sensing system operation is important for the assay consistency, reliability in operation, and user-friendliness. In the multi-cardiac marker sensing system, MEMS technique was implemented for the automatic flow control unit (Fig. 30.3ab). An electronically controllable micro-pump and seven microvalves were used to deliver the sample and reagents to the sensing module and the microchannel network. The automatic control of the micro-pump, micro-valves, fluorometer, and other electronic parts were done by a customized LabVIEW™ code with an easy and simple interface. Therefore, a MEMS based biosensing system was developed for simultaneous, quantitative measurement of the four cardiac markers (Fig. 30.3c). Using this automated sensing system, for all four

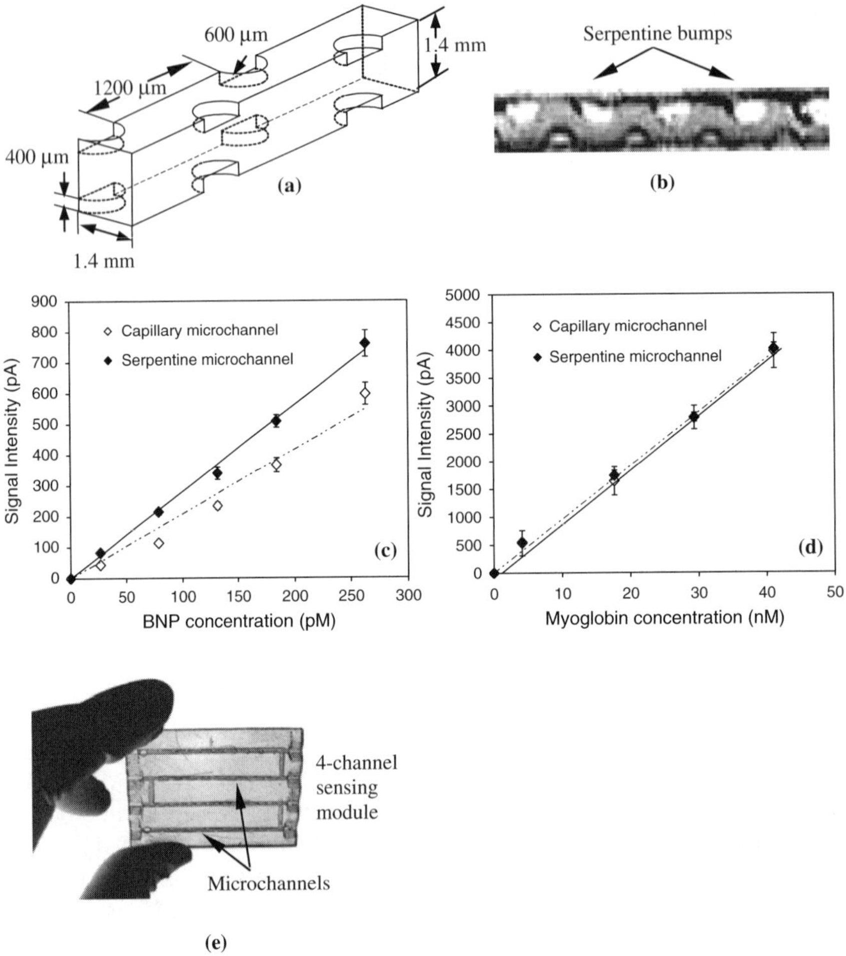

Fig. 30.2 (a) Schematic diagram of the serpentine microchannel and (b) its actual side view. [Structure: a series of half-circular bumps at 600 μm radius, 400 μm height and 1200 μm spacing between bumps and 1.4×1.4 mm square cross-section]; the sensing performance of (c) BNP and (d) MG sensors using capillary microchannel (◇) and serpentine microchannel (◆). [Experimental conditions: 3-cm sensor, 3/4 minutes for sample and AF647-2°Mab incubation, flow velocity at 1.2 cm/sec, NGPR, 5nmNGP-SAM2nm in 1-butanol , automatic sensing.]; and (e) a four-channel serpentine sensing module.

sensors, the S/N ratio in average was doubled from 18 to 35, with signal intensities similar to those by the manual operation. In addition, with the micro-fabricated flow network, the sample volume required for each assay decreased from 1 mL to 300 μL.

In order for our sensing system to be used for a rapid diagnosis of diseases, especially for AMI, a shorter assay time is highly desired. Here, the sensing performance of four cardiac marker sensors in our serpentine sensing module

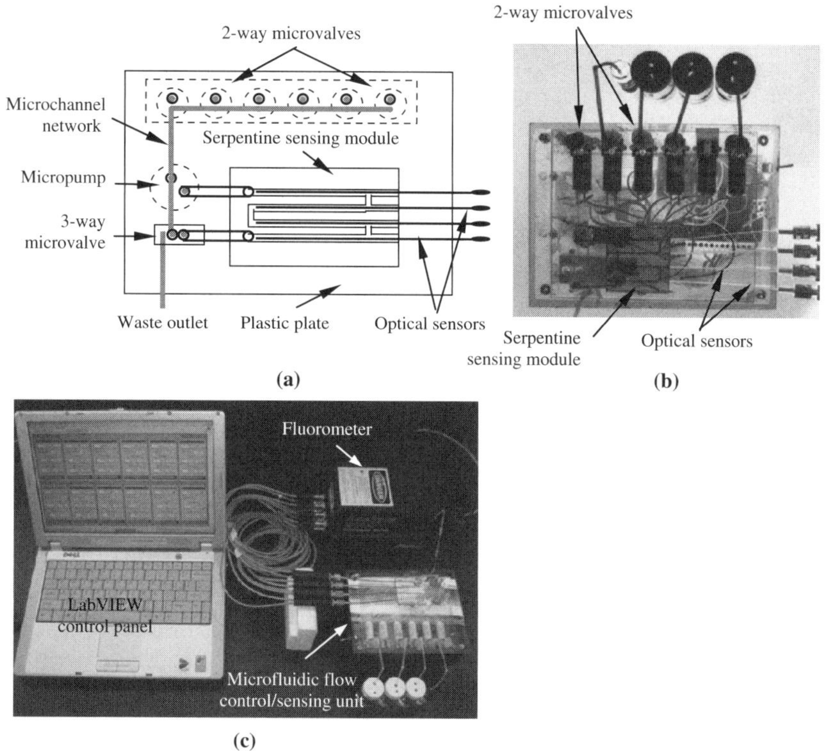

Fig. 30.3 Automatic sensing system: (a) Schematic diagram and (b) the top view of the actual microfluidic sensing unit: imbedded microchannel network, micro-pump, micro-valves, serpentine sensing module; (c) The entire sensing system including the laptop computer with LabVIEW control panel, microfluidic flow control/sensing unit with serpentine sensing module, fluorometer with four detection channels.

(Fig. 30.2e), was studied with changes in the incubation times for the sample and the AF647-2°Mab, by the new assay protocol with NGPR. Plasma samples with four markers at their lower concentration limits (e.g., 0.1 ng/ml BNP, 0.7 ng/ml cTnI, 70 ng/ml MG, and 700 ng/ml CRP) were incubated for 1, 2 or 3 min, at a constant incubation time of 4 min for the AF647-2°Mab mixture (Fig. 30.4a). Results of BNP and cTnI sensors were shown in one figure (Fig. 30.4a1), while MG and CRP in another figure (Fig. 30.4a2), because of their similar sensing ranges. For cTnI (Fig. 30.4a1, $\leq$) and MG (Fig. 30.4a2, (), the signal intensities increased sharply only after 2 min. From the results of four sensors, 3 min seems the optimal reaction time for the assay. Although the signals for all sensors may increase after 3 min, the signal intensities at 3 min were all high with high S/N ratios. Similarly, the effect of the AF647-2°Mab incubation time (1, 2, 3, or 4 min) on sensing performance was studied with a constant sample incubation time of 3 min. All four sensors showed a similar

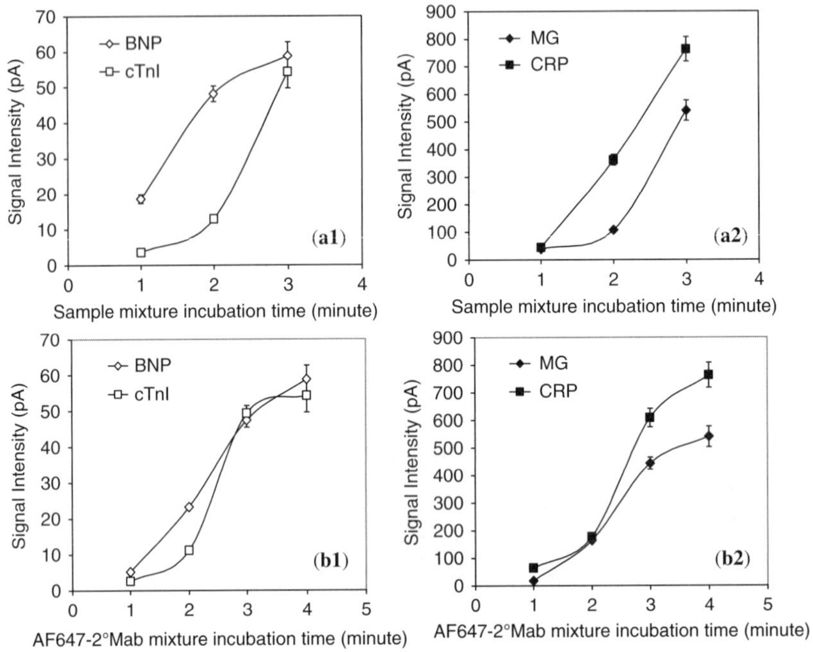

Fig. 30.4 The effects of reaction time for (a) sample and (b) AF647-2°Mab on the sensing performance. [Experimental conditions: four cardiac markers at their lower limit of sensing ranges; sensor size, 3 cm; flow velocity, 1.2 cm/sec; NGPR, 5nmNGP-SAM2nm in 1-butanol; serpentine sensing module; automatic sensing.]

signal profile with the increase of the AF647-2°Mab incubation time (Fig. 30.4b). At 3 min, the signal increase was slowly tapered, indicating 3 min is sufficient.

In a sum, with the application of NGPR and MEMS, the four cardiac marker biosensing can be completed within 9 minutes (3, 3, and 3 min for sample incubation, AF647-2°Mab incubation, and all other times such as sample/reagent delivery, sensor washing and regeneration, respectively) with a 3-cm sensor size, and a high S/N ratio of 35.

30.4 Conclusions

A MEMS based, multi-analyte, point-of-care biosensing system was developed to simultaneously quantify four important cardiac markers in blood plasma. To improve the sensitivity for analytes, fluorescence enhancing NGPR was applied. The sensitivities of BNP, cTnI, MG and CRP sensors increased by 60%, 50%, 180% and 230%, respectively. A serpentine sensing chamber was microfabricated to improve the analyte mass transport and the structure improved the sensitivity well, especially for the analyte with low concentration.

MEMS technology was also incorporated to the system for a reliable detection and user-friendly operation. The sensing consistency of the system (S/N ratio) was doubled, the assay time became 9 min and the sample volume decreased to 300 μL.

Our MEMS based, multi-analyte biosensing device can be used for quantifying disease-representing multi-biomarkers, rapidly, accurately, and user-friendly.

Acknowledgment Authors acknowledge the financial support from Kentucky Science and Engineering Foundation (KSEF-148-502-03-55) for fluorescence enhancement studies and National Science Foundation (BES-0330075) for cardiac marker biosensing. The Sigma Xi honor society is acknowledged for Bin Hong's Grants-in-Aid Research award for NGPs related studies and the Institute for Molecular Diversity and Drug Design (IMD[3]) at the University of Louisville for Bin Hong's Graduate Fellowship.

References

1. American Heart Association. Heart Disease and Stroke Statistics, Update, 10–12 (2005).
2. F. S. Apple, R. H. Christenson, R. Valdes, A. J. Andriak, K. Mascotti, and A. H.B. Wu, Simultaneous rapid measurement of whole blood myoglobin, creatine kinase MB, and cardiac troponin I by the triage cardiac panel for detection of myocardial infarction, *Clin. Chem.* **45**(2), 199–205 (1999).
3. A. S. Maisel, P. Krishnaswamy, H. C. Herrmann, and P. A. McCullough, Rapid measurement of B-type natriuretic peptide in the emergency diagnosis of heart failure, *New Engl. J. Med.* **347**, 161–167 (2002).
4. M. S. Sabatine , D. A. Morrow, C. P. Cannon, and E. Braunwald, Multimarker approach to risk stratification in non-ST elevation acute coronary syndromes: Simultaneous assessment of troponin I, c-reactive protein, and b-type natriuretic peptide, *Circ.* **105**, 1760–1763 (2002).
5. L. Tang, Y. J. Ren, B. Hong, and K. A. Kang, A fluorophore-mediated, fiber-optic, multi-analyte, immuno-sensing system for rapid diagnosis and prognosis of cardiovascular diseases, *J. Biomed. Optics* **11**, 021011 (2006).
6. B. Hong and K. A. Kang, Biocompatible, nanogold-particle fluorescence enhancer for fluorophore mediated, optical immunosensor, *Biosens. Bioelectron.* **21**(7), 1333–1338 (2006).
7. K. A. Kang and B. Hong, Biocompatible nano-metal particle fluorescence enhancers, *Crit. Rev. Eukar. Gene Expres.* **16**(1), 45–60 (2006).
8. L. Tang, 2005. Multi-analyte, fiber-optic immuno-biosensing system for rapid disease diagnosis: model systems for anticoagulants and cardiac markers. *Dissertation.* Chemical Engineering, University of Louisville, Louisville, KY.
9. Y. Sohn, J. H. Kai, C. H. Ahn, Protein array patterning on Cyclic Olefin Copolymer (COC) for disposable protein chip, *Sensor Lett.* **2**, 171–174 (2005).
10. L. Tang, H. J. Kwon, and K. A. Kang, Theoretical and experimental analysis of analyte transport in a fiber optic, protein C immuno-biosensor, *Biotech. Bioeng.* **88**, 869–879 (2004).

Chapter 31
Tumor-specific Nano-entities for Optical Detection and Hyperthermic Treatment of Breast Cancer

Hanzhu Jin[1], Bin Hong[1], Sham S. Kakar[2], and Kyung A. Kang[1]

Abstract The ultimate goal of this study is to develop a tumor-specific multi-functional, nano-entity that can be used for both cancer detection and treatment. Low heat (42~45°C) hyperthermia is an effective cancer treatment method with little side effect. Magnetic nanoparticles, such as Fe_3O_4, can be heated by alternating electromagnetic (AEM) fields at well selected frequencies, without heating normal tissue. Nanogold particles (NGPs) are effective optical absorbers and also excellent fluorescent enhancers. Therefore, coating gold on Fe_3O_4 particles can enhance the optical contrast as well as keeping the particle property for hyperthermia. Indocyanine green (ICG), a FDA approved fluorophore, has a very low quantum yield, and its fluorescence can be enhanced by linking ICG to gold-coated Fe_3O_4 nanoparticles. Luteinizing hormone releasing hormone (LHRH), which has high affinity to breast cancer, can be used for tumor-specific targeting. Our study results showed: Fe_3O_4 particles at a size range of 10~30 nm can be heated well by an AEM field at a rate of 18°C/wt%-minute; the fluorescence of ICG was extensively enhanced by NGPs; LHRH-coated gold nanoparticles provided as much cancer specificity as LHRH alone. Combining these properties in one entity, i.e., LHRH/ICG linked, gold-coated Fe_3O_4 nanoparticles, can be a tumor-specific nano-agent for optical detection and electro-magnetically induced hyperthermia for breast cancer.

31.1 Introduction

Low heat hyperthermia (42~45°C) is an effective cancer treatment method with very little side effect [1,2]. At this temperature range, the enzymes needed for tumor growth/ survival become deactivated and, with repeated low heat treatments, the tumor is gradually destroyed [3]. Magnetic nanoparticles have been considered to deliver heat to the tumor *via* an alternating electromagnetic

[1]Department of Chemical Engineering.
[2]Department of Medicine, University of Louisville, Louisville, KY 40292.

(AEM) field [4,5]. One of the most frequently used particles for this purpose is Fe_3O_4 nanoparticles [6,7], although previous applications may not have necessarily been tumor specific. Gold is very effective NIR absorber at a nano size and, therefore, coating gold on Fe_3O_4 particles is expected to enhance the NIR contrast, as well as keeping the heating property for hyperthermia [8,9]. In addtion, nanogold particles (NGPs) are chemically inert and their surface can easily be functionalized for other bio-entities.

Fluorophores can also be very effectively used as optical contrast agents [10]. One of the most widely used fluorophores in the NIR range for humans is Indocyanine green (ICG; excitation/emission maxima: 780/830 nm). Nevertheless, ICG has a very low quantum yield (only 0.012 in whole blood) [11]. Hong and Kang [12] have demonstrated that NGPs, when placed at an appropriate distance from a fluorophore, can significantly enhance the fluorescence emission. Therefore, the fluorescence of ICG may be enhanced by linking ICG on the gold-coated Fe_3O_4 particles *via* a spacer at a predetermined length.

The diameter of a usual capillary blood vessel is approximately 7 μm [13]. Nano-sized particles are much smaller than capillaries, and therefore are easily circulated in blood vessels. The surface of our proposed nano-entity can be functionalized with tumor-specific anti-receptors to reduce the systemic toxicity, when it is applied to the body. Luteinizing hormone releasing hormone (LHRH) is a peptide of 10 amino acids (pGlu-His-Trp-Ser-Tyr-D-Trp-Leu-Arg-Pro-Gly-NH_2) [14]. LHRH spontaneously reacts with the surface of nanometal particles by its N-terminal amine group *via* its self-assembling nature [15]. Many cancer types, including breast cancers, express receptors for LHRH [14, 16, 17]. In contrast, most visceral organs do not express LHRH receptors, or express only at a low level. Researchers have demonstrated that the breast cancer cells can be targeted through their high affinity LHRH receptors present on the cell membrane [15].

In this paper, optical/thermal properties of gold-coated Fe_3O_4 nanoparticles were investigated. A preliminary study of adding fluorescence property to this nano-entity, by linking ICG to the NGPs, was also conducted. Also, LHRH was linked to the surface of NGPs, and the binding affinity of the particles to the mouse gonadotrope cell line (LβT2), which expresses high levels of LHRH receptors, was measured and compared with that of free LHRH.

31.2 Materials and Methods

31.2.1 Measurement of the Thermal Properties of Gold-coated Fe_3O_4 Nanoparticles

Gold coated Fe_3O_4 nanoparticles (core size: 10~20 nm; gold layer thickness: 4~7 nm) were provided by Dr. Shi at the University of Cincinnati, OH. To test the effect of a gold layer on the heating performance of Fe_3O_4 nanoparticles,

Fe_3O_4 or gold-coated Fe_3O_4 nanoparticles were uniformly mixed with agar gel (2.15wt% agar in water: weight percentage) at a concentration of 1wt% of particles. The mixture containing the particles at a volume of 4 ml was placed into a glass test tube of 1.6 (diameter) × 10 (length) cm, and solidified at room temperature. The AEM field was generated at 450 KHz and 5 KW by an AEM generator (MKII-5; Taylor-Winfield induction Company; Brookfield, OH). The AEM applicator (induction heater coil) was made of a copper tube at 0.5 cm diameter that formed a coil of three turns, at a dimension of 2 cm (diameter) × 3 cm (height). The glass tube containing the sample was then placed inside the induction heater coil and the AEM field was applied for a predetermined period. The sample temperature was measured by a digital thermometer (Traceable®· All QA product; Belmont, NC) before and after the application of the field.

31.2.2 Measurement of the Optical Properties of Gold-coated Fe_3O_4 Nanoparticles

Optically breast-tissue-like, experimental models at a dimension of 24 × 14 × 5 cm were constructed following the procedure described by Jin and Kang [18]. The absorption and scattering properties of the model were adjusted by India ink (Design Higgins®; Sanford Co., Bellwood, IL) and skimmed milk (Kroger Co., Cincinnati, OH). Agar (Sigma-Aldrich, St. Louis, MO) was used for the mechanical property of the experimental breast model. Breast tumors usually have higher absorption than normal breast tissue [19]. Therefore, for the tumor model without the contrast agent, the absorption coefficient was adjusted to be 4 times higher than that of the breast model. For the particle accumulated tumor models, Fe_3O_4 or gold-coated Fe_3O_4 nanoparticles were added at a concentration of 0.1wt% to the same ingredient of the breast model. Each tumor model ingredient, while it was still in liquid phase, was injected into an empty oval, Vitamin E capsule shell (1.5 × 1.0 × 1.0 cm; National Vitamin Company, LLC; Las Vegas, NV). Once the filling of the capsule was solidified, the capsule was placed in the breast model solution at a predetermined depth. After the solidification of the breast model, the surface of the model was scanned using the optical fibers of the NIR time resolved spectroscopy (TRS) instrument, as described by Honar and Kang [20]. Obtained TRS spectra were converted to the frequency domain by Fourier transformation [21].

31.2.3 Cypate Fluorescence Signal Enhancement by Nanogold Particles

Cypate (M.W. = 705), an ICG derivative with a carboxylic group, was provided by Dr. Achilefus from the Department of Radiology, Washington University School of Medicine. Protein A (PA; ~1 nm) linked NGPs at sizes of 5 (5 nmNGP-PA) and

10 nm (10nmNGP-PA) were purchased from Ted Pella (Redding, CA). Streptavidin (SA: ~3 nm) linked NGPs at a size of 10 nm (10nmNGP-SA) were obtained from Sigma-Aldrich (St. Louis, MO). The thickness of the protein (PA or SA) layer on NGPs was estimated using the software, HyperChem 7.0 (Hypercube, Inc.; Gainesville, FL). For the reaction between the carboxylic acid of Cypate and the amine group of PA or SA, ethanol and *N,N′*-Dicyclohexylcarbodiimide (DCC; Sigma-Aldrich, St. Louis, MO) were used as a solvent and a catalyst, respectively. Cypate and DCC were dissolved in a minimal amount of ethanol, and then the mixture was immediately transferred to 1 mL NGP solution, stirring it at 4°C for 5 hours. After the reaction, Cypate linked NGPs were separated from the unreacted molecules by a dialysis tube [DispoDialyzer® (Molecular cut off: 25 KD); Spectrum Laboratories, Inc.; Rancho Dominguez, CA]. The concentration of Cypate in the final product was measured at 780 nm by UV/Vis Spectrophotometer (DU® 520, Beckman Coulter, Inc., Fullerton, CA). The Cypate linked NGPs were mixed with the experimental breast model ingredients and the mixture was injected into an empty Vitamin E capsule shell as a tumor model. The tumor model was placed in the experimental breast model at a predetermined depth as described in Section 2.2. The surface of the breast model was then scanned with the source (788 nm) and detector probes of NIR-TRS instrument to measure the fluorescence contrast generated by Cypate or Cypate linked NGPs. For fluorescence detection, a long pass filter (cut-off wavelength: 830 nm; B+W 093 IR 830 nm; Schneider Optics, Inc.; New York, USA) was placed between the detector fiber and the detector to allow passing the fluorescence generated by the fluorophore with minimal detection of source light (788 nm).

31.2.4 LHRH Linked NGPs for Breast Tumor Targeting

LHRH (MW = 1311.45; Sigma-Aldrich; St. Louis, MO) was coated on the surface of 10 nm sized NGPs. After the coating, a dialysis tube was used to remove unbound LHRH. The solution containing LHRH coated NGPs was adjusted to pH 9.0 using 0.1 mM sodium carbonate solution. Then, various concentrations of LHRH or LHRH linked NGPs were applied to the mouse gonadotrope cell line (LβT2) expressing LHRH receptors. After four hours, cells were lysed and assayed for luciferase activity using Luciferase Assay Kits (Promega Bioscience, Inc, San Luis Obispo, CA) [22].

31.3 Results and Discussion

31.3.1 Gold-coated Fe_3O_4 Nanoparticles as Optical/Thermal Agent

Our previous study results [23] showed that Fe_3O_4 nanoparticles at size 5~60 nm are heated well in the AEM field at a frequency of 450 KHz and a power of 5 KW, without heating any tissue components. For the same wt%, the particles at a size range of 10–30 nm were heated the best among the particle size tested. Fe_3O_4 nanoparticles are reasonably good near infrared (NIR) absorbers. But NGPs at a size range of 10∼250 nm are even stronger NIR absorbers, and gold has other advantageous properties for multi-functional uses. Gold coated Fe_3O_4 nanoparticles were, therefore, considered.

31.3.1.1 Effect of Gold Layer on Heating Performance of Gold-coated Fe_3O_4 Particles

The effect of the gold layer on the heating of Fe_3O_4 nanoparticles was studied. Either Fe_3O_4 (10∼20 nm) or gold-coated Fe_3O_4 nanoparticles (gold layer thickness: 4∼7 nm) in the range of 0.1 ∼ 1 wt% were added to 4 ml of agar gel. The samples were placed in a glass tube and the AEM field was applied at 450 KHz and 5 KW for 2 minutes. Figure 31.1 shows the temperature increase of the samples containing the Fe_3O_4 particles, with (?) and without (?) gold coating. For both samples, the heating was linearly proportional to the particle concentration. The temperature increases for both were approximately at a rate of 15∼18°C/wt% – particles per minute, indicating gold coating has little effect on the heating.

31.3.1.2 Effect of Gold Layer on NIR Contrast Enhancement by Gold-coated Fe_3O_4 Particles

As a next step, the optical contrast was studied for a tumor model (absorption 4 times of normal tissue), a model containing Fe_3O_4 particles (0.1wt%), or a

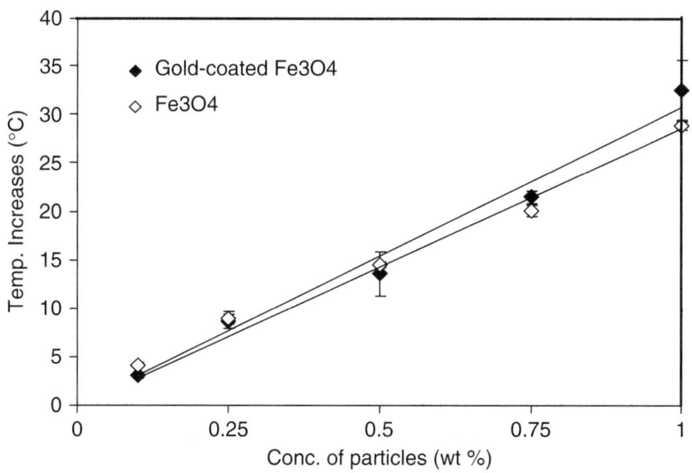

Fig. 31.1 The effect of a gold layer on the heating performance of Fe_3O_4 nanoparticles by AEM field. [Experimental conditions: Fe_3O_4 nanoparticle size: 10∼20 nm; gold layer thickness: 4∼7 nm; AEM frequency: 450 KHz; Power: 5 KW; heating time: 2 min.]

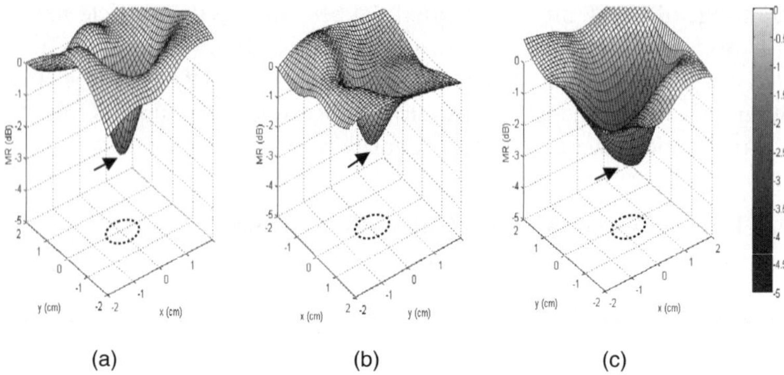

Fig. 31.2 Contrasts by (a) a tumor model (b) a model with 0.1wt% of Fe_3O_4, and (c) a model with 0.1 wt% of gold-coated Fe_3O_4, placed 1 cm deep in the breast model; by NIR-TRS at the wavelength of 788 nm in transmittance; modulation frequency analyzed at 100 MHz. The black dashed ellipsoids indicate the tumor model size and position. The arrows indicate a maximum absorption contrast.

model with gold coated Fe_3O_4 particles (0.1wt%). Since Fe_3O_4 or gold-coated Fe_3O_4 nanoparticles are mixed with the same ingredients of the breast model, the absorption contrast is created only by the particles. Each tumor model was placed 1 cm deep in the breast model and the optical contrast was measured by the NIR-TRS instrument, in transmittance. The measurements were performed at an interval of 1 cm on an area of 5.0×5.0 cm immediately above the tumor model. The tumor model [Fig. 31.2(a)] generated a maximum absorption contrast of 2.5 dB. The contrast by Fe_3O_4 nanoparticles was 2 dB [Fig. 31.2(b)], indicating that Fe_3O_4 is an effective contrast agent. For the gold-coated Fe_3O_4, a maximum contrast observed was 3 dB [Fig. 31.2(c)], which is greater than those by the Fe_3O_4 nanoparticles or by the tumor model.

31.3.2 Fluorescence Contrast Enhancement by NGP Linked Cypate

Cypate (an ICG derivative) was considered to be linked on the gold-coated Fe_3O_4 nanoparticles to add a highly effective fluorescent contrast property to our entity. As an initial attempt to enhance the fluorescence of Cypate, Cypate was linked to 5 or 10 nm NGPs (eventually gold-coated Fe_3O_4 particles) *via* Protein A (PA: 1 nm) or Streptavidin (SA: 3 nm), and their fluorescence was compared to that of Cypate only (Fig. 31.3). Cypate concentration for this experiment was 30 μM. For all cases, the fluorescence signal was enhanced by $300 \sim 900$ times of that by Cypate alone. The fluorescence signal by Cypate linked *via* PA spacer and NGP size of 5 nm showed the best enhancement by 900 times.

Cypate or Cypate linked NGPs was then mixed with the breast model ingredients at a Cypate concentration of 5 μM, and the mixture was filled into

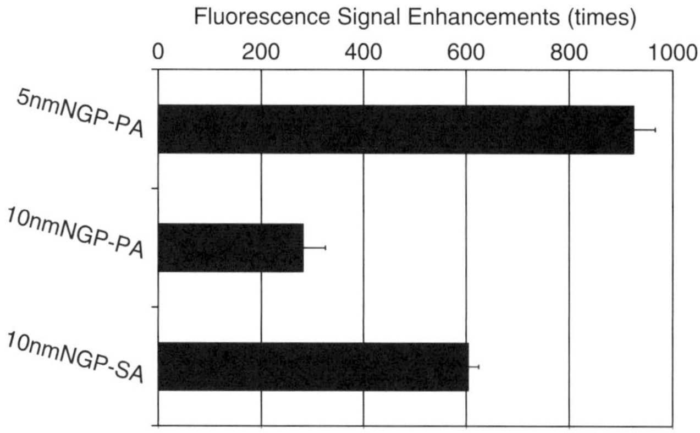

Fig. 31.3 Fluorescence signal enhancement by Cypate linked PA coated 5 nm gold (5nmNGP-PA), PA coated 10 nm gold (10 nmNGP-PA), or SA linked 10 nm gold (10nmNGP-SA) compared to Cypate alone.

the Vitamin E capsules as tumor model. The tumor model was, then, placed at 1 cm depth in the breast model, and the breast model was scanned using the NIR-TRS instrument at 788 nm excitation wavelength in reflectance. Figure 31.4 shows the fluorescence contrast enhancement of Cypate *via* Protein A linked 5 nm gold (5nmNGP-PA), Protein A linked 10 nm gold (10 nmNGP-PA), or Streptavidin linked 10 nm gold (10nmNGP-SA). With a constant spacer length by PA, the fluorescence enhancement was similar for both 5 and 10 nm NGPs, approximately 1.3 times enhancement. For the NGP size at 10 nm, using SA spacer provided more effective enhancement (2.2 times) than PA spacer (1.3 times).

Compared to the result of free NGP-fluorophore solution (Fig. 31.3), the results in the experimental breast model showed much less enhancement. One

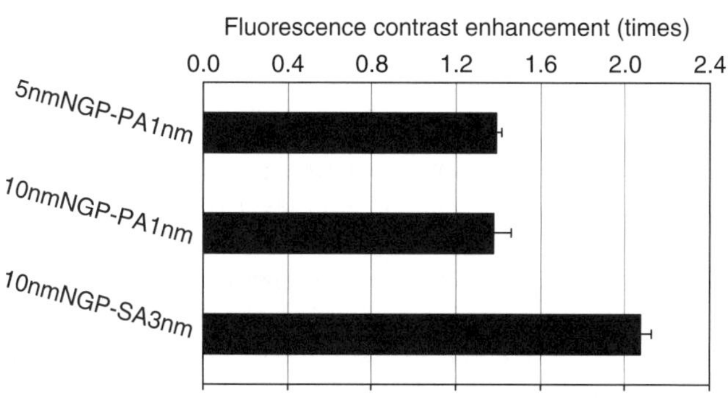

Fig. 31.4 Fluorescence contrast enhancement by Cypate linked PA coated 5 nm gold (5 nmNGP-PA), PA coated 10 nm gold (10 nmNGP-PA), or SA coated 10 nm gold (10 nmNGP-SA) compared to the contrast by Cypate alone.

possible reason could be that Cypate linked NGPs may be interacting with the ingredients of the experimental breast model. Another reason could be that the fluorescence light generated by Cypate is highly scattered by a turbid media like our experimental breast model, and the scattered fluorescence signal is measured by a single photon counting system (NIR-TRS instrument). Further investigation is on going. The range of the fluorescent contrast by Cypate linked NGPs was 10~17 dB, which is much higher than the absorption contrast (~3 dB). This result demonstrates that the feasibility of linking Cypate on the gold-coated Fe_3O_4 nanoparticles to enhance Cypate fluorescence. Currently, further investigation for optimizing NGP size and spacer length to maximize Cypate fluorescence enhancement is being performed.

31.3.3 Affinity of LHRH to Receptor with/without Linking to NGPs

As an initial test for using LHRH as a tumor targeting agent, LHRH was linked to NGPs or Fe_3O_4 nanoparticles (eventually LHRH will be linked to gold-coated Fe_3O_4 nanoparticles). The binding affinity of LHRH linked NGPs was studied. The mouse gonadotrope cell line (LβT2) expressing LHRH receptors was plated on 6-well plates. 24 hours after plating, cells were transfected with reporter gene construct CRE-Luciferase (1 μg/well), as described by Kakar et al [22]. After another 24 hours, the medium was replaced with serum free medium and the cells were incubated for 60 minutes, followed by treatment of cells with various concentrations of free LHRH or LHRH linked NGPs, for four hours. LHRH-linked NGPs showed a similar binding affinity (about 0.1 nM) to the native LHRH peptide, suggesting that the LHRH conjugated NGP retains its binding affinity.

31.4 Conclusions

From the studies performed on optical/thermal properties of gold-coated Fe_3O_4 nanoparticles, it is concluded that: the gold layer on iron oxide nanoparticles enhanced the NIR absorption at 788 nm; gold-coated Fe_3O_4 particles showed the same heating performance as non-coated ones in AEM field, indicating that the gold layer has almost no negative effect on the heating performance of the nanoparticles; Cypate linked NGPs *via* streptavidin spacer (~3 nm) have shown twice higher fluorescence contrast than ICG alone in our experimental breast model system. The binding affinity of LHRH linked NGPs was found to be similar to the LHRH alone, indicating that LHRH retains its binding affinity after being bound to NGPs.

Future studies include linking LHRH and Cypate on gold-coated Fe_3O_4 nanoparticles as a prototype of a multi-functional nano-entity and then testing the optical/thermal properties and binding affinity of conjugated LHRH.

Aknowledgment The authors thank the U.S. Army Medical Research and Materiel Command (DAMD17-03-1-0572) for partial financial support. The authors also thank Dr. Shi's group at the University of Cincinnati, OH for supplying the gold-coated nanoparticles, and Dr. Achilefu's group at Washington University, MO for supplying Cypate.

References

1. F. Kristian Storm, *Hyperthermia in Cancer Therapy*, (G. K. Hall Medical Publishers, Boston, MA, 1983).
2. S. Sharma, S.P. Sandhu, F. D. Patel, S. Ghoshal, B. D. Gupta, and N. S. Yadav, Cervix cancer and hyperthermia: Side-effects of local hyperthermia: results of a prospectively randomized clinical study, *Int. J. Hyperthermia*, 6 (2), 279–285, (1990).
3. T. Ohtsubo, H. Igawa, T. Saito, H. matsumoto, H. Park, C. W. Song, E. Kano, and H. Saito, Enhancement of cell killing by induction of apoptosis after treatment with mild hyperthermia at 42°C and cisplatin, *Radiation Research*, 156, 103–109, (2001).
4. A. Jordan, R. Scholz, P. Wust, H. Fahling, and R. Felix, Magnetic fluid hyperthermia (MFH): cancer treatment with AC magnetic field induced excitation of biocompatible superparamagnetic nanoparticles *J. Magn. Magn. Mater.* 201, 413–419, (1999).
5. P. Tartaj, M.P. Morales, S. Veintemillas-Verdaguer, T. Gonz'alez-Carreno, and C.J. Serna, The preparation of magnetic nanoparticles for applications in biomedicine, *J. Phys. D: Appl. Phys.* 36, R182–197, (2003).
6. D. Bahadur, and J. Giri, Biomaterials and magnetism, *Sadhana*, 28 (3 and 4), 639–656, (2003).
7. S. Mornet, S. Vasseur, F. Grasset, and E. Duguet, Magnetic nanoparticle design for medical diagnosis and therapy, *J. Mater. Chem.*, 14, 2161–2175, (2004).
8. S. J. Oldenburg, J. B. Jackson, S. L. Westcott, and N. J. Halas, Infrared extinction properties of gold nanoshells, *Applied Physics Letters*, 78 (19), 2897–2899, (1999).
9. C. H. Chou, C. D. Chen, and C. R. Wang, Highly Efficient, Wavelength-Tunable, Gold Nanoparticle Based photothermal Nanoconvertors, *J. Phys. Chem. B*, 109, 11135–11138, (2005).
10. S. Achilefu, R. B. Dorshow, J. E. Bugaj, and R. Rajagopalan, Novel receptor-targeted fluorescent contrast agents for in vivo tumor imaging. *Invest Radiol*, 35:479–485, (2000).
11. C. D. Geddes, A. Parfenov, D. Roll, M. J. Uddin, and J. R. Lakowicz, Fluorescence spectral properties of indocyanine green on a roughened platinum electrode: Metal-enhanced fluorescence, *Journal of Fluorescence*, 13 (6), 453–457, (2003).
12. B. Hong and K.A. Kang, Biocompatible, nanogold-particle fluorescence enhancer for fluorophore mediated, optical immunosensor, *Biosensors and Bioelectronics*, 21(7), 1333–1338, (2006).
13. R. Eckert, D. Randall, and G. Augustin, *Animal Physiology*, 3rd Edition, W. H. Freeman and Company, New York, 435–473, (1988).
14. S. S. Kakar, L. C. Musgrove, D. C. Devor, J. C. Sellers, and J. D. Neill, Cloning, sequencing, and expression of human gonadotropin releasing hormone (GnRH) receptor. *Biochem. Biophys. Res. Commun.* 189, 289–295, (1992).
15. M. Preuss, W.G. Schmidt, and F. Bechstedt, Coulombic amino group-metal bonding: Adsorption of adenine on Cu (110), *Physical Review Letters*, 94, 236102–4, (2005).

16. S. S. Kakar, W. E. Grizzle, and J. D. Neill, The nucleotide sequences of human GnRH receptors in breast and ovarian tumors are identical with that found in pituitary, *Mol. Cell. Endocrinol.* 106, 145–149, (1994).
17. S. S. Kakar, M. T. Malik, S. J. Winters, and W. Mazhawidza, Gonadotropin-releasing hormone receptors: structure, expression, and signaling transduction. *Vitam Horm.* 69, 151–207, (2004).
18. H. Jin and K. A. Kang, Fluorescent mediated detection of Heterogeneity in a highly scattering media, *Adv. Exper. Med. Bio.*, 566, 167–172, (2005).
19. T. L. Troy, B. W. Pogue, E. D. Genety, S. B. Poplack, O. L. Osterburg, and K. D. Paulsen, Spectroscopic diffuse optical tomography for the quantitative assessments of hemoglobin concentration and oxygen saturation in human breast tissue, *Appl. Opt.,* 38(25), 5480–5490, (1999).
20. A.L. Honar and K.A. Kang, Effect of the source and detector configuration on the detectability of breast cancer, *Comp. Biochem.Physio - Part A: Molecular & Integrative Physiology,* 132(1), 9–15, (2002).
21. D. F. Bruley, Pulse reduction code written for process identification (personal communication), (1974).
22. S. S. Kakar, S. J., Winters, W. Zacharias, D. M. Miller, and S. Flynn, Identification of distinct gene expression profiles associated with treatment of LbetaT2 cells with gonadotropin-releasing hormone agonist using microarray analysis. *Gene.* 308, 67–77, (2003).
23. H. Jin and K. A. Kang, Application of Novel Metal Nanoparticles as Optical/Thermal Agents in Optical Mammography and Hyperthermic Treatment for Breast Cancer, Proceedings of the 33rd ISOTT Annual Meeting, August 28-September 2, Brisbane, Australia, Manuscript Submitted, (2005).

Chapter 32
LHRH Receptor Targeted Therapy for Breast Cancer

S.S. Kakar[1], H. Jin[2], B. Hong[2], J.W. Eaton[1], and Kyung A. Kang [2]

Abstract Breast cancer remains the most common cancer among women, with an estimated 212,920 new cases and 40,970 deaths in the United States in 2006. The present work extends the studies of nanoparticles targeted to the luteinizing hormone-releasing hormone (LHRH) receptor which is overexpressed in breast, ovarian, endometrial and prostate cancer cells. In contrast, LHRH receptors are not expressed, or expressed at a low level in most visceral organs. In our studies, we conjugated Fe_3O_4 nanoparticles (20–30 nm) with [D-Trp^6]LHRH (Triptorelin), a decapeptide analog of LHRH currently used for treatment of sex-hormone-dependent tumors. Conjugation of [D-Trp^6]LHRH to Fe_3O_4 particles retained its binding affinity and biological activity for the LHRH receptor. Treatment of two separate breast tumor cell lines (MCF-7 and MDA-MB231) with these conjugated nanoparticles resulted in 95–98% cell death and loss of viability within 24 h whereas no change in cell proliferation or cell apoptosis was observed in cells treated with equal amounts of either [D-Trp^6]LHRH or unconjugated Fe_3O_4 nanoparticles. These studies provide critical and important information regarding use of LHRH receptor targeted therapy for breast cancer.

32.1 Introduction

Cancer is a complex disease that affects millions of people worldwide. Currently, one in four deaths in the United States is due to cancer. This disease affects diverse tissues and organs including colon, prostate, lung and breast. Breast cancer is the most common cancer among women with an estimated 212,920 new cases and 40,970 deaths in 2006 [1]. Although the death rate from breast tumors has been reduced by the introduction of breast

[1]Department of Medicine and James Graham Brown Cancer Center, University of Louisville, Louisville. KY 40202.
[2]Department of Chemical Engineering, J.B Speed School of Engineering, University of Louisville, Louisville. KY 40202.

K.A. Kang et al. (eds.), *Oxygen Transport to Tissue XXIX*

screening mammography and adjuvant therapies, more efficacious treatment modalities are needed. The primary treatment modality for breast cancer is cytoreductive surgery followed by adjuvant chemotherapy, radiotherapy, or both [2]. This strategy is successful in the majority of patients, however it is always accompanied by cytotoxicity to normal organs and tissues. Successful chemotherapy is also hindered by intrinsic or acquired resistance of breast cancer cells.

An alternative to standard chemotherapies involves more direct targeting of cancer cells using agents specifically directed to binding sites on cancer cells. Several different targets have been explored, including carbohydrates, lectins, receptor ligands and antibodies. Usually, ligands which recognize these targets are coupled with low molecular weight anti-neoplastic drugs. However, a new class of potential anticancer agents has appeared in recent years – nanoparticles – which hold promise for improved cancer detection and treatment. Some of these applications include iron-based nanoparticles which may permit magnetic drug targeting, hyperthermia, magnetic field-assisted radio nucleotide therapy and magnetic resonance imaging (MRI) contrast enhancement. To meet application requirements, nanoparticles are generally coated with various functional surface layers to increase residence time in circulation but there has been little work aimed at the strategies for preferential delivery of these nanoparticles to tumors.

In the present work, we exploited LHRH receptors for targeting of nanoparticles because these receptors are overexpressed in a variety of tumors including breast, ovarian, endometrial, prostate, and melanoma, and not expressed in a detectable level in most visceral organs. We are targeting this receptor using [D-Trp^6]LHRH (Triptorelin), a decapeptide analog of LHRH. In addition, LHRH is inexpensive compared to the most frequent used ligands and humanized monoclonal antibodies.

Luteinizing hormone-releasing hormone (LHRH), also known as gonadotropin-releasing hormone (GnRH), is a hypothalamic decapeptide (p-Glu-His-Trp-Ser-Tyr-Gly-Leu-Arg-Pro-Gly-NH2). Based on studies from our laboratory and others, it became clear that high affinity LHRH receptors are overexpressed in most of the tumors analyzed to date and treatment of tumor cells with LHRH agonists and antagonists results in reduction in tumor cell growth and proliferation in vitro (see ref [3] for review). However, such antiproliferative affects of LHRH analogs are moderate (15–20% inhibition after three to four days of treatment), making LHRH analogs not particularly useful for the treatment of breast cancer. Our investigations are based on the idea that we might be able to employ nanospheres displaying analogs of LHRH to effect selective delivery of these nanospheres to breast tumors, thereby sparing normal, non-cancerous cells from unnecessary exposure. In support of this general concept, Schally and his colleagues developed cytotoxic analogs of LHRH containing doxorubicin or derivatives of doxorubicin and showed inhibition of proliferation of various tumor cell lines and tumor growth in nude mice [4].

Recently, Dharap et al. [5] and Zhou et al. [6] have investigated breast tumor specific targeting using LHRH peptide to deliver anti-cancer drugs for cancer treatment or magnetic nanoparticles for enhancing magnetic resonance imaging (MRI) contrast [7]. Their results showed that the use of LHRH peptide as targeting moiety substantially enhances the uptake of the anti-cancer drugs or magnetite nanoparticles (20 nm) in tumors, with much less accumulation in liver or kidney compared to magnetic particles alone. In their studies, magnetite nanoparticles were conjugated with LHRH and intravenously injected to female mice bearing carcinogen initiated tumors. After 20 hr, the distribution of the nanoparticles was examined in tumor, liver and kidney. No studies were performed to analyze the effects on tumor growth or regression by these investigator [6]. Most of the un-conjugated particles (55.5%) were found in the liver, while those conjugated with LHRH were found primarily in the tumors (59.1%) and in pulmonary metastases of the tumors (20.3%), confirming the high specificity of targeting achieved by LHRH derivitization.

32.2 Materials and Methods

32.2.1 Cross Linking of an LHRH Analog to Fe_3O_4 Particles

The clinically used LHRH analog [D-Trp^6]LHRH (Triptorelin) has an amino acid sequence of p-Glu-His-Trp-Ser-Tyr-D-Trp-Leu-Arg-Gly-NH2. It can be coupled to the surface of appropriately prepared magnetite nanoparticles via its N-terminal amine group. To conjugate [D-Trp^6]LHRH (MW = 1311.45) to nanoparticles, 10 mg of magnetite particles (20-30 nm from Alfa Aesar, Ward Hill, MA) were treated with 3 ml of 28% NH_4OH solution in a glass bottle coated with silicone for 2 h. The hydroxylated particles were then centrifuged and washed three times with H_2O followed by three washes with methanol. The particles were dried at 65°C for 1 h. Two ml of toluene was added to the particles and sonicated for 1–2 min and sparged with N_2 for 5 min. Ten μl of (3-aminopropyl)trimethoxysilane (APTS) was added to the mixture and incubated at 40°C for 2 h. The mixture was centrifuged and particles were washed three times with toluene. Particles were resuspended in 10 ml H_2O. To this, 15.6 mg of dicyclohexylcarbodiimide solubilized in 50 μl of ethanol was added. To this mixture, 1 mg of [D-Trp^6]LHRH solubilized in 3 ml of PBS was added, mixed and incubated for 2 h. The mixture was centrifuged to collect the [D-Trp^6]LHRH conjugated particles and washed [6]. Unconjugated [D-Trp^6]LHRH was measured in the supernatant. The conjugated particles were resuspended in 1% BSA/PBS buffer solution to a final volume of 5.0 ml (Fig. 32.1). The amount of

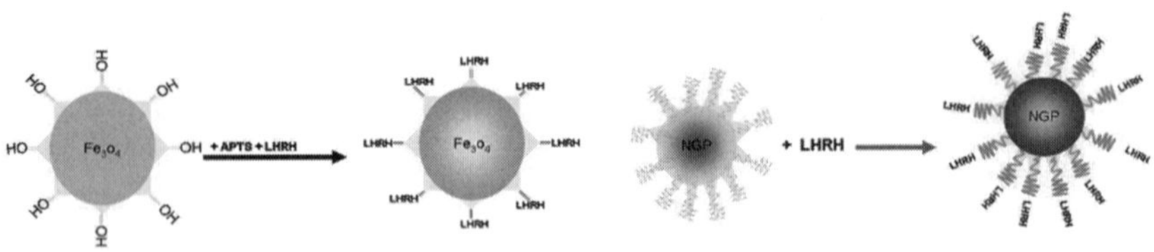

Fig. 32.1 Scheme of the reaction between hydroxylated Fe_3O_4 particles (A) or gold (NGP) coated Fe_3O_4 particles (B) and [D-Trp^6] LHRH.

[D-Trp^6]LHRH conjugated to particles and amount of particles was determined as described by Zhou et al [6].

32.2.2 Cross Linking of an LHRH analog to Gold Coated Fe_3O_4 Particles

For control experiments, gold nanoparticles (NGP) were conjugated with [D-Trp^6]LHRH as described by Aslam et al.[8] Briefly, nanogold colloids coated with surfactant tannic acid was adjusted to pH 9.0 using 0.1 M sodium carbonate solution in the silicone coated glass bottle. One mg of [D-Trp^6]LHRH was dissolved in 1 ml of H_2O and immediately transferred to gold colloid, and the mixture was intensively vibrated at 4°C for one h. Tannic acid on the NGP surface was then replaced with [D-Trp^6]LHRH and [D-Trp^6]LHRH linked NGP (LHRH-NGP) were formed (Fig. 32.1). LHRH-NGP were collected by centrifugation at 10,000 RPM for 10 min. Particles were washed three times with H_2O and resuspended in 1% BSA PBS buffered solution to a final volume of 5.0 ml. The amount of [D-Trp^6]LHRH linked to NGP was calculated as described above.

32.2.3 Cell Culture

Breast tumor cell lines (MCF-7 and MDA-MB231) were obtained from American Type Tissue Culture (ATCC) (Rockville, MD). Mouse pituitary gonadotrope cell line LβT2 was obtained from Dr. Pamela Mellon, (Univeristy of California, SanDiego, CA). The cell lines were cultured according to the suppliers' recommendations in 10% fetal calf serum (FCS) at 37°C in an atmosphere of 5% CO_2 in humidified air. The cell lines were subcultured on routine basis every 3–4 days.

32.2.4 Determination of Binding Affinity of LHRH Conjugated Nanoparticles

The binding affinity of LHRH-NP to LHRH receptor and activation of LHRH receptor was determined as described previously [9,10]. For this purpose, we used a mouse pituitary gonadotrope tumor cell line (LβT2) that expresses high levels of high affinity LHRH receptors [10,11]. The cells were transfected with CRE-Luc reporter construct as described previously [10]. The reporter construct (CRE-luciferase) is used for the measurement of cAMP response element (CRE) activation. It contains four copies of a CRE enhancer element fused to luciferase cDNA (pCRE-Luc plasmid, Stratagene, La Jolla, CA). This reporter construct is commonly used to determine the activation of G-protein coupled receptors such as LHRH receptor in response to ligand binding resulting in change in intracellular cAMP levels leading to activation of lucifierase activity [12]. After overnight transfection of cells, the medium was replaced with serum free medium for 4 h. Cells were treated with various concentrations of [D-Trp^6]LHRH, LHRH-NP or NP. After 6 h of treatment, the cells were lysed and luciferase activity was assayed. IC_{50} value was calculated as described previously [10].

32.2.5 Treatment of Breast Tumor Cells with LHRH-NP and Determination of Cell Death

Breast tumor cells (MCF-7 and MDA-MB231) were plated in 6-well plates. After 24 h, the FCS-containing medium was replaced with serum-free medium. After 2 h, the cells were treated with various concentrations of LHRH analog [D-Trp^6]LHRH, LHRH-NP or NP for 15 min. The cells were rinsed twice with PBS and 1.0 ml of medium containing 1% FCS was added to each well. The cells were incubated at 37°C for overnight and examined under an Olympus microscope for cell survival and morphological changes. To quantitate the cell number, we fixed the cells with 10% formaldehyde and then stained the cells with 1% crystal violet. Following thorough rinsing in PBS, the cells were examined microscopically.

32.2.6 Cell Viability Assay

To determine the effect of LHRH-NP on breast tumor cell proliferation and cell survival, we performed cell viability assays. We plated breast tumor cells (MCF-7 and MDA-MB231) that express LHRH receptors [13,14] in 96-well opaque plates. After 24 h, the medium was replaced with serum-free medium. Cells were incubated at 37°C for 2 h and then treated with various concentrations of free [D-Trp^6]LHRH, LHRH-NP or nanoparticles (NP) for 15 min. The amount of LHRH bound to the nanoparticles was adjusted to a concentration equivalent to free [D-Trp^6]LHRH and final concentration of nanoparticles was 0.002–0.02 mg/ml. After 15 min of treatment, the medium was aspirated and cells were rinsed twice with PBS. One hundred μl of medium containing 1% FBS was added to each well and the cells were incubated at 37°C in the incubator. After 48 h of treatment, cells were equilibrated to room temperature for 30 min, 100 μl of CellTiter-Glo reagent from the kit (Promega, Madison, WI) was added to cells. The cells were incubated for 10 min and luminescence was recorded by Luminometer.

32.2.7 Cell Membrane Integrity Assay

To determine the cytotoxic effect of LHRH-NP on breast tumor cells, we used CytoTox-One Membrane Integrity Assay kit from Promega. This assay is based on the release of lactate dehydrogenase (LDH) from cells with damaged membranes. Cells were plated in 96-well opaque plates and treated with [D-Trp^6]LHRH, LHRH-NP or NP as described above. To measure LDH release in the medium, after 48 h of treatment of cells, 100 μl of CytoTox-One reagent from the kit was added to cells containing 100 μl of growth medium and incubated at room temperature for 10 min. To each sample 50 μl of stop solution was added and mixed. Fluorescence was measured using spectrofluorometer (Molecular Devices) with an excitation wavelength of 560 nm and an emission wavelength of 590 nm.

32.2.8 Determination of Release of Iron from Fe_3O_4 Particles at Low pH

Finally, we investigated the possibility that nanoparticle-induced cytotoxicity is mediated by iron released from the magnetite particles in the acidic and cysteine-rich interior of the lysosomal compartment. To test the hypothesis that LHRH conjugated nanoparticles release iron in lysosomes, we incubated 1 mg of LHRH conjugated or non-conjugated magnetite beads suspended in 500 μl of 20 mM Tris and 2 mM cysteine. In one case the pH was adjusted to 7.4 whilst in the other the pH was lowered to 4.5 to mimic intralysosomal pH. The supernatant was sampled at 1, 4 and 24 hours and 'free' iron was determined using the ferene S reaction [15].

32.3 Results and Discussion

32.3.1 LHRH Conjugated Nanoparticles Bind to LHRH Receptors with High Affinity and Activate the Production of cAMP

Fe_3O_4 or gold nanoparticles conjugated with the LHRH analog [D-Trp6]LHRH showed high binding affinity. To determine the binding affinity and activation of LHRH receptor by LHRH conjugated nanoparticles, we transfected LβT2 cells with pCRE plasmid followed by treatment of cells with [D-Trp^6]LHRH, LHRH-NP or NP. Treatment of cells with [D-Trp^6]LHRH or LHRH-NP showed activation of LHRH receptor resulting in an increase in production of intracellular cAMP which was measured by activation of CRE to drive luciferase. Using this latter as a surrogate marker for LHRH receptor occupancy, the binding affinity (IC_{50} value) for [D-Trp^6]LHRH conjugated magnetite particles was found to be similar to free [D-Trp^6]LHRH (0.1 nM), suggesting that the LHRH analog bound to magnetite nanoparticles retained its normal affinity for the receptor and its biological activity. Similar results were obtained when gold coated particles (LHRH-NGP) were used for conjugation, suggesting that conjugation of [D-Trp^6] to Fe_3O_4 or gold coated Fe_3O_4 did not lose its binding affinity for the LHRH receptor or its biological activity to induce intracellular signaling. Binding affinity of [D-Trp^6]LHRH or conjugated particles was similar to our earlier studies [9,10].

32.3.2 Effect of LHRH-NP on Breast Tumor Cell Survival

To determine the effect of LHRH-NP particles on breast tumor growth and survival, we treated the MCF-7 and MDA-MB231 cells with various concentration of free [D-Trp^6]LHRH, LHRH-NP or NP as described in materials and methods. Cells after treatment were examined microscopically. As shown in Fig. 32.2, treatment of both MCF-7 cells with LHRH-nanoparticles caused substantial dose-dependent cell death compared to cells treated with vehicle, free [D-Trp^6]LHRH or uncoated nanoparticles. Maximum cell death (95–98%) was observed with conjugated LHRH-nanoparticle concentration of 0.5 μM to 1 μM (particle concentration = 0.01–0.02 mg/ml). Similar results were obtained when MDA-MB231 cells were used. Staining of cells with crystal violet showed similar results confirming that LHRH conjugated nanoparticles induce cell death in breast tumor cells (Figs. 32.3 and 32.4). In contrast HEK293 cells that do not express LHRH receptor [16] showed no change in cell proliferation or cell survival when treated with free [D-Trp^6]LHRH, LHRH-NP or NP. Therefore, cell death induced by LHRH-NP is specific and is achieved through LHRH receptors present on breast tumor cells.

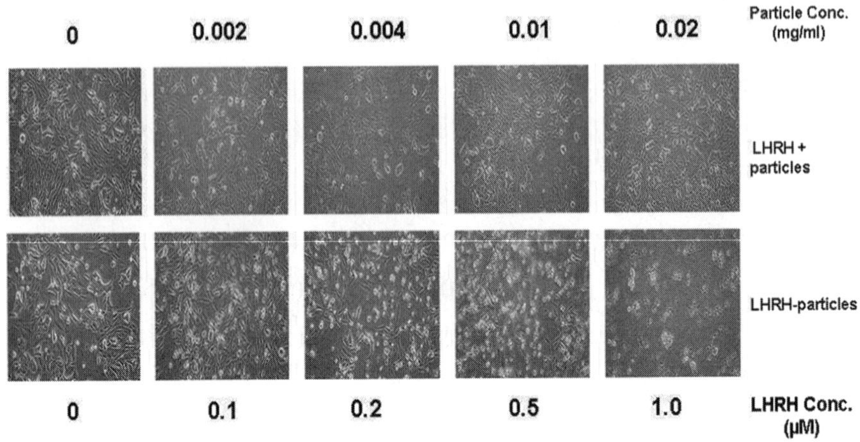

Fig. 32.2 Treatment of breast tumor cells (MCF7) with Fe_3O_4-LHRH nanoparticles. MCF7 cells were treated with [D-rp^6]LHRH, LHRH-NP, NP. After 24 h of treatment cells were examined under a microscope to examine death and morphological changes. Nanoparticles concentration (mg/ml) and corresponding LHRH concentration are indicated.

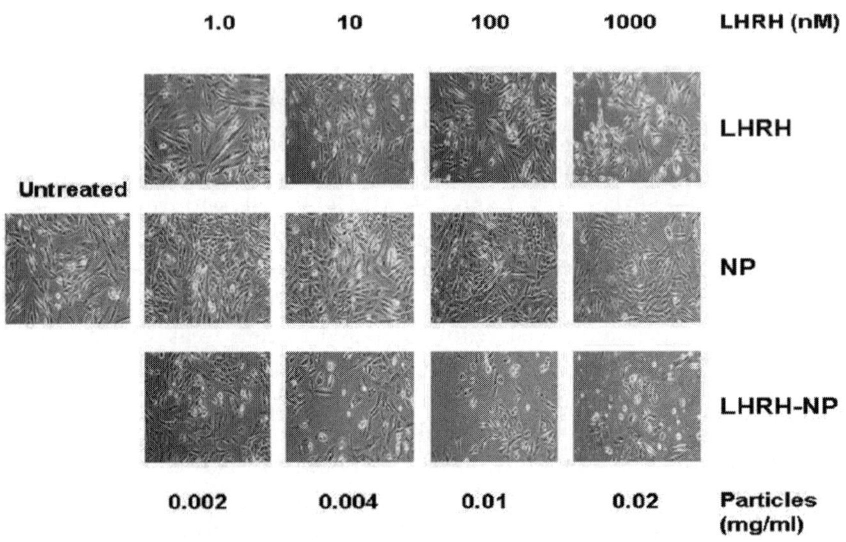

Fig. 32.3 Effect of treatment of MCF7 cells with LHRH-NP on cell survival. MCF7 cells were treated with [D-Trp^6]LHRH, LHRH-NP or NP. After 24 h of treatment cells were stained with crystal violet and examined under a microscope and photographed. Nanoparticles concentration (mg/ml) and corresponding LHRH concentration are indicated.

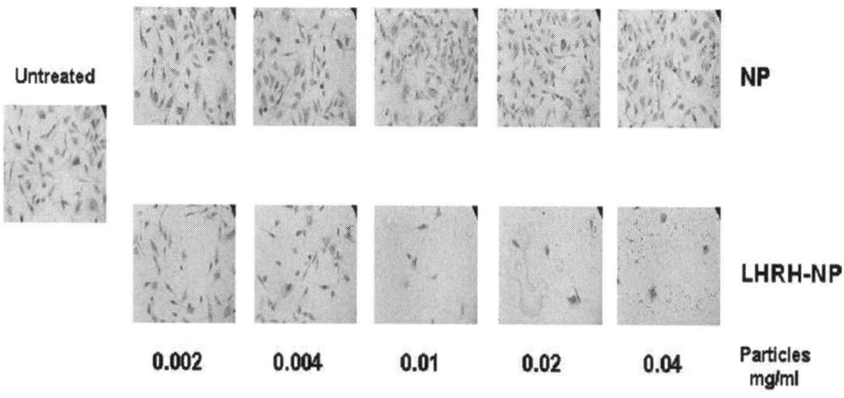

Fig. 32.4 Effect of treatment of MDA-MB231 cells with LHRH-NP on cell survival. MDA-MB231 cells were treated with [D-Trp^6]LHRH, LHRH-NP or NP. After 24 h of treatment of cell were stained with crystal violet and examined under a microscope and photographed. Nanoparticles concentration and corresponding LHRH concentration are indicated.

32.3.3 Effect of LHRH-NP on Breast Tumor Cell Proliferation

To quantitate the effect of LHRH-NP on breast tumor cell proliferation, we performed cell viability assays. This method is based on quantitation of intracellular ATP which reflects the metabolic activity of control and treated cells. Using this surrogate measure of cell viability, we found that both MCF-7 and MDA-MB231 cells are efficiently killed (95–98%) when treated with LHRH-NP at a final concentration of [D-Trp^6]LHRH of 0.5 μM or 1 μM. In contrast very few cells were killed when treated with free [D-Trp^6]LHRH or NP alone. No change in cell proliferation or cell survival was observed when gold coated LHRH conjugated Fe_3O_4 particles were used for treatment, suggesting that loss of cell viability (cell death) achieved by LHRH-conjugated Fe_3O_4 particles is due to cellular cytotoxicity induced by Fe_3O_4.

32.3.4 Effect of LHRH-NP on Breast Tumor Cellular Toxicity

Binding of LHRH or its analog results in desensitization of LHRH receptors and internalization of the ligand-receptor complex (endocytosis) (see ref 3 for review). In lysosomes, LHRH is released from the receptor and most of the LHRH receptors undergo proteolytic degradation (although some receptors may be recycled to the plasma membrane. Once within the lysosomal compartment, Fe_3O_4 might be released, thereby, inducing cellular cytotoxicity. To determine the extent of cytotoxicity caused by LHRH-NP, we performed cytotoxicity assays using the CytoTox-One system from Promega. This method is based on the measurement of lactate dehydrogenase (LDH) release from cells which reflects loss of plama membrane integrity. After treatment of cells with LHRH-NP for 48 h, cytotoxic assay was performed as described in materials and methods. There was a dose-dependent cell death as indicated by LDH release. Such effects of LHRH-NP were observed to be dose dependent. No cytotoxic effects were observed when cells were treated with free [D-Trp^6]LHRH or NP alone.

32.3.5 LHRH-Conjugated Fe_3O_4 Nanoparticles Release Iron at Low pH

Finally, we are entertaining the possibility that nanoparticle-induced cytotoxicity is mediated by iron released from the magnetite particles in the acidic and cysteine-rich interior of the lysosomal compartment. In brief, we earlier found that in normal cells almost all redox active iron is located within lysosomes [17,18] and that the lysosomal iron derives from the continuous 'autophagocytosis' of ferruginous intracellular material such as mitochondria and ferritin. Furthermore, this iron pool seems most important in mediating oxidant-induced cell death (which can involve reactive oxygen species produced by normal mitochondrial metabolism). Most importantly, iron- and oxidant-mediated cell death involves lysosomal rupture as the first event, followed by activation of the classical apoptotic cascade [19]. To indirectly test the hypothesis that LHRH conjugated nanoparticles might release iron within the lysosomes, we incubated 1 mg of LHRH conjugated or non-conjugated magnetite beads suspended in 500 μl of 20 mM Tris and 2 mM cysteine. In one case (A) the pH was adjusted to 7.4 whilst in (B) the pH was lowered to 4.5 to mimic intralysosomal pH. The supernatant was sampled at 1, 4 and 24 hours and 'free' iron was determined using the ferene S reaction [15]. Note the absence of iron release at neutral pH (or, at least, the failure of 'loose' iron to appear in the supernatant). In contrast, Fe_3O_4 nanoparticles incubated at low pH showed substantial free iron in the supernatant (Table 32.1),

Table 32.1 Release of free iron from LHRH-NP or NP

	SUPERNATANT Fe CONTENT (nmol)		
SAMPLES	1 h	4 h	24 h
Nano particles @ pH 7.4	0*	0*	0*
LHRH-nanospheres @ 7.4	0*	0*	0*
Nano particles @ pH 4.5	3.06	5.46	8.66
LHRH-nanospheres @ pH 4.5	1.51	2.92	5.56

* Below limits of detection

suggesting that the release of free iron from LHRH conjugated and unconjugated Fe_3O_4 particles at acidic pH which may promote cell death.

32.4 Conclusions

Our results clearly define that conjugation of the LHRH analog [D-Trp^6]LHRH to Fe_3O_4 nanoparticles retains normal LHRH receptor binding affinity and biological activity. Our studies also demonstrate that treatment of breast tumor cells (MCF-7 and MDA-MB 231) with LHRH-NP specifically induces cytotoxicity leading to cell death and loss of cell viability. Incubation of LHRH-conjugated and unconjugated particles release free iron at pH 4.5 (intralysosomal pH). The effect of LHRH conjugated Fe_3O_4 nanoparticles on tumor growth and metastasis in vivo remains to be tested.

Acknowledgment This work was supported in part by a grant from the U.S. Army Medical Research and Material Command (W81XWH-06-1-0662), and funds from James Graham Brown Cancer Center, University of Louisville. The authors are thankful to Dr. Mohammad T. Malik and Ms. Alison L. Burton for their technical help.

References

1. A. Jemal, R. Siegel, E. Ward et al. Cancer statistics, 2006. CA: a Cancer J Clin 2006; **56**(2): 106–130
2. A.M. Bajo, A.V. Schally, G, Halmos et al. Targeted doxorubicin containing luteinizing hormone-releasing hormone analogue AN-152 inhibits the growth of doxorubicin-resistant MX-1 human breast cancers. Clin Cancer Res 2003; **9**: 3742–3748
3. S.S. Kakar, M.T. Malik, S.J. Winters et al. Gonadotropin-releasing hormone receptors: structure, expression, and signaling transduction. Vit Horm 2004; **69**: 151–207
4. A. Nagy, A.V. Schally. Targeting of cytotoxic luteinizing hormone-releasing hormone analogs to breast, ovarian, endometrial, and prostate cancers. Biol Reprod 2005; **73**(5): 851–859
5. S.S. Dharap, Y. Wang, P. Chandna et al. Tumor-specific targeting of an anticancer drug delivery system by LHRH peptide. Proc Nat Acad Sci USA 2005; **102**(36): 12962–12967
6. J. Zhou, C. Leuschner, C. Kumar et al. Sub-cellular accumulation of magnetic nanoparticles in breast tumors and metastases. Biomaterials 2006; **27**(9): 2001–2008
7. C. Leuschner, C. Kumar, M. Urbina et al. The use of ligand conjugated supermagnetic iron oxide nanoparticles (SPION) for early detection of metastasis. In INST Naotech Tech Proc 1005; **1**: 5–6
8. M. Islam, L. Fu, M. Su et al. Novel one-step synthesis of amine-stabilized aqueous colloidal gold nanoparticles. J Material Chem 2004; **14;** 1795–1797
9. S.S. Kakar, L.C. Musgrove, D.C. Dever et al. Cloning, sequencing, and expression of human gonadotropin-releasing hormone (GnRH) receptor, Biochem Biophys Res Commun 1992; **189**: 289–295
10. 10. S.S. Kakar, S.J. Winters, W. Zacharias et al. Identification of distinct gene expression profiles associated with treatment of LbetaT2 cells with gonadotropin-releasing hormone agonist using microarray analysis. Gene 2003; **308**: 67–77

11. P. Thomas, P.L. Mellon, J. Turgeon et al. The L beta T2 clonal gonadotrope: a model for single cell studies of endocrine cell secretion. Endocrinology 1996; **137**: 2979–2989
12. G.Y. Bedecarrats, K.D. Linher, J.A. Janovick et al. Four naturally occurring mutations in the human GnRH receptor affect ligand binding and receptor function. Mol Cell Endocrin 2003; **205**(1–2), 51–64
13. A. Mangia, S. Tommasi, S.J. Reshkin et al. Gonadotropin releasing hormone receptor expression in primary breast cancer: comparison of imunohistochemical, radioligand and Western blot analyses, Oncol Rep 2002; **9**(5): 1127–1132
14. T. Moriya, T. Suzuki, M. Pilichowska et al. Immunohistochemical expression of gonadotropin releasing hormone receptor in human breast carcinoma. Pathol Int 2001; **51**(5): 333–337
15. M. W. Qian, J.W. Eaton. Iron translocation by free fatty acids, Am J Path 1991; **139**(6): 1425–1434
16. S.S. Kakar. Inhibition of growth and proliferation of EcRG293 cell line expressing high-affinity Gonadotropin-releasing hormone (GnRH) receptor under the control of an inducible promoter by GnRH agonist (D-Lys6)GnRH and antagonist (Antide). Cancer Res 1998; **58**(20): 4558–4560
17. Z. Yu, H.L. Persson, J.W. Eaton et al. Intralysosomal iron: a major determinant of oxidant-induced cell death. Free Rad Biol Med 2003; **34**(10): 1243–1252
18. H.L. Persson, Z. Yu, O. Tirosh et al. Prevention of oxidant-induced cell death by lysosomotropic iron chelators. Free Rad Biol Med 2004; **34**(10): 1295–1305
19. U.T. Brunk, J. Neuzil, J.W. Eaton. Lysosomal involvement in apoptosis, Redox Rep 2001; **6**(2): 91–97

Part VIII
Translational and Clinical Studies

Chapter 33
Saturation of Hemoglobin in Intracranial Arteries is Similar in Patients with Hemodynamically Relevant and Irrelevant Stenosis of the Internal Carotid Artery

U. Jensen[1], S. Wolff[2], K. Alfke[2], K. Börsch[1], O. Jansen[2], and R. Stingele[1]

Abstract The aim of this study was to establish if patients with hemodynamically relevant or irrelevant stenoses of the extracranial internal carotid artery have different intracranial arterial oxygen saturation as measured by transcranial pulse oximetry using near infrared spectroscopy.

Patients with unilateral stenosis $\geq$70% according to North American Symptomatic Carotid Endarterectomy Trial (NASCET) were included. Hemodynamic relevance was assessed using ultrasound criteria. Transcranial spectroscopy recordings were taken before and after surgical or interventional treatment of the stenosis. Optodes were placed bilaterally on the intact fronto-parietal aspect of the skull. Oxygen saturation and diversion angle alpha from the hemoglobin plane were measured.

There were no significant differences regarding arterial oxygen saturation between the two groups. Oxygen saturation ranged from 0.910 ± 0.08 to 0.957 ± 0.028 in the subgroups (all values as mean $\pm$ S.E.). These values are consistent with previous studies and theoretical values. In smokers we found a significantly shifted diversion angle from the hemoglobin plane to the negative side. This indicates the presence of an absorber other than oxy- and desoxyhemoglobin in the optical field.

We conclude that transcranial pulse oximetry cannot distinguish between patients with hemodynamically relevant and irrelevant stenosis of the internal carotid artery. However it seems to be capable of distinguishing smokers from non-smokers.

33.1 Introduction

Transcranial pulse oximetry (TCO) using near infrared spectroscopy (NIRS) is a method used to detect desaturation of hemoglobin in the arterial vessels of the brain. It was demonstrated that it detects failure of collateral blood flow in

[1]Department of Neurology, University of Kiel, Schittenhelmstr. 10, 24105 Kiel, Germany.
[2]Section of Neuroradiology, University of Kiel, Schittenhelmstr. 10,24105 Kiel, Germany.
Corresponding author: U. Jensen, e-mail address: Ulfrjensen@web.de, Tel.: + 49-431-5978550; Fax: + 49-431-5978502

patients with impaired blood supply [1]. We tested the possiblity to assess the hemodynamical relevance of stenosis of the internal carotid artery (ICA). Additionally we investigated the influence of smoking on the near infrared pulse oximetry.

33.2 Material and Methods

33.2.1 Near Infrared Spectroscopy

With each cycle of the heart there is a pulsatile change of blood volume in the intracranial arteries. Venous vessels and capillaries do not change their volume. This leads to a pulsatile change in infrared absorption since hemoglobin is a strictly intravascular near infrared chromophore [2]. The pulsatile changes of blood volume lead to pulsatile changes of NIR-absorbance that were measured at 3 wavelengths with 10 Hz acquisition frequency. A plot of the absorption values against each other is given in Fig. 33.1. Each datapoint represents a triple of absorption values at one point in time. In such a three wavelength-plot, absorption changes in time along a straight line. The straight lines in Fig. 33.1 represent the theoretical absorption values with saturation 1 and 0, respectively. These theoretical lines span a plane in the 3-D plot (termed hemoglobin plane) that contains all absorption values that can be explained by any arterial oxygen

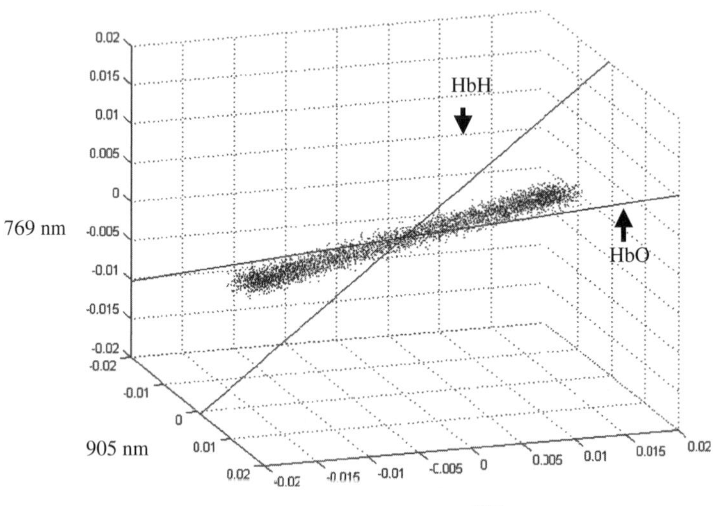

Fig. 33.1 Oxygen saturation of 0.89 recovered from a recording of a patient. Note the two vectors for desoxy- (HbH) and oxyhemoglobin (HbO) spanning the hemoglobin plane.

saturation. The line resulting from the linear fit of the measured absorption values is analyzed in two ways: First, the projection of the line onto the hemoglobin plane is used to calculate the arterial oxygen saturation, as described before (1). Second, the angle formed by the measured line with the hemoglobin plane (alpha) is calculated to quantify how well the observed absorption values are explained by oxy- and deoxy-hemoglobin in the optical field. If pulsatile absorption changes are due to these absorbers only the angle alpha is 0. An angle alpha $\neq 0^\circ$ means that other absorbers than oxy- or desoxyhemoglobin are present intraarterially.

As a first step the received scattered clouds of datapoints had to be reduced to the main vector. This was done by principal component analysis [3]. The second step required a coordinate system transformation. Graphically it included stretching and rescaling of the hemoglobin plane [4]. The last step was a recovery of the oxygen saturation by trigonometric calculation using Eq. (33.1) in which β represents the angle formed by the main resulting vector and the new x-axis of the coordinate system.

$$\text{Saturation} = (1 + \tan(\beta))^{-1} \tag{33.1}$$

For details regarding the different steps and the technique itself see literature on TCO using NIRS [1,2].

Additionally a goodness of fit (GOF) using multiple correlation was calculated for every recording [5]. It ranges between 0 and 1 and is an indicator for the quality of the recording. A goodness of fit of 1 indicates a perfect measurement with no diversion of the datapoints from the main vector formed in 3D-space during the recording. A GOF of 0 indicates a bad measurement with no correlation of the scattered datapoints over time.

33.2.2 Apparatus

TCO using NIRS (OXYMON spectroscope, University of Nijmegen, three wavelengths: 769, 850, 905 nm, 10 Hz sampling frequency, 2 channels) was used. TCO was performed before and after angioplasty with stent or surgery. Optodes were placed over both hemispheres simultanously. Source-detector distance was 55 mm. Patients were asked not to speak, move, and to breathe normally and external conditions were kept constant during the recording. Each recording consisted of an arterial saturation and an angle alpha. The received signals were bandpass-filtered.

33.2.3 Subjects

The 21 patients with unilateral extracranial stenosis of the ICA were examined. Inclusion criteria were a symptomatic stenosis $\geq 70\%$ according to NASCET. Stenosis were categorized prospectively in hemodynamically relevant (group 1) and irrelevant (group 2) stenosis according to ultrasound criteria. Features assessed by ultrasound were bloodflow velocity in the ipsilateral middle cerebral artery and the presence of collateral bloodflow. After the treatment patients were again evaluated by ultrasound to rule out restenosis or insufficient treatment. In addition patients were categorized as smokers and non-smokers to assess the possibility to detect the presence of carboxyhemoglobin in the optical field. Participants were informed in oral and written form about the nature of the study. Informed consent was obtained from all participants and the experiment was approved by the local Ethical Committee.

33.3 Results

Only measurements with a GOF ≥ 0.9 were taken into account. All patients were treated successfully. Values are given as mean ± S.E. In group 1, 9 patients (♂ = 9, age: 67.22 ± 6.83 years) were included before and 6 patients (♀ = 6, age: 66.17 ± 6.46 years) after treatment. In group 2, 10 patients (♂ = 9, ♀ = 1, age: 67.8 ± 9.21 years) were included before and 10 patients (same group) after treatment. 16 patients (♂ = 15, ♀ = 1, age: 67.19 ± 8.09 years) were included before and after treatment. 6 smokers and 13 non-smokers could be identified.

No statistically significant differences were found between group 1 and group 2, before and after the treatment and between the affected and unaffected hemisphere ($p > 0.05$, all comparisons with Wilcoxon-Mann-Whitney U-test, see Table 33.1).

All smokers' hemispheres ($n = 22$) had a smaller angle alpha (–0.291 ± 3.7°) compared to non-smokers' hemispheres ($n = 48$, 1.38 ± 3.73°, $p < 0.05$). This difference was present in all hemispheres (−1.325 ± 3.55°, 2.6 ± 3.06°, $p < 0.001$), affected (−2 ± 2.27°, 2.8 ± 1.89°, $p < 0.001$) and not-affected hemispheres (-0.645 ± 4.64°, 2.4 ± 3.99°, $p < 0.05$) in smokers ($n = 12$, $n = 6$) and non-smokers ($n = 26$, $n = 13$) but only before the treatment. After the treatment this difference disappeared ($p > 0.05$, all comparisons with Wilcoxon-Mann-Whitney U-test).

33.4 Discussion

Our results show that TCO is not able to distinguish hemodynamically relevant from irrelevant stenosis of the ICA if this dichotomization is performed using ultrasound criteria. We conclude that oxygen saturation in patients with hemodynamically relevant and irrelevant stenosis of the ICA is similar. Previous studies using this method showed decreased oxygen saturation in intracranial arteries in situations with bad collateral blood supply [1]. Therefore TCO can be regarded as a method sensitive to failure of collateral blood supply. The values of intracranial oxygen saturation are consistent with theoretical values and values found in previous studies in healthy volunteers and patients with stenosis

Table 33.1 Results of the subgroups for oxygen saturation and angle alpha†

	Group 1 before		Group 2 before		Group 1 after		Group 2 after	
	affected‡	normal¶	affected	normal	affected	Normal	affected	Normal
sO_2	0.94±0.05	0.91±0.08	0.95±0.05	0.92±0.06	0.94±0.04	0.94±0.04	0.95±0.04	0.96±0.03
alpha	1.33±2.67	2.39±4.56	1.24±3.44	0.59±4.15	0.83±3.48	1.4±2.97	-0.6±3.67	0.21±4.89

†: angle alpha in degrees
‡: affected side
¶: not affected side

of the ICA but intact collateral blood supply [1,6,7]. In the patients presented here collateral blood supply was still sufficient to allow for normal intraarterial situations. Smokers have a smaller angle alpha. This can be explained by larger amounts of carboxyhemoglobin in the blood of smokers [8]. The probable cause of the normalisation after the treatment is the drop of concentration of carboxyhemoglobin due to less tobacco consumption during hospitalisation.

References

1. R. Stingele, H. Schnippering, E. Keller, T. Steiner, and W. Hacke, Transcranial pulse oximetry using fast near infrared spectroscopy can detect failure of collateral blood supply in humans, Comp. Biochem. and Physiol. Part A 134, 539–543 (2003).
2. J. W. Severinghaus and Y. Honda, Pulse oximetry, Int. Anesthesiol. Clin. 25, 205–214 (1987).
3. N. A. Gershenfeld, The Nature of Mathematical Modeling (Cambridge University Press, Cambridge, 1999).
4. G. Strang: Introduction to Linear Algebra (Wellesley Cambridge Press, 1998).
5. L. Sachs In: Angewandte Statistik, edited by L. Sachs (11. ed., Springer, Heidelberg, 2003), pp. 571–580.
6. A. S. Popel, Theory of oxygen transport to tissue, Crit. Rev. Biomed. Eng. 17, 257–321 (1989).
7. M. Sharan, M. D. Jones Jr., R. C. Koehler, R. J. Traystman, and A. S. Popel, A compartmental model for oxygen transport in brain microcirculation, Ann. Biomed. Eng. 17, 13–38 (1998).
8. A. Deller, R. Stenz, K. Forstner, M. N. Schreiber, F. Konrad, and T. Fosel, Carbomonoxyhemoglobin and Methemoglobin in patients with and without a smoking history during ambulatory anesthesia, Anesthesiol. sIntensivmed. Notfallmed. Schmerzther. 26 (4), 186–190 (1991).

Chapter 34
A Three-tiered Approach for Calibration of a Biosensor to Detect Estrogen Mimics

Sarah A. Andres, D. Alan Kerr II, Stefanie B. Bumpus, Traci L. Kruer, Joshua W. Thieman, Irina A. Smolenkova, and James L. Wittliff[1]

Abstract A three-tiered approach was developed to determine the influence of a chemically-diverse group of compounds exhibiting estrogen mimicry using recombinant human estrogen receptor (rhER) activity to calibrate a receptor protein-based biosensor. In the initial tier, a ligand competition array was developed to evaluate compounds inhibiting [3H]estradiol-17β binding to rhER. Each of six different concentrations of [3H]estradiol-17β was mixed with increasing concentrations of an unlabeled putative mimic. Each of these mixtures was incubated with a constant amount of rhERα and then receptor-bound [[3H]estradiol-17β was measured. This array protocol analyzes ligand binding affinities of hERα with a potential inhibitor over the entire range of receptor protein saturation.

When either hERα or hERβ binds to an estrogenic ligand, the receptor monomer forms both homo- and hetero-dimers. Then the ligand-receptor dimer complex activates transcription by associating with an estrogen response element (ERE), which is a specific DNA sequence located upstream of estrogen-responsive genes. The second tier for ligand evaluation utilized an electrophoretic mobility shift assay (EMSA), which was performed with an ERE sequence labeled with [α[32]P]dATP and incubated with rhER in the presence or absence of unlabeled ligand. ERE-hER complexes were separated by electrophoresis and analyzed using phosphor imaging technology.

To assess biological effects of an estrogen mimic on expression of an ER-target gene, a yeast cell-based bioassay was constructed with recombinant DNA technology using *Saccharomyces cerevisiae*. Each of these engineered yeast cells contained a rhERα expression plasmid (YEpE12) and a separate reporter plasmid (YRG2) containing an ERE sequence upstream of a β-galactosidase reporter gene. Incubation of these yeast cells with an estrogenic compound allows formation of ligand-hERα complexes, which

[1]Sarah A. Andres, D. Alan Kerr II, Stefanie B. Bumpus, Traci L. Kruer, Joshua W. Thieman, Irina A. Smolenkova, and James L. Wittliff, Department of Biochemistry & Molecular Biology, University of Louisville, Louisville, KY 40292 USA.

K.A. Kang et al. (eds.), *Oxygen Transport to Tissue XXIX*

recognize the ERE sequence regulating β-galactosidase expression. Estrogenic compounds, which were evaluated as calibrators for ligand-based and ERE-based biosensors, elicit varying responses in each of the three tiers of the protocol.

34.1 Introduction

Estrogen plays a crucial role beginning with conception and continuing through normal development, as well as in many health-related events, e.g., prevention of cardiovascular diseases, osteoporosis and treatment of cancer [1,2]. Many substances in the environment, such as phytoestrogens and mycoestrogens, as well as man-made therapeutic estrogens and certain industrial pollutants, may act as estrogen mimics [3,4]. Exposure to endocrine disrupting compounds has recently emerged as a major public health concern due to potentially hazardous effects through their interaction with steroid hormone receptors and their target pathways. Determination of the extent of estrogen mimicry by environmental compounds is essential to estimate risk/benefit ratios in both human and animal populations.

An evanescent biosensor (Endotect™) was developed jointly with IA, Inc./Threefold Sensors (Ann Arbor, MI) for screening compounds exhibiting potential estrogen mimicry effects [5–7]. This biosensor utilizes laser-based fiber optics with Cy5-labeled recombinant human estrogen receptor-α (rhERα) as a probe for either ligand or estrogen response element-based fibers. Calibration of the biosensor by independent methodologies, e.g, ligand binding and ERE recognition, as well as biological response is necessary to insure accuracy of mimic detection. The long-term objective is to utilize the biosensor to identify molecules with either hormone-stimulating or disrupting properties before human exposure so that rational assessment of risk/benefit ratios may be evaluated.

34.2 Methods

34.2.1 Expression of Human Estrogen Receptor-α

Saccharomyces cerevisiae strain BJ3505, transformed with expression vector YEpE12 containing the hERα gene, was expressed as an ubiquitin fusion under the control of a CUP1 promoter [8,9]. Lysates were separated into pellet and extract containing hERα by centrifugation at 40,000 rpm for 45 min at 4°C and purified by various modes of chromatography.

34.2.2 *Ligand Competition Array*

A ligand competition array performed in both the presence and absence of a compound suspected of estrogen mimicry was employed [10]. Each of six different concentrations of [[3H]estradiol-17β was mixed with increasing concentrations of a putative mimic, and the mixture was incubated with rhERα for 16 hr at 4°C. Then receptor-bound [[3H]estradiol-17β was measured in a scintillation counter. Each curve describing the binding isotherm (Fig. 34.1, B-H) represents results from assays in which an unlabeled candidate mimic were added to increasing concentrations (a-f) of [^{3}H]estradiol-17β. Curve A represents total binding of [^{3}H]estradiol-17β to both receptor and non-receptor proteins in the absence of unlabeled candidate competitor. Curve I represents radiolabeled-ligand binding in the presence of excess unlabeled diethylstilbestrol, which serves as a measure of low affinity, non-specific binding (i.e., the control). By solving differences in binding curves (A-H), specific binding capacities and affinities of [^{3}H]estradiol-17b were calculated in the presence of unlabeled test compound from an array of binding isotherms. This allowed identification of the activity of an estrogen mimic, as well as detection of mimic-induced alterations in affinity constants of [^{3}H]estradiol-17b binding using Lundon One-Site® and GraphPad Prism® software (calculations I & II). Calculation III designates experimental results used to directly generate ligand competition curves with Lundon Compete® and GraphPad Prism® software. The IC_{50} value of an unlabeled test compound was used to estimate relative affinity according to the following equation:

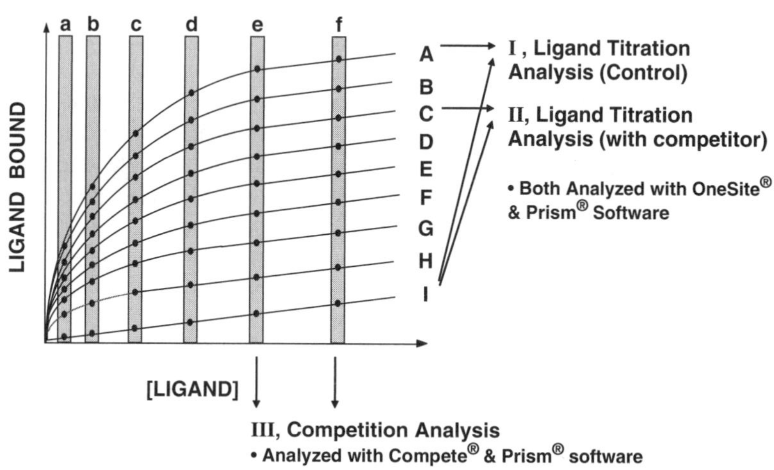

Fig. 34.1 Schematic of the ligand competition array protocol for analyzing competition of a candidate compound for the human estrogen receptor. Adapted from Raffelsberger and Wittliff [10].

$$Kd_{unknown} = (Kd_{reference} \times IC50_{unknown})/IC50_{reference} \quad (34.1)$$

34.2.3 Electrophoretic Mobility Shift Assays

Double-stranded ERE sequences (i.e., VitA2, pS2, h-fos, jun and cathepsin D), were radiolabeled using DNA Polymerase I Large (Klenow) fragment in 1X Klenow buffer (Promega), 5% β-mercaptoethanol (Sigma), 750 μM dCTP, dGTP and dTTP (Promega) and 1.25 μM [α[32P]dATP (800 Ci/mmol, Perkin Elmer) [8,9]. Each reaction was incubated at 37°C for 30 min, followed by gravity-flow separation through a NICK column (Amersham). Poly (dI-dC) (50 ng, Amersham), 10 mM KCl & 1% glycerol were added to 40 mM Tris-HCl buffer, pH 8.0, containing 500 μM PMSF and 10 μM monothioglycerol [8,9]. rhER preparations were incubated with a candidate estrogen mimic for 30 min, 4°C. Labeled ERE was then added to each reaction and incubated overnight at 4°C. Electrophoresis was performed as described previously using non-denaturing PAGE with 0.5x TBE running buffer [8,9]. Gels were dried and exposed to phosphor screens (Perkin Elmer) overnight, and bands representing [^{32}P]ERE-protein complexes and free [^{32}P]ERE were visualized, quantified and analyzed using a Cyclone Storage Phosphor System with OptiQuant® software (Perkin Elmer).

34.2.4 Cell-based Bioassay

Saccharomyces cerevisiae (BJ3505) containing both a plasmid encoding hERα (YEpE12) and a reporter plasmid (YRG2) containing an ERE driving the expression of β-galactosidase (Fig. 34.2) were grown in YNBD medium without tryptophan in an orbital shaker at 30°C overnight in a 100 ml culture [1,11]. The following morning rhER expression was induced with 100 μM $CuSO_4$, and the culture was divided into 5 ml mini-cultures, each of which was treated with a test compounds for 4 hr. Mini-cultures were then centrifuged, and each pellet was resuspended in Z-Buffer (50 mM Na_2HPO_4/NaH_2PO_4 buffer pH 7.0, 10 mM KCl, 1 mM $MgSO_4$ and 50 mM 2-mercaptoethanol), and the cell density was measured in a spectrophotometer at 600 nm. Cells were then lysed by addition of 0.4% N-lauroyl sarcosine in Z buffer. β-Galactosidase activity was determined after addition of 2.6 mg/ml ONPG. When sufficient yellow product had formed, reactions were terminated by the addition of 625 mM Na_2CO_3. Reactions were analyzed in a spectrophotometer at wavelengths of 420, 550 and 600 nm. Miller units were calculated with the following equation:

$$U = \{1000 \times [(OD420) - (1.75 \times OD550)]\}/[(t) \times (v)(OD600)] \quad (34.2)$$

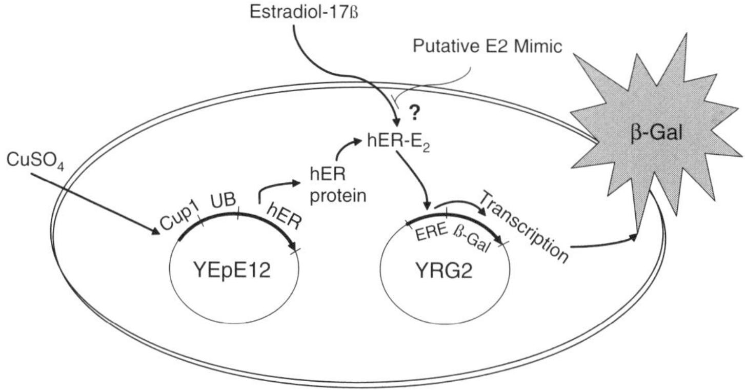

Fig. 34.2 Schematic of the yeast cell-based bioassay for analysis of an estrogen mimic. Adapted from Wittliff and Raffelsberger, 1995.

Where: t = time of reaction (min)
v = volume of culture in assay (ml)
OD600 = cell density at the start of assay
OD420 = combination of absorbance by ONP & light scattering by cell debris
OD550 = light scattering by cell debris

34.3 Results and Discussion

When a compound is selected as a calibrator candidate for the biosensor, it is evaluated initially in the ligand competition array. This allows analyses of ligand binding capacities over the entire range of steroid hormone receptor saturation. A series of ligand binding isotherms is developed for each concentration of competitor candidate and analyzed independently for specific binding to rhERα. Then competition curves are created at each concentration of [^{3}H]estradiol-17β (Fig. 34.3) over a broad range receptor saturation. As shown in Fig. 34.3, the Ki value determined at 0.35 nM [^{3}H]estradiol-17β (below saturation) was comparable to that observed at 2.5 nM [^{3}H]estradiol-17β, which represents a saturating concentration of labeled ligand. Each competition curve is analyzed independently and binding affinities for the competitor candidate are calculated using data obtained from each curve (Fig. 34.4). These ligand-binding affinities of calibrator estrogen mimics are then correlated with sensor-gram data obtained from the biosensor using the ligand-based optic fiber.

The next tier for evaluating a calibrator candidate utilizes both ligand- and DNA-binding characteristics of rhERα. Briefly, ligand association with rhERα allows the complex to interact with ERE sequences appearing upstream of estrogen responsive genes encoded in DNA. Assessment of a compound's

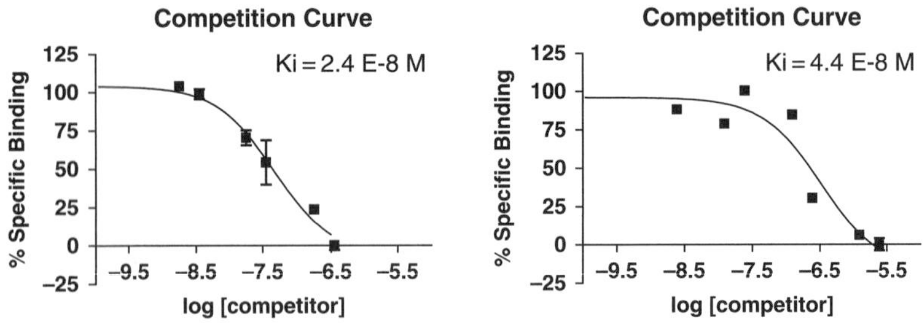

Fig. 34.3 Representative competition curves performed at different levels of rhERα saturation: 0.35 nM [^{3}H]Estradiol-17β (*left*) and 2.5 nM [^{3}H]Estradiol-17β (*right*).

influence on these binding activities is required to calibrate the biosensor and ERE-based optic fiber. Figure 34.5 shows alteration in rhERα-ERE complex migration due to association of the receptor protein with tamoxifen, an man-made, pharmacologically-active estrogen mimic [1,2]. While certain estrogen mimics induce a downward shift in the migration profile, indicating a more compact conformation of the rhER-ERE complex, other mimics induce an upward shift indicative of a less compact conformation. Results such as those shown in Fig. 34.5A and B may be quantified using the densitometric profiles, which also reflect rhER-ERE complex differences with and without a competitor candidate. Titration results of estrogen mimic-hERα complexes with an ERE sequence, when compared to that of estradiol-liganded ER-ERE

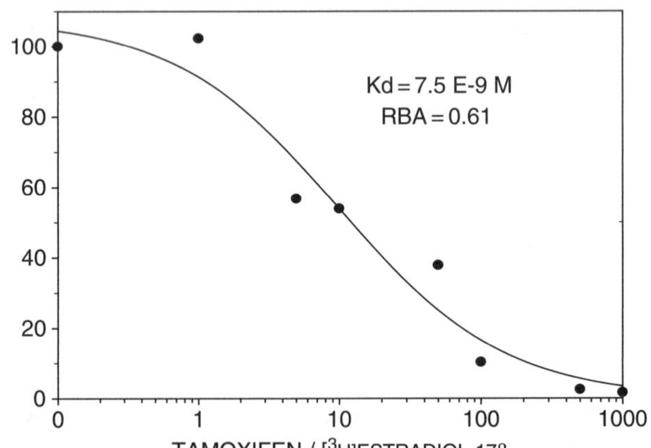

Fig. 34.4 Representative graph depicting the results from eight independent competition analyses of tamoxifen inhibition of [^{3}H]Estradiol-17β to rhERα. Each point represents the collective results from an experiment performed with the ligand competition array. Note that the affinity calculated from this analysis is slightly different from that of the individual analyses.

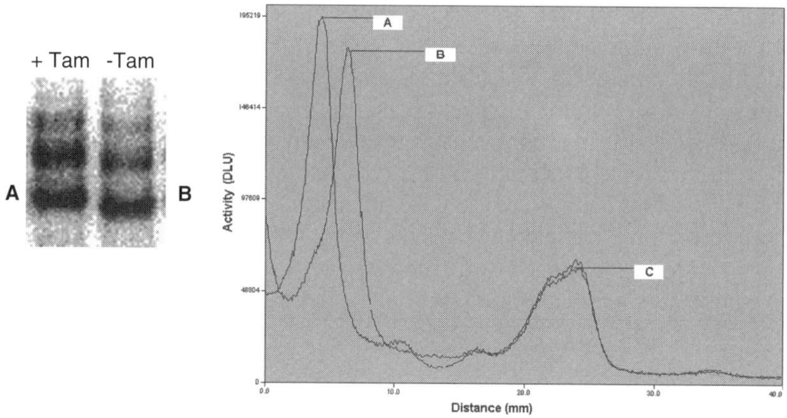

Fig. 34.5 Representative EMSA results illustrating the different migration of the ER-ERE complex in the presence (A) and absence (B) of tamoxifen. Migration is quantified in the densitometric profile on the right. Peak A represents the ER-ERE complex in the presence of tamoxifen, while peak B represents the complex without ligand. Peak C (not shown on gel) is utilized as an alignment control between lanes on the EMSA gel.

complexes, allows estimation of relative binding affinities for calibration of the ERE-based optic fiber of the biosensor.

The cell-based bioassay encompasses the activities of the entire estrogen receptor transcriptional complex, including ligand binding to the rhERα protein, its association with the ERE sequence, recruitment of transcription factors, as well as the expression of a target gene [1,2]. The yeast cell expression system, described in Fig. 34.2, provides transcriptional machinery that is highly conserved between mammals and *Saccharomyces*, and therefore is considered relevant to the

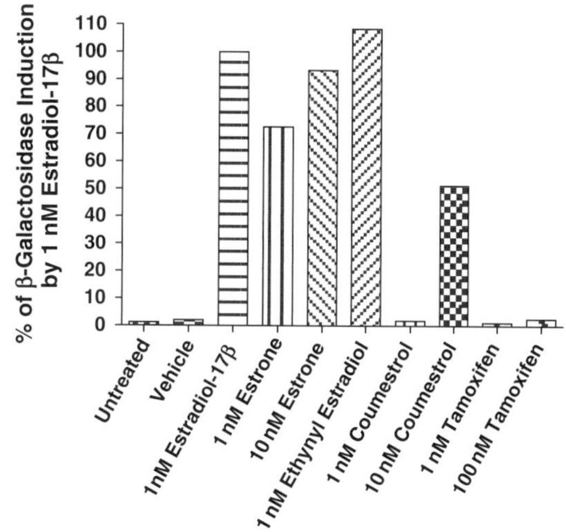

Fig. 34.6 Representative analysis of five known estrogenic compounds using the cell-based bioassay.

examination of the human estrogen response mechanism [9,11]. Addition of known estrogenic compounds to the growth media allows formation of liganded-hERα complexes, which subsequently associate with the ERE sequence regulating expression of the β-galactosidase reporter gene. Figure 34.6 illustrates the relative effects of five known estrogenic compounds [1–4] compared to the activity of estradiol-17β. Results shown correlate with the known properties in vivo of these compounds to induce gene transcription in mammals [1]. This cell-based bioassay provides essential results for calibration of the biosensor and confirms the results obtained with the estrogen mimic candidate from of the ligand competition array and the EMSA experiments.

34.4 Summary and Conclusions

A three-tiered approach using recombinant human ER-based analyses was employed for calibration of a fiber-optic biosensor that will be used to identify estrogen mimics in environmental samples. Each tier evaluates a specific mode of ER action, such as ligand binding, ERE sequence recognition and induction of transcriptional activity. Using ligand competition arrays, apparent dissociation constants of mimic candidates were generated and utilized to calibrate ligand-based fiber measurements. EMSA results using rhERα in the presence and absence of mimic candidate identified ligands that altered rhER-ERE complex migration, which may be correlated with biosensor measurements using the ERE-based fiber. Results from the cell-based bioassay were used as a confirmation of estrogen mimicry derived from the ligand competition array and EMSA, as well as determined alterations in target gene expression induced by a suspected estrogen mimic. Collectively, this three-tiered approach provides a broad spectrum of analyses for biosensor calibration to ensure its utility in detecting and characterizing compounds with estrogenic activities.

Acknowledgment Supported in part by grants NIEHS/SBIR #1R43-ES10076-01, #2R44-ES10076-02, USAID CFDA No.98.009 & the Phi Beta Psi Charity Trust. SAA & DAK are recipients of IPIBS Fellowships from the University of Louisville. SBB is a recipient of a Fellowship from NCI 5R25CA44789.

References

1. J. L. Wittliff and W. Raffelsberger, Mechanisms of Signal Transduction: Sex Hormones, Their Receptors, and Clinical Utility, J. Clin. Ligand Assay 18(4), 211–235 (1995).
2. J. L. Wittliff, R. Pasic, K. I. Bland, In: The Breast: Comprehensive Management of Benign and Malignant Diseases, edited by K. I. Bland and E. M. Copeland III (W.B. Saunders Co. , Philadelphia, 1998), pp. 458–498.
3. T. Colburn, C. Clement, Chemically-Induced Alterations in Sexual and Functional Development: The Wildlife/Human Connection (Princeton Sci. Publ. Co., 1992), pp. 403.

4. B. Gutendorf and J. Westendorf, Comparison of an Array of in vitro Assays for the Assessment of the Estrogenic Potential of Natural and Synthetic Estrogens, Phytoestrogens and Xenoestrogens, Toxicology 166(1–2), 79–89 (2001).
5. R. H. Smith, W. J. Lemon, J. L. Erb, J. R. Erb-Downward, J. G. Downward, O. E. Ulrich, and J. L. Wittliff, Development of Kinetic Ligand-binding Assays Using a Fiber Optic Sensor, Clin. Chem. 45(9), 1683–1685 (1999).
6. E. A. E. Garber, J. L. Erb, J. G. Downward, E. M. Priuska, J. L. Wittliff, W. Feng, J. Magner, and G. L. Larsen, Biosensor, ELISA, and Frog Embryo Teratogenesis Assay: Xenopus (FETAX) Analysis of Water Associated with Frog Malformations in Minnesota, Proc. Soc. Photo-Optical Instrumentation Engineers (SPIE) 4206, 147–158 (2001).
7. J. L. Erb, E. A. E. Garber, J. G. Downward IV, and E. M. Priuska, Data from an Estrogen Receptor-based Biosensor Correlates with Evidence of Frog Malformation and Demonstrates a Differential Response of hERα & β to Beneficial and Harmful Estrogenic Compounds, In: Proc. 2nd Intl. Conf. Pharmaceuticas & Endocrine Disrupting Chemicals in Water, p. 203–217, Westerville, OH, The National Ground Water Association (2001).
8. J. L. Wittliff, L. L. Wenz, J. Dong, Z. Nawaz, and T. R. Butt, Expression and Characterization of an Active Estrogen Receptor as a Ubiquitin Fusion Protein from Escherichia coli, J. Biol. Chem. 265(35), 22016–22025 (1990).
9. K. Graumann, J. L. Wittliff, W. Raffelsberger, L. Miles, A. Jungbauer, and T.R. Butt, Structural and Functional Analysis of N-terminal Point Mutants of the Human Estrogen Receptor, J. Steroid Biochem. Mol. Biol. 57(5–6), 292–300 (1996).
10. W. Raffelsberger and J. L. Wittliff, A Novel Approach for Comparing Ligand Binding Results from Titration and Competition Analyses to Study Hormone Mimics, J. Clin. Ligand Assay 20(4), 275–280 (1997).
11. C. R. Lyttle, P. Damian-Matsumura, H. Juul, and T. R. Butt, Human Estrogen Receptor Regulation in a Yeast Model System and Studies on Receptor Agonists and Antagonists, J. Steroid Biochem. Mol. Biol. 42(7), 77–685 (1992).

Chapter 35
Biosensors for Detecting Estrogen-like Molecules and Protein Biomarkers

James L. Wittliff[1], Sarah A. Andres[1], Traci L. Kruer[1], D. Alan Kerr II[1], Irina A. Smolenkova[1], and Judith L. Erb[2]

Abstract A novel evanescent-based biosensor (Endotect™, ThreeFold Sensors, Inc.) was developed with laser-based fiber optics using fluorescent dye-labeled recombinant human estrogen receptor-α (rhERα) and hERβ as probes. A three-tiered approach evaluating various steps in the formation of the estrogen-receptor complex and its subsequent activity was developed for instrument calibration to detect estrogen mimics in biological samples, water and soil. Using this approach, binding affinities and activities of certain known estrogen mimics were determined for their use as calibrator molecules. Results indicated rhERα and rhERβ may be employed as probes to distinguish estrogen mimics with a broad range of affinities. In addition, application of the biosensor for detecting DNA-binding proteins in human tissue extracts was demonstrated. The later studies suggest the biosensor may be used as a clinical laboratory tool for assessing tumor marker proteins.

35.1 Introduction

Estrogen mimics used in clinical management of breast cancer and osteoporosis, e.g. Tamoxifen and Raloxifene respectively, mediate their therapeutic effects through direct interaction with estrogen receptor proteins. Determination of the extent of estrogen mimicry by new generations of these drugs as well as estrogenic compounds encountered in the environment (endocrine disruptor compounds, EDCs) is essential to estimate risk/benefit ratios.

The effects of putative EDCs on the reproductive functions of human and animal species has become an area of intense concern [1]. The Environmental Protection Agency (EPA) has been mandated by Congress to develop a plan for testing some 87,000 chemicals known to be present in the environment to

[1]Department of Biochemistry & Molecular Biology, Institute for Molecular Diversity & Drug Design, University of Louisville, Louisville, KY USA.
[2]IA, Inc./ThreeFold Sensors, Ann Arbor, MI, USA.

determine their potential for producing endocrine disruption. In response to this mandate, the Endocrine Disruptor Screening and Testing Advisory Committee (EDSTAC) was created by the EPA and has issued recommendations for a multi-tier screening program to be executed on 15,000 chemicals beginning in 1999. Tier 1 screening involves assessment of the effects of test compounds on estrogen receptor mediated functions. Additional EDC activities may be due to cancer therapeutics, such as tamoxifen, which was designed to exert endocrine disrupting effects, useful in treating certain reproductive cancers [2]. A thorough understanding of the mechanisms by which each type of effect occurs is required in order that informed regulatory decisions may be made, and useful new drugs may be developed for treating reproductive cancers.

Evidence supporting an aggressive testing program for endocrine disruptors suggests that the reproductive functioning of both animal and human species is being adversely affected and that environmental exposure to chemicals is likely to be involved [1]. In response to this need, a novel biosensor (Endotect^{TM}, ThreeFold Sensors, Inc.) was developed to identify compounds in biological samples which contained EDCs, such as estrogen mimics [3–5].

The physical principles of operation of the sensor instrument are explained in detail in previous publications [3–5]. The biosensor technology utilizes an evanescent field generated on the surface of an optical fiber held in a cartridge. The fiber surface possesses covalently-linked molecules (e.g., steroids, ERE sequences, antibodies), to which the molecules of interest are attracted. In the case of the biosensor described, this is either Cy5-labeled recombinant hERα or hERβ proteins. When the fluorophore-labeled molecules are attracted to the ligand on the fiber surface, the fluorophore will be excited by the evanescent field upon binding. Because of the nature of the evanescent field [3–5], fluorophore-labeled molecules in the surrounding solution are not excited, thus kinetic analyses becomes possible.

35.2 Methods

35.2.1 Biosensor

The essential feature of the fiber-optic evanescent biosensor is confinement of fluorescence sensing to the immediate surface of an optical fiber. It takes advantage of the evanescent field produced by total internal reflection of light propagating within the fiber, as described by Hirchfield [6]. Figure 35.1A illustrates the ligand-based optic fiber, in which estrone-1-glucuronide (E1g) has been chemically attached to the fiber surface. Recombinant hER, acting as the probe for the biosensor, was fluorescently labeled by incubation with Cy5 (Amersham Pharmacia Biotech) for 30 min, and separation of Cy5-hER from free Cy5 was performed using a 10DG gravity flow column (Biorad). Cy5-labeled hER is used as a probe in solution, which flows through the cartridge around the optical fiber. When bound to the ligand E1g on the fiber surface, the evanescent field excites fluorescence of Cy5-hER. In solution, E1g exhibits an affinity for rhERα with an apparent Kd value of 2.5 × 10[–9 M[3]. Figure 35.1B illustrates the estrogen response element (ERE)-based optic fiber, in which Vitellogenin A2 ERE (5′-GATCCGTCAGGTCACAGTGACCTGATG-3′) is bound to the optic fiber in a system similar to that of the ligand-based fiber.

A three-tiered approach was developed with hERα and hERβ consisting of 1) ligand titration and competition arrays [7], 2) gel mobility shift and super-shift assays [8,9] and 3) yeast cell-based bioassays [10] with various known estrogen mimics for instrument calibration to detect suspected estrogen mimics in biological samples, water and soil. The therapeutic tamoxifen associated with recombinant hERα (Fig. 35.2) with lower affinity (Kd = 0.8–2 × 10^{-7} M) than that of hERβ (Kd = 3–7 × 10^{-8} M). Using this approach, binding affinities of other calibrator substances were determined: 4-Hydroxytamoxifen- 2–4 × 10^{-10} M; Clomiphene- 2–5 × 10^{-7} M; Ethynylestradiol- 2–5 × 10^{-10} M; Nafoxidine- 1–3 × 10^{-8} M; and Raloxifene- 3–8 × 10^{-10} M.

For calibration of the ERE-based fiber, the second tier approach was employed using electrophoretic mobility shift assays (EMSA) to identify hER or other proteins that may bind to ERE sequences located upstream of estrogen receptor target genes. Double-stranded ERE sequences (i.e., VitA2, pS2, h-fos, jun, cath D), were labeled with [a [32] P]dATP (Perkin Elmer) [8,9]. EMSA reactions were performed in 40 mM Tris-HCl buffer, pH 8.0, containing 500

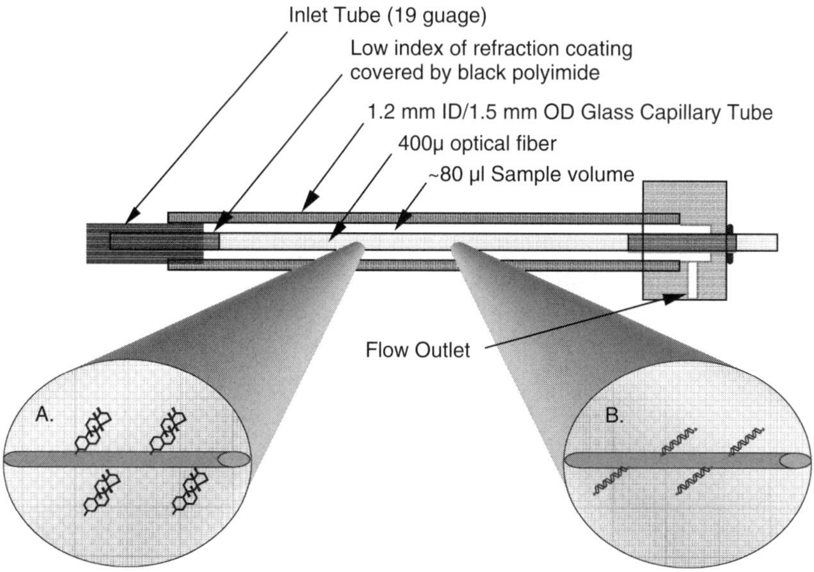

Fig. 35.1 Schematic of biosensor cartridge illustrating the ligand-based (A) and ERE-based (B) optic fibers.

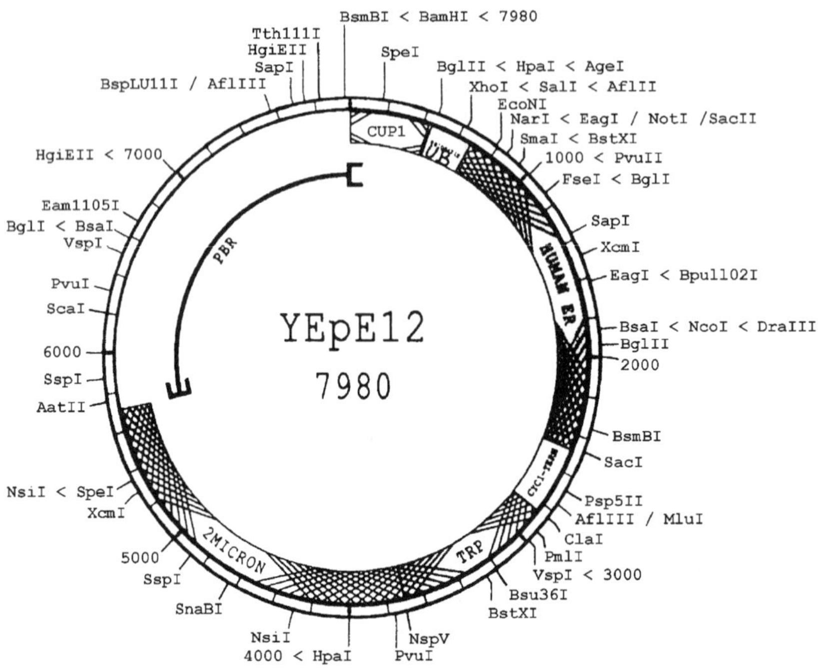

Fig. 35.2 Schematic of plasmid expressing recombinant human estrogen receptor-α (YEpE12) in yeast cells. Adapted from Wittliff et al. 1993 [11].

μM PMSF and 10 μM monothioglycerol [8,9]. rhER and tissue extract preparations were incubated with labeled ERE overnight at 4°C. Electrophoresis was performed as described previously [8,9] using 0.5x TBE running buffer. Gels were dried and exposed to phosphor screens (Perkin Elmer) overnight, and bands representing [[32] P]ERE-protein complexes and free [[32] P]ERE were analyzed using a Cyclone Storage Phosphor System with OptiQuant® software (Perkin Elmer).

These results and others using the three-tiered approach for calibration indicate recombinant hERα and hERβ isoforms may be employed as biosensor probes to distinguish therapeutic estrogen mimics with a broad range of affinities. Currently we are investigating the detection of these and other estrogen mimics using both the ligand-based and the ERE-based fibers with the biosensor with a focus on ascertaining their endocrine-activating and endocrine-disrupting activities.

35.3 Results and Discussion

The biosensor utilizes either a ligand-based or an ERE-based optic fiber to detect estrogen-like compounds in various solutions [3–5]. Figure 35.3A illustrates the association of Cy5-labeled recombinant hER with the ligand-based fiber producing an increase in fluorescence as detected by the sensor. The ligand-based fiber allows detection of estrogen mimics present in a test solution, which compete for the Cy5-labeled recombinant hER and create a decreased fluorescence. Figure 35.3B illustrates the competition of a test solution for the ERE fiber. The upper curve represents loading of unliganded Cy5-labeled hERα, while the lower curve represents the reaction of Cy5-labeled hERα with an estrogen mimic in a test solution creating diminished fluorescence detected by the sensor. The sensogram results are converted to relative binding affinities using the results of known estrogens and mimics from the three-tiered calibration approach (Table 35.1).

Using the ERE-based fiber-optic biosensor, protein molecules recognizing hormone response element sequences, such as hERα and hERβ may be detected in extracts. The sensograms in Fig. 35.4A indicate the presence of ERE-binding proteins since addition of extracts prepared from two human reference samples decreased Cy5-hERα binding to the ERE fiber. Curves a and b represent two reactions showing association of uncompeted Cy5-hERα (control) with the ERE sequence bound to the optical fiber. Curve c represents a reaction in which an extract of a reference myometrium specimen exhibited 36% inhibition

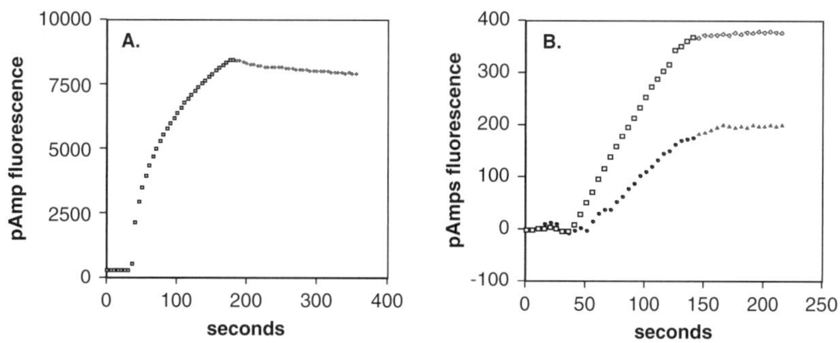

Fig. 35.3 Representative sensograms of Cy5-labeled recombinant human estrogen receptor-α associating with a ligand-based optic fiber (A) and ERE-based fiber (B) with (upper curve) and without (lower curve) competition.

Table 35.1 Comparison of Kd values of binding of various estrogen mimics to rhERα calculated from biosensor data and those obtained by standard radioligand binding methods

Competitor	K_d obtained from biosensor	K_d obtained from radioligand binding
estradiol-17ß	2×10–10 M	1.1×10–10 M
estrone	2×10–8 M	1.8×10–8 M
estriol	2×10–8 M	3.1×10–8 M
DES	2×10–10 M	1.8×10–10 M
zearalenone	2×10–8 M	3.6×10–8 M
tamoxifen	2×10–9 M	1.1×10–9 M

Note: All first digits of the sensor data are 2, because solutions used to determine K_d consisted of concentrations of test compounds which were 2 times some power of 10.

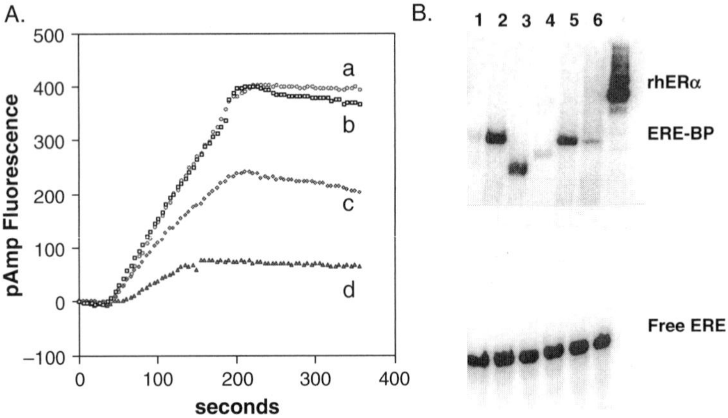

Fig. 35.4 (A) Representative sensogram demonstrating the presence of ERE- binding proteins in extracts of human reference samples of breast tissue. Curves a and b represent loading of Cy5-labeled hERα alone on the ERE-based optical fiber. Curves c and d represent loading of recombinant hERα in the presence of protein extracts of human reference specimens of myometrium (c) and breast cancer (d), showing the decrease in binding indicative of ERE-binding proteins in the tissue preparations. (B) EMSA gel pattern showing migration and distribution of various ERE-binding proteins (ERE-BP) in extracts of five different (lanes 1–5) breast cancers compared to that of recombinant human ERα (rhERα), confirming the results from biosensor measurements.

compared to that of the control. Curve d represents a reaction in which an extract of a reference breast cancer specimen exhibited 68% inhibition.

The degree of inhibition by a tissue extract determined from a sensogram (e.g., Fig. 35.4A) is being calibrated against the results of the same sample obtained by EMSA (e.g., Fig. 35.4B). Preliminary evaluation suggests a good qualitative relationship between measurements taken by the two methods. However, considerably more analyses of various tissue extracts will be required to establish a quantitative relationship. This is due to the physically different conditions of the ERE-binding reactions, i.e., soluble ERE sequences in the EMSA compared to covalently-bound ERE sequences on the optical fiber. Experiments addressing the latter relationship are in progress with the goal to use the biosensor to detect as well as quantify these new biomarker proteins.

Further examination of these activities in human tissue extracts by EMSA revealed the presence of several species of ERE-binding proteins that appear to be unrelated to human estrogen receptor. As shown in Fig. 35.4B, reference tissue extracts exhibited varied abundance of these novel ERE-binding proteins that migrated more rapidly than that of intact recombinant human ERα. Furthermore, certain cytosols contained ERE-binding protein species with different migration properties (e.g., lanes 13). Levels of expression of ERE-binding proteins were determined by scanning each lane of the EMSA gels and using Optiquant™ imaging software for quantification after normalization for the total protein content [8,9]. There was no apparent correlation between the level of ERE-binding protein and the amount of estrogen receptor in a reference tissue extract, suggesting these proteins recognizing ERE sequences are unrelated to estrogen receptor proteins. Although preliminary, results using super-shift assays of human tissue extracts incubated with monoclonal antibodies prepared against specific epitopes located in various functional domains of rhERα and rhERβ indicated these ERE-binding proteins are not related to known human estrogen receptor isoforms [12].

35.4 Summary and Conclusions

As demonstrated, the biosensor was employed in a laboratory setting to analyze compounds known to express estrogen-like properties, as well as in a practical setting to detect the presence of estrogen mimics with suspected endocrine disrupting activities in lake water, in which deformed frogs were identified [4,5]. Analyses are easily and rapidly performed on extracts to determine the presence of hormone mimics and EDCs in various samples, such as industrial waste, foodstuffs, hospital waste, and ground water. Biosensor instrumentation, optic fiber composition and fluorescent probe selection may be modified readily to analyze other types of molecular interactions including kinetics of ligand-receptor association and dissociation of candidate drugs. Furthermore, our research demonstrated that hormone response element-based fibers may be used as a discovery tool for detecting novel DNA-binding proteins with potential to serve as cancer biomarkers.

The long-term goal of our research is directed toward an understanding of the mechanisms of action by which chemical compounds and ligands exert their influences on receptor-mediated signal transduction. This includes the development and testing of an instrument and methods for screening environmental chemicals for endocrine disrupting potential [1], as a result of their participation in hormone-controlled pathways. Estrogens and certain of their mimics (e.g., phytoestrogens and therapeutic estrogen mimics) appear to influence the rates of association and dissociation of estrogen receptors-α and -β with their cognate accessory proteins (co-activators and co-repressors) as well as the composition of the transactivation complex in the ERE and promoter regions of responsive genes [1]. A new generation of instrumentation is being developed to address these complex issues using a FRET-based evanescent fiber optic sensor (J.L Erb, J. Downward & J.L. Wittliff, unpublished) that provides real time kinetic measurements reflecting association and dissociation of receptors with co-regulatory proteins, response elements and ligands.

Acknowledgment Supported in part by grants NIEHS/SBIR #1R43 ES010076-01 (JLE & JLW) & #2R44-ES010076-02 (JLE & JLW) and grants from Phi Beta Psi Sorority Charity Trust (JLW). SAA, TLK and DAK II are recipients of fellowships from the Integrated Programs in Biomedical Sciences, University of Louisville.

References

1. ICCVAM/NICEATM Final Report, Expert Panel Evaluation of the Validation Status of in vitro Test Methods for Detecting Endocrine Disruptors: Estrogen Receptor and Androgen Receptor Binding and Transcriptional Activation Assays, 2002 (http:// iccvam.niehs.nih.gov).
2. J. L. Wittliff, R. Pasic, K. I. Bland: Steroid and Peptide Hormone Receptors: Methods, Quality Control and Clinical Use, in Bland KI, Copeland III EM (eds): The Breast: Comprehensive Management of Benign and Malignant Diseases. Philadelphia, PA, W. B. Saunders Co, 458–498, 1998.
3. R. H. Smith, W. J. Lemon, J. L. Erb, J. R. Erb-Downward, J. G. Downward, O. E. Ulrich, and J. L. Wittliff, Development of Kinetic Ligand-binding Assays Using a Fiber Optic Sensor, Clin. Chem. 45(9), 1683–1685 (1999).
4. E. A. E. Garber, J. L. Erb, J. G. Downward, E. M. Priuska, J. L. Wittliff, W. Feng, J. Magner, and G. L. Larsen, Biosensor, ELISA, and Frog Embryo Teratogenesis Assay: Xenopus (FETAX) Analysis of Water Associated with Frog Malformations in Minnesota, Proc. Soc. Photo-Optical Instrumentation Engineers (SPIE) 4206, 147–158 (2001).
5. J. L. Erb, E. A. E. Garber, J. G. Downward IV, and E. M. Priuska, Data from an Estrogen Receptor-based Biosensor Correlates with Evidence of Frog Malformation and Demonstrates a Differential Response of hERα & β to Beneficial and Harmful Estrogenic Compounds, In: Proc. 2nd Intl. Conf. Pharmaceuticas & Endocrine Disrupting Chemicals in Water, p. 203–217, Westerville, OH, The National Ground Water Association (2001).
6. T. E. Hirchfield, Fluorescent Immunoassay Employing Optical Fiber in a Capillary Tube, U.S. Patent No. 4,447,546 (1984).
7. W. Raffelsberger and J. L. Wittliff, A Novel Approach for Comparing Ligand Binding Results from Titration and Competition Analyses to Study Hormone Mimics, J. Clin. Ligand Assay 20(4), 275–280 (1997).
8. J. L. Wittliff, L. L. Wenz, J. Dong, Z. Nawaz, and T. R. Butt, Expression and Characterization of an Active Estrogen Receptor as a Ubiquitin Fusion Protein from Escherichia coli, J. Biol. Chem. 265(35), 22016–22025 (1990).
9. K. Graumann, J. L. Wittliff, W. Raffelsberger, L. Miles, A. Jungbauer, and T.R. Butt, Structural and Functional Analysis of N-terminal Point Mutants of the Human Estrogen Receptor, J. Steroid Biochem. Mol. Biol. 57(5–6), 292–300 (1996).
10. C. R. Lyttle, P. Damian-Matsumura, H. Juul, and T. R. Butt, Human Estrogen Receptor Regulation in a Yeast Model System and Studies on Receptor Agonists and Antagonists, J. Steroid Biochem. Mol. Biol. 42(7), 77–685 (1992).
11. J. L. Wittliff, P. Folk, J. Dong, C. Schaupp, and T. R. Butt, Characteristics of the Human Estrogen Receptor Protein Produced in Microbial Expression Systems. In: V. K. Moudgil (ed.), Steroid Hormone Receptors: Basic and Clinical Aspects, pp. 473–501, Birkhauser Boston (1993).
12. T. L. Kruer, I. A. Smolenkova and J. L. Wittliff, Expression of Novel ERE-Binding Proteins in Breast and Uterine Cells, American Society for Biochemistry & Molecular Biology, Abstract # 1381 (2006).

Part IX
Modeling and Analysis of Metabolism and Transport

This section was contributed by the Center for Modeling Integrated Metabolic Systems (MIMS) of the Case Western Reserve University, Cleveland, OH

Chapter 36
Muscle Oxygen Uptake Differs from Consumption Dynamics During Transients in Exercise

Nicola Lai[1,3], Nakisha Syed[1], Gerald M. Saidel[1,3], and Marco E. Cabrera[1,2,3]

Abstract Relating external to internal respiration during exercise requires quantitative modeling analysis for reliable inferences with respect to metabolic rate. Often, oxygen transport and metabolism based on steady-state mass balances (Fick principle) and passive diffusion between blood and tissue are applied to link pulmonary to cellular respiration. Indeed, when the work rate does not change rapidly, a quasi-steady-state analysis based on the Fick principle is sufficient to estimate the rate of O_2 consumption in working muscle. During exercise when the work rate changes quickly, however, non-invasive *in vivo* measurements to estimate muscle O_2 consumption are not sufficient to characterize cellular respiration of working muscle. To interpret transient changes of venous O_2 concentration, blood flow, and O_2 consumption in working muscle, a mathematical model of O_2 transport and consumption based on dynamic mass balances is required. In this study, a comparison is made of the differences between simulations of O_2 uptake and O_2 consumption within working skeletal muscle based on a dynamic model and quasi-steady-state approximations. The conditions are specified under which the quasi-steady-state approximation becomes invalid.

36.1 Introduction

During exercise, oxygen transport and metabolism within muscle in healthy and disease states (e.g., heart failure and diabetes) can be studied under various experimental protocols with non-invasive measurements. Measurement methods include pulmonary O_2 uptake by indirect calorimetry, muscle oxygenation by near-infrared spectroscopy and microvascular oxygenation by phosphorescence quenching. Muscle O_2 consumption (UO_{2m}) during a fast

[1]Department of Biomedical Engineering and [2]Pediatrics, Center for Modeling Integrated Metabolism Systems and [3]Rainbow Babies and Children's Hospital, Case Western Reserve University, Cleveland, OH 44106, USA.

transient change (e.g., exercise) cannot be directly evaluated by measuring oxygen uptake in the lungs (VO_{2p}). Consequently, a mathematical model is needed to relate these variables. Often, a quasi-steady-state model (e.g., Fick principle) is used to analyze dynamic responses at the onset of exercise in order to quantify relationships between the oxygen uptake, blood flow, and O_2 concentration dynamics in the capillary bed (or index of extraction) within muscle [1–3]. Although this method is strictly applicable only under steady-state conditions [4], it could provide a reasonable approximation under some conditions [5]. More generally, interpretation of non-invasive measurements related to O_2 transport and metabolism in tissue can be made using dynamic mass balances if the tissue volume can be determined. Nevertheless, this has limitations also. At the microvascular level, the volume of tissue supplied with oxygen by the blood vessels is uncertain. Furthermore, in a macroscopic tissue volume, the heterogeneous spatial distribution and temporal variation of blood flow and O_2 concentration in tissue can have a significant effect on the interpretation of measurements. In this study, simulations using a quasi-steady-state model [3] and those using a multi-compartment dynamic model [6] are compared to experimental measurements during exercise. The effect of changes in muscle blood flow on the dynamic responses of venous oxygen concentration was investigated during exercise assuming specific dynamics of muscle O_2 consumption.

36.2 Methods

We consider oxygen transport and consumption in muscle to occur in a system of perfectly mixed blood and tissue compartments (Fig. 36.1) as developed previously [6]. In the blood compartment, the oxygen concentration $C(t)$ changes with time depending on flow Q_m(t) through the capillary bed of volume

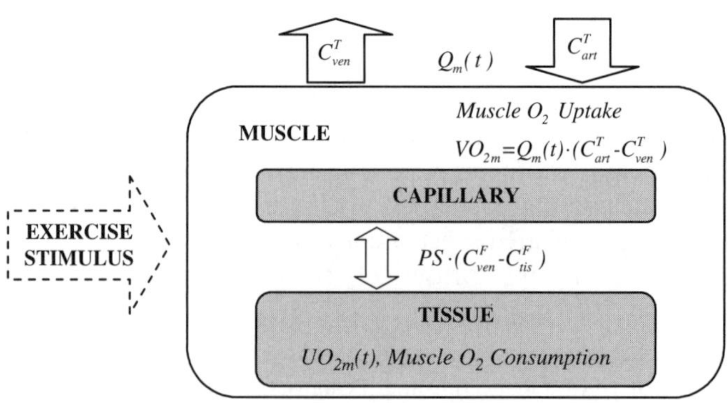

Fig. 36.1 Oxygen consumption and transport in skeletal muscle.

V_{cap} and diffusion between blood and tissue with rate coefficient *PS* according to the dynamic mass balance:

$$V_{cap}\frac{dC^T_{ven}}{dt} = Q_m(t)\left(C^T_{art} - C^T_{ven}\right) - PS\left(C^F_{ven} - C^F_{tis}\right) \tag{36.1}$$

where C^T_j and C^F_j represent the total and free oxygen concentrations in compartment *j*, which are related by nonlinear equations [6]. In the tissue compartment, the oxygen concentration $C(t)$ changes with time depending on the rates of diffusion between blood and tissue cells and oxygen consumption in muscle $UO_{2m}(t)$:

$$V_{tis}\frac{dC^T_{tis}}{dt} = PS\left(C^F_{ven} - C^F_{tis}\right) - UO_{2m}(t) \tag{36.2}$$

At steady state, these equations can be combined to yield the Fick principle for oxygen uptake of skeletal muscle:

$$VO_{2m} = Q_m\left(C^T_{art} - C^T_{ven}\right) = PS\left(C^F_{ven} - C^F_{tis}\right) = UO_{2m} \tag{36.3}$$

In response to a step increase in work rate, the oxygen consumption has typically an exponential response [3]:

$$UO_{2m}(t) = UO^{BL}_{2m} + \Delta UO_{2m}\left[1 - \exp^{t_0 - t/\tau_{UO_{2m}}}\right] \tag{36.4}$$

Also, the blood flow has two phases (I, II) with different amplitudes and time constants [3]:

$$\begin{aligned} Q_m(t) = Q^{BL}_m &+ \Delta Q_{m,1}\left[1 - \exp^{(t_0 - t)/\tau_{Q_{m,I}}}\right] \\ &+ \Delta Q_{m,2}\left[1 - \exp^{(t_0 + TD_2 - t)/\tau_{Q_{m,II}}}\right] \end{aligned} \tag{36.5}$$

To simulate the oxygen concentration dynamics of skeletal muscle to a step change in work rate from baseline (*BL*), we must specify initial conditions at t_0: $C^F_{ven}(t_0) = C^{F,BL}_{ven}$; $C^{F,BL}_{tis}(t_0) = C^{F,BL}_{tis}$. Responses of venous oxygen concentration were simulated between two steady states assuming different values of muscle volume (V_m) engaged during exercise with different dynamic changes of blood flow in muscle. The differential equations of the model were solved numerically using a robust algorithm for stiff systems [7].

For comparison with the results of Ferreira [3], we used the same initial conditions, muscle blood flows ($\tau_{Qm,I}$ and $\tau_{Qm,II}$), and oxygen consumption time constant ($\tau_{UO_{2m}} = 30$s). Except for the parameter values in Table 36.1, values of other model parameters were obtained from previous studies [6].

Following Ferreira [3] we chose values for the arterial oxygen concentration, C^T_{art}, and free oxygen concentration $C^{F,BL}_{ven}$. The free oxygen concentration intissue was determined based on $PO_2 = 25$ mmHg. The value of *PS* was computed from Eq. (36.3) at steady state.

Table 36.1 Initial conditions and model parameters

$C_{ven}^{F,BL}$ [mM]	$C_{tis}^{F,BL}$ [mM]	$C_{art}^{F,BL}$ [mM]	PS [L min^{-1}]
$3.48 \cdot 10^{-2}$	$3.37 \cdot 10^{-2}$	$1.05 \cdot 10^{-1}$	2026

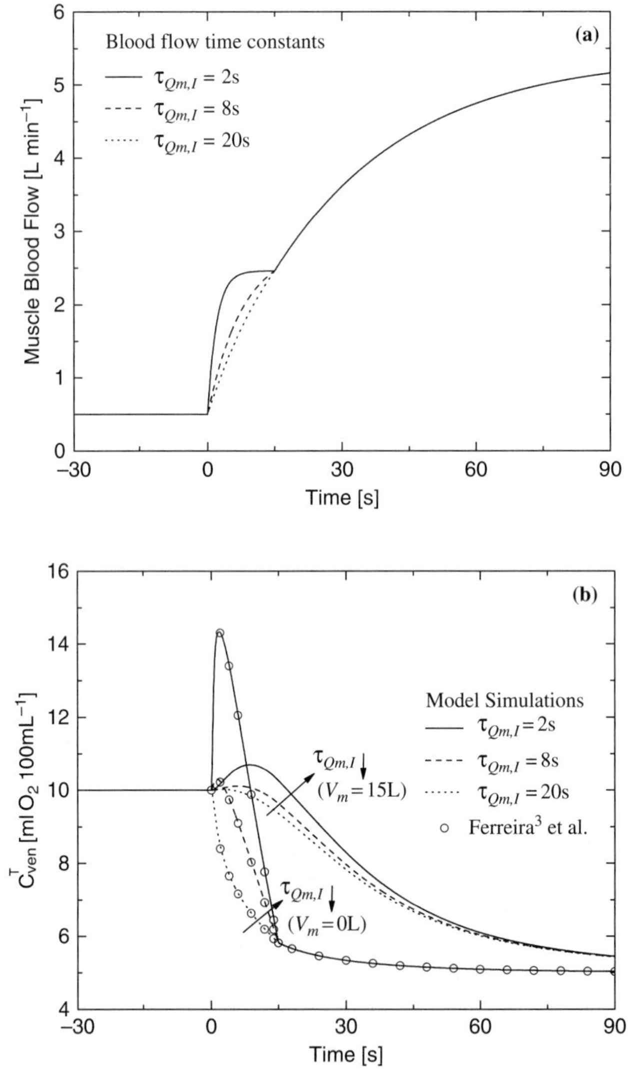

Fig. 36.2 (a) Blood flow dynamic for different time constants $\tau_{Qm,I}$, ($\tau_{\text{Qm,II}} = 30$ s); (b) Effect of blood flow dynamic on the dynamic response of the venous O_2 concentration obtained with quasi-steady-state and dynamic model.

36.3 Results

The effects of various blood flow time profiles on venous oxygen concentration for a specific oxygen consumption dynamics are simulated.

Figure 36.2 shows dynamic changes in phase I of muscle blood flow (Fig. 36.2a) and corresponding dynamic responses of venous oxygen concentration (Fig. 36.2b). The dynamic response of the venous oxygen concentration is faster

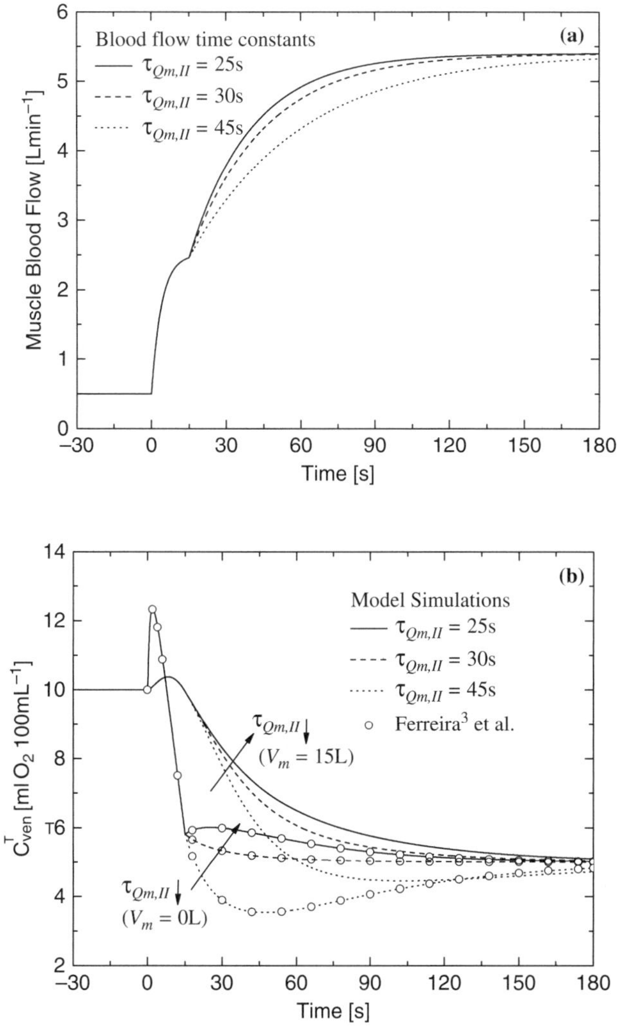

Fig. 36.3 (a) Blood flow dynamic for different time constants $\tau_{Qm,II}$, ($\tau_{\mathrm{Qm,I}} = 4$ s); (b) Effect of blood flow dynamic on the dynamic response of the venous O_2 concentration obtained with quasi-steady-state and dynamic model.

when the transient term (i.e., rate of oxygen change) is negligible ($V_m \approx 0$L) than when it is significant ($V_m = 15$L). The overshoot of the venous oxygen concentration is greater with a smaller muscle volume and with a shorter time constant of phase I ($\tau_{Qm,I}$). Even when the time constant of phase I is long, the dynamics of the venous oxygen concentration depends on the muscle volume.

Figure 36.3 shows dynamic changes in phase II of muscle blood flow (Fig. 36.3a) and corresponding dynamic responses of venous oxygen concentration (Fig. 36.3b). The dynamic response of the venous oxygen concentration is faster when the transient term is negligible ($V_m \approx 0$L) than when the transient term is significant ($V_m = 15$L). The undershoot of the venous oxygen concentration is greater with a smaller muscle volume and with a longer time constant of phase II ($\tau_{\mathrm{Qm,II}}$). Even when the time constant of phase II is small, the dynamics of the venous oxygen concentration depends on the muscle volume.

36.4 Discussion

Model simulations of the time response of venous oxygen concentration in working skeletal muscle were obtained with a dynamic computational model for quasi-steady-state (e.g., negligible muscle volume) and transient conditions. The simulations under quasi-steady-state conditions are equivalent to those of Ferreira [3], who applied a steady-state (Fick principle) analysis.

With a larger muscle volume, a change of muscle blood flow has less effect on the time profile of the venous oxygen concentration. In any case, an overshoot in venous oxygen concentration can occur during phase I of muscle blood flow as reported in human exercise studies [8]. Corresponding to simulations in venous oxygen concentration during phase II of muscle blood flow, experiments with rat muscle contractions show a similar undershoot response [9]. In these studies with diabetic rats where the disease induces a mismatch between oxygen delivery and oxygen consumption, an undershoot can occur in microvascular O_2 pressure at the onset of exercise.

Based on the Fick principle, the red blood cell flux (or oxygen delivery) and microvascular O_2 pressure measurements are used to compute oxygen consumption in the diabetic state [2]. This simplified analysis shows a mismatch between oxygen delivery and oxygen consumption, which accounts for the observed undershoot in microvascular O_2 pressure. Correct interpretation of this mismatch requires quantitative analysis with a more general dynamic model to determine the effect of muscle volume in the transient term of the oxygen balance. Furthermore, a dynamic model is essential to analyze the response of microvascular oxygen pressure in muscle at the onset of contraction in heart failure [10] and aging [11], where O_2 delivery is impaired.

Generally, venous oxygen concentration during exercise depends on the rate of oxygen mass accumulation within muscle that results from a dynamic interplay of convection, diffusion, and metabolism. Consequently, the dynamics of muscle oxygen uptake and muscle oxygen consumption differ during transient changes in exercise [6]. In this regard the extent of muscle involvement (i.e., muscle volume), which is often uncertain (especially at the microvascular level), plays a relevant role during exercise.

The more general dynamic model applied in this study consists of spatially lumped, dynamic mass-balance equations. A special case of this dynamic model is the quasi-steady-state model, which is commonly used to analyze oxygen exchange in capillary blood and in tissue of the working muscle. Although this dynamic model of oxygen consumption in skeletal muscle [6] is sufficient for some purposes, modifications are needed to reflect more physiological conditions. For example, the product of permeability and surface area should be a function of blood flow due to the capillary recruitment occurring at the onset of exercise [12]. Furthermore, this dynamic model assumes an exponential function to describe cellular oxygen consumption [13]. While this simple expression is sufficient to make some inferences about dynamic responses, key metabolic processes should be incorporated into the model to provide a mechanistic basis for oxygen consumption dynamics. For this purpose, future models should incorporate substrates and enzymes participating in mitochondrial oxidative phosphorylation during exercise.

The dynamics of the oxygen concentration in blood depend on the spatial distribution and temporal variation of the variables such as blood flow and hemoglobin oxygen saturation that affect convective and diffusive transport of oxygen in the microcirculation. Although these effects are not directly measurable during muscle contraction [14], more general models have been applied to account for heterogeneities of blood flow and oxygen consumption of the muscle [15–18].

In conclusion, physiological relations between oxygen transport and consumption within skeletal muscle during exercise require a model based on dynamic mass balances for oxygen in blood and tissue. Such a model can be used together with non-invasive or minimally invasive experiments to study capillary oxygen exchange during an exercise stimulus where active muscle, convection and diffusion have a significant effect. This could contribute to quantifying changes associated with aging in healthy subjects, as well as with potential pathological alterations of oxygen transport and metabolism in unhealthy subjects suffering from diabetes and heart failure [19].

Acknowledgment Supported by grant (P50 GM-66309) from the National Institute of General Medical Sciences (NIH).

References

1. Barstow, T.J., Lamarra, N., and Whipp, B.J., 1990, Modulation of muscle and pulmonary O_2 uptakes by circulatory dynamics during exercise, *J. Appl. Physiol.* **68**, 979–989.
2. Behnke, B.J., Barstow, T.J., Kindig, C.A., McDonough, P., Musch, T.I, and Poole, D.C., 2002b, Dynamics of oxygen uptake following exercise onset in rat skeletal muscle, *Respir. Physiol. Neurobiol.* **133**, 229–239.

3. Ferreira, L.F., Poole, D.C., and Barstow, T.J., 2005, Muscle blood flow-O_2 uptake interaction and their relation to on-exercise dynamics of O_2 exchange, *Respir. Physiol. Neurobiol.* **147**, 91–103.
4. Stringer, W.W., Whipp, B.J., Wasserman, K., Porszasz, J., Christenson, P., and French, W.J., 2005, Non-linear cardiac output dynamics during ramp-incremental cycle ergometry, *Eur. J. Appl. Physiol.* **93**, 634–639.
5. Kemp, G., 2005, Kinetics of muscle oxygen use, oxygen content and blood flow during exercise, *J. Applied Physiology* **99**, 2463–2469.
6. Lai, N., Dash, R.K., Nasca, M.M., Saidel, G.M., and Cabrera, M.E., 2006, Relating pulmonary oxygen uptake to muscle oxygen consumption at exercise onset: in vivo and in silico studies, *Eur. J. Appl. Physiol.* **97**(4), 380–94.
7. Hindmarsh, A.C., 1983, A systematized collection of ode solvers, *Scientific computing* 55–64, DLSODE, http://www.netlib.org/odepack/.
8. Grassi, B., Poole, D.C., Richardson, R.S., Knight, D.R., Erickson, B.K., and Wagner, P.D., 1996, Muscle O_2 uptake kinetics in humans: implications for metabolic control, *J. Appl. Physiol.* **80**, 988–998.
9. Behnke, B.J., Kindig, C.A., McDonough, P., Poole, D.C., and Sexton, W.L., 2002a, Dynamics of microvascular oxygen pressure during rest-contraction transition in skeletal muscle of diabetic rats, *Am. J. Physiol. Heart Circ. Physiol.* **283**, H926–H932.
10. Diederich, E.R., Behnke, B.J., McDonough, P., Kindig, C.A., Barstow, T.J., Poole D.C., and Musch, T.I., 2002, Dynamics of microvascular oxygen partial pressure in contracting skeletal muscle of rats with chronic heart failure, *Cardiovasc. Res.* **56**, 479–486.
11. Behnke, B.J., Delp, M.D., Dougherty, P.J., Musch, T.I, and Poole, D.C., 2005, Effects of aging on microvascular oxygen pressures in rat skeletal muscle, *Respir. Physiol. Neurobiol.* **146**, 259–268.
12. Caldwell, J.H., Martin, G.V., Raymond, G.M, and Bassingthwaighte, J.B., 1994, Regional myocardial flow and capillary permeability-surface area products are nearly proportional, *Am. J. Physiol. Heart Circ. Physiol.* **267**, H654–H666.
13. Binzoni, T., Colier, W., Hiltbrand, E., Hoofd, L., and Cerretelli, P., 1999, Muscle O_2 consumption by NIRS: a theoretical model, *J. Appl. Physiol.* **87**, 683–688.
14. Pittman, R.N., 2000, Oxygen supply to contracting skeletal muscle at the microcirculatory level: diffusion vs. convection, *Acta Physiol. Scand.* **168**, 593–602.
15. Beard, D.A., 2001, Computational framework for generating transport models from databases of microvascular anatomy, *Ann. Biomed. Eng.* **29**, 837–843.
16. Beard, D.A., Schenkman, K.A., and Feigl, E.O., 2003, Myocardial oxygenation in isolated hearts predicted by an anatomically realistic microvascular transport model, *Am. J. Physiol. Heart Circ. Physiol.* **285**, H1826–H1836.
17. Dash R.K., and Bassingthwaighte, J.B., 2006, Simultaneous blood–tissue exchange of oxygen, carbon dioxide, bicarbonate and hydrogen Ion. *Ann. Biomed. Eng.* **34**, 1129–1148.
18. Popel, A.S., 1989, Theory of oxygen transport to tissue, *Crit. Rev. Biomed. Eng.* **17**, 257–321.
19. Poole, D.C., Behnke, B.J., and Padilla, D.J., 2005, Dynamics of muscle microcirculatory oxygen exchange, *Med. Sci. Sports Exerc.* **37**, 1559–1566.

Chapter 37
Modeling Oxygenation and Selective Delivery of Drug Carriers Post-Myocardial Infarction

Bin Wang[2], Robert C. Scott[1], Christopher B. Pattillo[1], Balabhaskar PrabhakarPandian[2], Shankar Sundaram[2], and Mohammad F. Kiani[1]

Abstract An anatomically realistic mathematical model of oxygen transport in cardiac tissue was developed to help in deciding what angiogenic strategies should be used to rebuild the vasculature post myocardial infarction (MI). Model predictions closely match experimental measurements from a previous study, and can be used to predict distributions of oxygen concentration in normal and infarcted rat hearts. Furthermore, the model can accurately predict tissue oxygen levels in infarcted tissue treated with pro-angiogenic compounds.

Immunoliposome (IL) targeting to areas of inflammation after MI could provide the means by which pro-angiogenic compounds can be selectively targeted to the infarcted region. The adhesion of model drug carriers and immunoliposomes coated with antibody to P-selectin was quantified in a MI rat model. Anti-P-selectin coated model drug carriers showed a 140% and 180% increase in adhesion in the boarder zone of the MI 1 and 4 hours post-MI, respectively. Circulating for 24 hrs, radiolabeled anti-P-selectin immunoliposomes showed an 83% and 92% increase in targeting to infarcted myocardium when injected 0 and 4 hrs post-MI, respectively. Targeting to upregulated adhesion molecules on the endothelium provides a promising strategy for selectively delivering compounds to the infarct region of the myocardium using our liposomal based drug delivery vehicle.

37.1 Introduction

Chronic cardiac failure (CCF) following myocardial infarction (MI) is a major health problem of epidemic proportions. A transmural MI involves a loss of necrotic cardiomyocytes and a proteolysis of extracellular matrix, vasculature and nerves. Subsequent tissue repair restores structural integrity at the infarct

[1]Department of Mechanical Engineering, Temple University, Philadelphia, PA 19122.
[2]CFD Research Corporation, Biomedical Technology Division, Huntsville, AL 35805.
Corresponding author: Mohammad F. Kiani, Department of Mechanical Engineering, 1947 North 12th Street, Philadelphia, PA 19122. Phone: 1-215-204-4644, Fax: 1-215-204-4956, e-mail: mkiani@temple.edu

K.A. Kang et al. (eds.), *Oxygen Transport to Tissue XXIX*

site; it does not involve a significant regeneration of cardiomyocytes. Of necessity, rebuilt myocardium must include a vascular network able to nourish it under diverse metabolic demands. This raises two questions: how and when to regrow a neovasculature at the infarct site.

When to regrow neovasculature? We have applied recently developed techniques in the field of tumor biology to quantify microvascular morphology and oxygenation at the infarct site and determined the optimal time for growing new tissue [1]. A mathematical model of oxygenation in myocardial tissue is being presented here to help resolve this question. A provisional neovasculature appears within 7–14 days after MI, which is not able to deliver sufficient oxygen to avoid regions of hypoxia. At 3–4 weeks post-MI, a dense collagen network has developed which leaves little space to accommodate the newly growing cardiomyocytes [2]. It appears the optimal time to regrow myocardium and its supporting vascular network may be within the first 14 days post-MI.

How to regrow neovasculature? Targeted delivery of pro-angiogenic compounds (e.g. VEGF) to the infarct site, using a new targeted drug delivery approach developed in our laboratory [3], can significantly enhance angiogenesis in post-infarct tissue. The inflammatory cascade is upregulated during the development of myocardial infarction [4, 5]. The endothelium becomes activated and increases its expression of receptors that bind ligands on the leukocytes. The up-regulated expression of endothelial cell adhesion molecules in the scar provides a potential avenue for targeting drugs to the infarct tissue similar to that developed in our laboratory to target drugs and/or other molecules to inflamed tissue [3, 6–8]. In this manner, the MI induced up-regulation of an endothelial cell adhesion molecule(s) within the diseased tissue is used as a target to deliver therapeutic agents (drugs, genes, etc.) selectively to the scar, see Fig. 37.1.

VEGF has been shown to augment perfusion through neovascularization post-MI [9–11]. However, systemic administration of VEGF has many observed and potential side effects [12, 13]. Many of these side effects can be circumvented through a targeted drug delivery approach in which the drugs can be preferentially targeted to the scar via the upregulation of various adhesion molecules in the scar tissue.

A targeted drug delivery system was developed to target drugs (e.g. pro-angiogenic compounds) to infarct myocardium. Liposomes coated with targeting ligands to adhesion molecules (immunoliposomes) upregulated on the

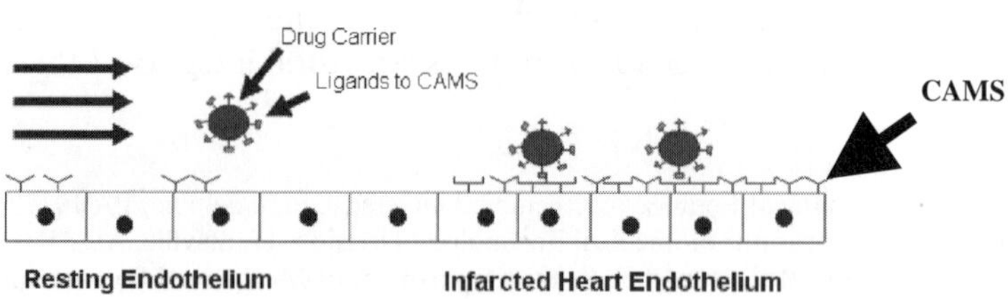

Fig. 37.1 The proposed scheme for delivering drugs to infarcted tissue.

surface of endothelium in the infarct regions were selectively delivered to the infarct regions.

Based on the experimental study [1], a mathematical model was developed and validated, to study oxygen transport from a capillary network in normal areas of infarcted heart, MI area, and losartan treated MI heart. This mathematical model can be used to guide the regrowth of the neovasculature using targeted drug delivery system.

37.2 Methods

37.2.1 Animal Model of Myocardial Infarction

As described previously [1], a rat model of myocardial infarction was used. In brief, following induction of anesthesia with isoflurane, an anterior transmural MI was created by ligation of the left coronary artery with silk ligature. A series of novel techniques from tumor biology were adapted to quantify the components and functionality (i.e., ability to deliver oxygen) of the scar vascular network at 1–4 weeks following MI. This combination of fluorescent and immunohistological stains was then used to define the distribution of distances from cells to the nearest anatomical or perfused vessel.

37.2.2 Mathematical Model of Myocardial Oxygenation in MI Rat Heart

A two-dimensional model of the physiological oxygen transport in tissue was developed using the finite volume Computational Fluid Dynamics code, CFD-ACE+ (ESI-CFD, Huntsville, AL) software on a standard desktop personal computer. It enables steady-state as well as transient analysis in realistic geometries. The CFD code uses fast and efficient numerical methods tailored to solve physiological transport problems in realistic geometries. The method was based on a continuous representation of the tissue around the vessels. All elements (myocyte, vessel wall, interstitial space) in this continuous system were assumed to represent a homogeneous combination of tissue types in this method. This provides a fast and efficient way of determining the effect of changes in oxygen concentration and vessel density on a macroscale in both normal and diseased myocardium. In addition, this was useful in quantitatively analyzing our experimental data. This was done by comparing the model with the experimental data derived previously from the EF5/Cy3 intensity cross section images of infarcted rat heart.

37.2.2.1 Vascular Geometry

The microvascular network geometry of the myocardium is derived from the perfused vessel image taken from the cross-section of heart tissue in rats obtained in previous experimental studies [1]. The outline of every vessel was obtained at the exact anatomic position and size in the rat heart. Vascular geometries were also obtained by randomly eliminating capillaries from an experimentally obtained normal heart microvascular network such that its capillary density would match that of experimentally obtained MI rat heart. The results from these simulated networks were compared with those from experimentally obtained vascular geometries. By randomly eliminating the capillaries, we can test what role vascular distribution per se plays in oxygenation of MI tissue, and we can use this vascular geometry to determine what level of vessel density is needed in the tissue to maintain normal function in cardiomyocytes.

37.2.2.2 Mathematical Model Validation

To determine if our mathematical model can also successfully predict improvements in tissue oxygenation by various pharmaceutical interventions, we validated this mathematical model using the vascular geometry from losartan treated MI rats to simulate the resulting changes in tissue oxygenation. We have shown that losartan, an AT1 receptor antagonist, significantly improves perfusion and reduces tissue hypoxia [1].

37.2.2.3 Transport Equations

Assuming no perfusion and fixed concentration of available oxygen at the vessel interface, the mass conservation equation for oxygen transport in the tissue reduces to

$$\frac{\partial C_{o_2}}{\partial t} = Do_2 \nabla^2 Co_2 - Vo_2(consumption) \qquad (37.1)$$

where t = time, Do_2= oxygen diffusivity in tissue, Co_2 = molar concentration of oxygen, and Vo_2 is the metabolic sink term.

Oxygen consumption was assumed to be homogeneously distributed within the homogeneous tissue according to Michaelis-Menten Kinetics

$$Vo_2 = Vo_{2(M)} * \frac{C_{o_2}}{(C_{o_2} + K_M)} \qquad (37.2)$$

where $VO_{2(\mathrm{M})}$ is defined as the maximal rate of oxygen consumption, CO_2 is oxygen concentration in the tissue and K_M is the effective Michaelis-Menten constant [14]. Matrix-based, stiff kinetics solvers were employed for the reaction kinetics. A second-order central differencing method was used for spatial interpolation of concentration variables and Euler time integration was employed. The model results were then compared to the experimental data previously derived from the EF5/Cy3 image (which shows the hypoxia levels) intensities of the infarcted rat heart.

37.2.2.4 Boundary and Initial Conditions

The vessels served as an inlet with the condition of fixed initial oxygen concentration and a zero flux at the external tissue boundary was imposed based on consideration of symmetry. Therefore, the only flow of oxygen into the tissue components was by diffusion due to the gradient of the oxygen concentration.

The model is used to predict the long-term, steady state distribution of oxygen in the tissue. As an initial condition, oxygen concentration everywhere in the tissue was set at zero; steady state tissue oxygen concentration was then calculated based on the balance of the vascular source and tissue metabolic sinks.

37.2.3 Targeted Delivery of Model Drug Carriers and Immunoliposomes

Model drug carriers (2μm red and blue fluorescent polystyrene microspheres) were used to show that particles can be targeted to MI tissue. We use model drug carriers to show "proof of concept" because they are cheaper and easier to use [15, 16]. The particles were coated with protein A via passive adsorption by incubating the particles in a 0.1M NaHCO3, pH 9.2 buffer containing 300 mg/ml protein A at room temperature for over an hour. The particles were then washed, incubated in a blocking buffer (Hank's balanced saline solution with 1% human serum albumin), washed and incubated with mAbs to anti-P-selectin diluted in blocking buffer. After 1 hour incubation, the mAb coated particles were washed and stored in the blocking buffer prior to use. The mAbs to anti-P-selectin were purchased commercially (e.g. R6D Systems; Minneapolis, MN). Particles were separated from solutions by centrifugation. After a predetermined amount of time post-MI, antibody coated model drug carriers (2×10^8 microspheres of each color) were injected *via* tail vein and allowed to circulate in the blood stream for 1 minute. The animal was euthanized and the heart was quickly removed and washed with a saline solution. The heart was quickly frozen and sectioned (10 μm thick). Images were taken in both the border zone (directly adjacent to the necrotic band in left ventricle wall) and non-infarcted myocardium (taken from right ventricle wall) with two different fluorescent filters (one for red fluorescing antibody coated model drug carriers and one for green fluorescing IgG coated model drug carriers).

Clinically relevant drug carriers (radiolabeled immunoliposomes) bearing mAbs to anti-P-selectin [3] were used to show that drug carriers can be targeted to MI area via upregulated adhesion molecules on the surface of scar endothelial cells. These liposomes are small (50–100 nm) and long circulating with a surface chemistry that is well-suited for immune tagging. The initial characterization of liposomal formulation, surface ligand density, and drug release profile were also performed. After warm 50mM H2CO3 (carbonate) buffer @ pH 7.0 at 40 °C, 3 mL of this buffer were added. Then solution was vortexed for 3 minutes at 40 °C and extruded @ 800 nm × 5, then 400 nm × 5 followed finally by 200 nm × 10. Phosphate assay was performed. 2-IT was added at 20 fold to Ab (buffer used is now pH 8.0) for 60 minutes @ room temperature (RT) following this use a molar ratio of 40:1 maleimide:Ab [17]. This solution was centrifuged at 14K for 10 minutes in a microcone vial then flipped in the vial and spun at 4k for 5 minutes. Then resuspend the Ab in the carbonate buffer @ pH 7.0. It was immediately added to liposomes and incubated at 4 °C overnight, on a shaker set on low speed. The success of the mAb coupling procedure was tested by quantifying the adhesion of immunoliposomes to endothelial cell monolayers in a parallel plate flow chamber as described [8, 18].

Radiolabeled immunoliposomes, 0.1 mL at a concentration of 10 mM, were injected *via* a tail vein and allowed to circulate for a predetermined amount of time. Afterwards, blood was flushed from the animal by injecting saline with a syringe into the left ventricle of the animal and making a small opening with micro-scissors in the right ventricle. The organs, blood, and heart tissue were removed as previously described. The tissue was processed as previously described for measurement in the scintillation counter.

37.2.4 Statistical Analysis

One way analysis of variance with planned contrasts was used to determine significant differences among experimental groups. Kolmogorov-Smirnov Test (StatGraphics Plus, Manugistics Inc.) was used to compare frequency distributions. $P<0.05$ was considered to be statistically significant.

37.3 Results

As shown in Fig. 37.2, throughout the 1–4 weeks post-MI, the progressive and significant increase in measured tissue hypoxia observed experimentally was successfully predicted by our mathematical model. From our previous study [1], the optimal time to regrow and/or rebuild myocardium and its supporting vascular network is before 14 days post-MI. Our mathematical model indicates that 2 weeks post-MI 29% of the myocardium is severely hypoxic ($PO_2<1.25$ mmHg) and that a 220% increase in vessel density is required to

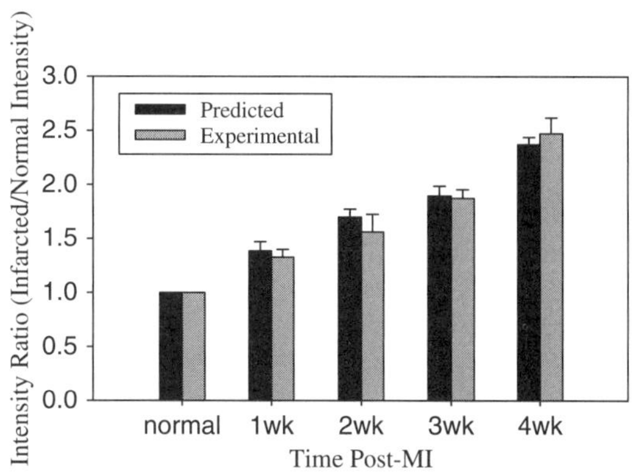

Fig. 37.2 The mathematical model developed in this study can successfully predict experimentally measured hypoxia levels in 1–4 weeks post-MI tissue.

ensure that no areas of the myocardium are severely hypoxic. Previously we have shown that losartan, an AT1 receptor antagonist, significantly improves perfusion and reduces tissue hypoxia [1]. Our model successfully predicted the changes in oxygenation induced by losartan treatment in infarcted tissue (Fig. 37.3).

We have investigated the adhesion of polystyrene microspheres (used as model drug carriers) coated with a mAb to anti-P-selectin to the vascular endothelium of rat scar and adjacent normal tissue. Figure 37.4 shows that in the scar a large number of anti-P-selectin conjugated microspheres adhere to the scar microvasculature while no IgG coated microspheres adhere to the same region. These results are summarized in Fig. 37.5 indicating a 3-fold increase in adhesion of anti-P-selectin coated microspheres in the infarcted region compared to normal myocardium.

The accumulation of radiolabeled anti-P-selectin coated immunoliposomes in the infarcted tissue was compared with the accumulation of the radiolabeled anti-P-selectin coated immunoliposomes in the remaining non-infarcted tissue

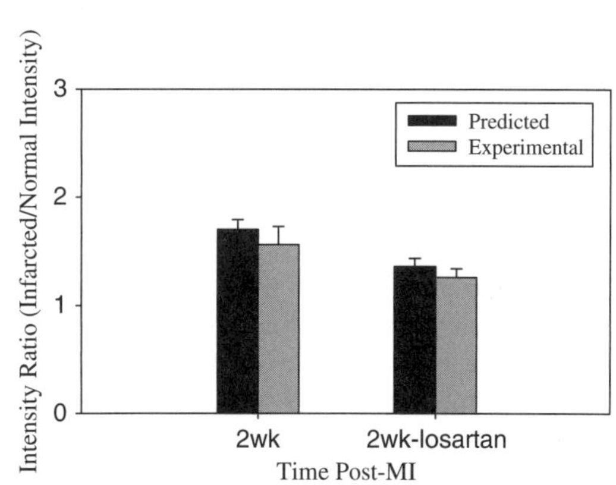

Fig. 37.3 The mathematical model developed here can successfully predict changes in tissue oxygenation in losartan treated animals.

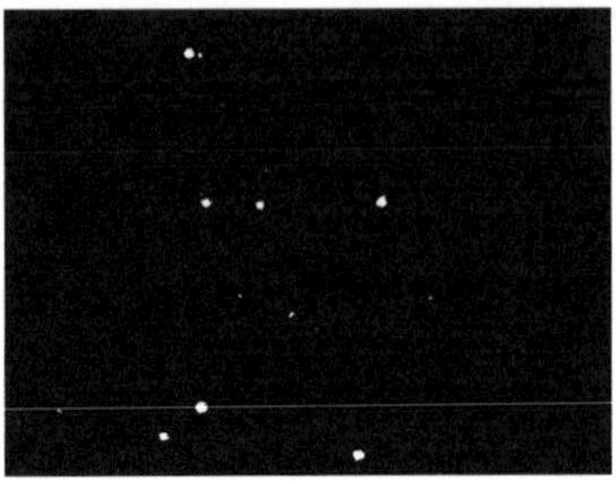

Fig. 37.4 Anti-P-selectin coated model drug carriers adhere preferentially to the infarcted myocardium (white dots) as compared to IgG coated model drug carriers in the same tissue 4 hours post-MI.

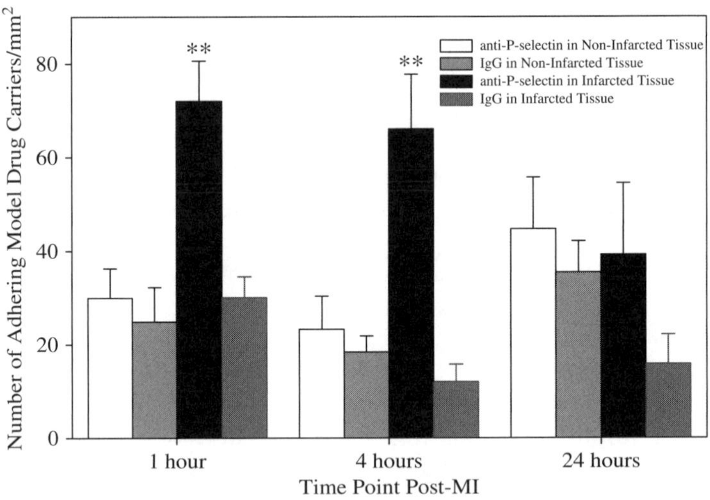

Fig. 37.5 Anti-P-selectin coated model drug carriers were found to adhere preferentially to myocardium in the border zone near the infarcted area at 1 and 4 hours post-MI (140% and 180%, respectively).

at different time periods post-MI. Anti-P-selectin coated immunoliposomes injected immediately post-MI and allowed to circulate for 24 hours showed a significant ($P = 0.01$) increase in adhesion (83%) to the infarct region as compared to the non-infarcted myocardium (Fig. 37.6). Anti-P-selectin coated immunoliposomes injected 4 hours post-MI and allowed to circulate for 24 hours showed a significant ($P = 0.04$) increase in adhesion (92%) to the infarcted region as compared to the non-infarcted myocardium (Fig. 37.6). These findings indicated that anti-P-selectin mAb can be used to selectively target infarct tissue post-MI using our liposomal drug carrier. In a smaller group of animals ($n = 3$), anti-P-selectin coated immunoliposomes were injected immediately post-MI and allowed to circulate for only 4 hours. This experiment showed a significant ($P = 0.03$) but small increase in adhesion (34%) to the infarcted region as compared to the non-infarcted myocardium

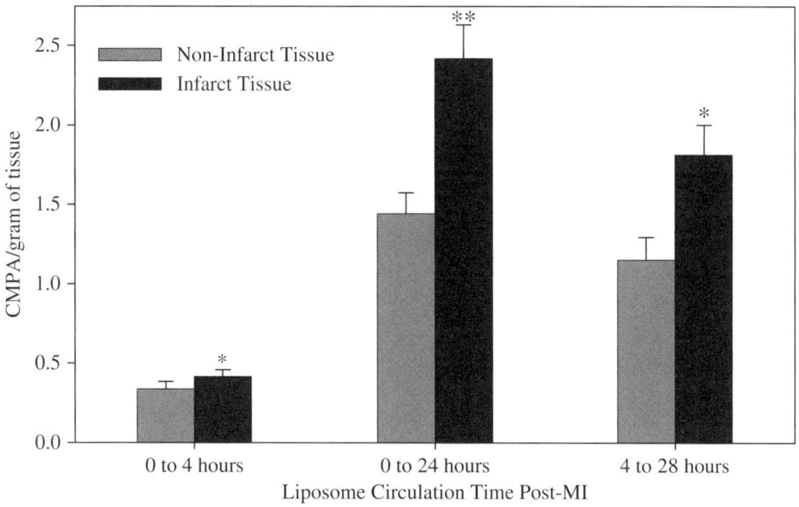

Fig. 37.6 Anti-P-selectin coated immunoliposomes (IL) were found to preferentially accumulate in the infarcted myocardium at various time points: 0 to 4 hrs (injected 0 hr post-MI, measured 4 hrs post-MI), 0 to 24 hours (injected 0 hr post-MI, measured 24 hrs post-MI), 4 to 28 hours (injected 4 hrs post-MI, measured 28 hrs post-MI).

(Fig.37. 6). These observations indicate that this innovative technique can be used to target drug carriers to select tissue via the up-regulation of adhesion molecules expressed on endothelial cells in response to MI.

37.4 Discussion

We have developed a series of experimental and mathematical techniques to characterize vascularity, perfusion, and levels of hypoxia of the scar vascular network up to 4 weeks post-MI in rats [1]. Utilizing microvascular anatomy of cardiac tissue based on available morphometric images, our model can be used to predict distributions of oxygen concentration in normal and infarcted rat hearts, as well as in infarcted tissue treated with pro-angiogenic compounds such as losartan. From the minimum oxygen concentration myocytes need to maintain their normal function, we can calculate the number of new perfused vessels needed in the heart to avoid tissue hypoxia, guiding our work of rebuilding vascular networks and myocardium. Our findings indicate that the optimal time to regrow and/or rebuild myocardium and its supporting vascular network is before 14 days post-MI.

Selectively targeting pro-angiogenic compounds in the infarcted myocardium represents an innovative approach for rebuilding damaged tissue. Our findings indicate that anti-P-selectin coated model drug carriers and liposomes can be preferentially targeted to infarcted regions in the myocardium post-MI. Utilizing the liposomal drug carriers, the upregulation of P-selectin due to inflammation and hypoxia can be used to deliver various compounds (e.g. pro-angiogenic agents such as VEGF) to diseased tissue. We observed a significant accumulation of clinically relevant anti-P-selectin coated immunoliposomes in the infarct region within the first 24 hours post-MI. These findings could have significant clinical implications in that delivering a 100–200% increased dose of a drug to the diseased area with no increase in the dose to normal tissue could yield a large benefit to the patient without an increase in side effects of the drug.

The development of a minimally invasive treatment for regenerating lost vasculature after a myocardial infarction would be very beneficial in our attempt at long term treatment of heart disease. This treatment could provide a neovasculature for other treatments (e.g. stem cells) which are limited by the lost blood flow after the onset of an MI. We are currently developing a methodology by which pro-angiogenic compounds can be selectively delivered to post-MI tissue.

Acknowledgment We thank Dr. Andrew Issekutz for providing us with the rat anti-P-selectin antibody. Bin Wang is a pre-doctoral fellow of the American Heart Association. Mohammad F. Kiani is an Established Investigator of the American Heart Association.

References

1. B. Wang, R. Ansari, Y. Sun, A.E. Postlethwaite, K.T. Weber, and M.F. Kiani. The scar neovasculature after myocardial infarction in rats. Am. J. Physiol Heart Circ. Physiol. **289,** H108 (2005).
2. Y. Sun, J.P. Cleutjens, A.A. Diaz-Arias, and K.T. Weber. Cardiac angiotensin converting enzyme and myocardial fibrosis in the rat. Cardiovasc. Res. **28,** 1423 (1994).
3. C.B. Pattillo, F. Sari-Sarraf, R. Nallamothu, B.M. Moore, G.C. Wood, and M.F. Kiani. Targeting of the antivascular drug combretastatin to irradiated tumors results in tumor growth delay. Pharm. Res. **22,** 1117 (2005).
4. B. Sun, H. Fan, T. Honda, R. Fujimaki, A. Lafond-Walker, Y. Masui, C.J. Lowenstein, and L.C. Becker. J. Activation of NF kappa B and expression of ICAM-1 in ischemic-reperfused canine myocardium. Mol. Cell Cardiol. **33,** 109 (2001).
5. Y. Sun, M.F. Kiani, A.E. Postlethwaite, and K.T. Weber. Infarct scar as living tissue. Basic Res. Cardiol. **97,** 343 (2002).
6. H. Yuan, M.W. Gaber, T. McColgan, M.D. Naimark, M.F. Kiani, and T.E. Merchant. Radiation-induced permeability and leukocyte adhesion in the rat blood-brain barrier: modulation with anti-ICAM-1 antibodies. Brain Res. **969,** 59 (2003).
7. K.T. Weber, I.C. Gerling, M.F. Kiani, R.V. Guntaka, Y. Sun, R.A. Ahokas, A.E. Postlethwaite, and K.J. Warrington. Aldosteronism in heart failure: a proinflammatory/fibrogenic cardiac phenotype. Search for biomarkers and potential drug targets. Curr. Drug Targets. **4,** 505 (2003).
8. M.F. Kiani, H. Yuan, L. Smith, M.W. Gaber, and D.J. Goetz. Targeting microparticles to select tissue via radiation-induced upregulation of endothelial cell adhesion molecules. Pharm. Res. **19,** 1317 (2002).

9. K. Suzuki, B. Murtuza, R.T. Smolenski, I.A. Sammut, N. Suzuki, Y. Kaneda, and M.H. Yacoub. Cell transplantation for the treatment of acute myocardial infarction using vascular endothelial growth factor-expressing skeletal myoblasts. Circulation **104,** I207 (2001).
10. S.B. Freedman and J.M. Isner. Therapeutic angiogenesis for coronary artery disease. Annals of Internal Medicine **136,** 54 (2002).
11. D.A. Engler. Use of vascular endothelial growth factor for therapeutic angiogenesis. Circulation **94,** 1496 (1996).
12. S.E. Epstein, R. Kornowski, S. Fuchs, and H.F. Dvorak. Angiogenesis therapy: amidst the hype, the neglected potential for serious side effects. Circulation **104,** 115 (2001).
13. R.J. Lee, M.L. Springer, W.E. Blanco-Bose, R. Shaw, P.C. Ursell, and H.M. Blau. VEGF gene delivery to myocardium: deleterious effects of unregulated expression. Circulation **102,** 898 (2000).
14. T.B. Bentley, H. Meng, and R.N. Pittman. Temperature dependence of oxygen diffusion and consumption in mammalian striated muscle. Am J Physiol **264,** H1825 (1993).
15. H. Yuan, D.J. Goetz, M.W. Gaber, A.C. Issekutz, T.E. Merchant, and M.F. Kiani. Radiation-induced up-regulation of adhesion molecules in brain microvasculature and their modulation by dexamethasone. Radiat. Res. **163,** 544 (2005).
16. E.E. Burch, P. Shinde, R.T. Camphausen, M.F. Kiani, and D.J. Goetz. The N-terminal peptide of PSGL-1 can mediate adhesion to trauma-activated endothelium via P-selectin in vivo. Blood **100,** 531 (2002).
17. M. Fleiner, P. Benzinger, T. Fichert, and U. Massing. Studies on protein-liposome coupling using novel thiol-reactive coupling lipids: influence of spacer length and polarity. Bioconjug. Chem. **12,** 470 (2001).
18. B. Prabhakarpandian, D.J. Goetz, R.A. Swerlick, X. Chen, and M.F. Kiani. Expression and functional significance of adhesion molecules on cultured endothelial cells in response to ionizing radiation. Microcirculation. **8,** 355 (2001).

Chapter 38
Hypobaric Hypoxia Reduces GLUT2 Transporter Content in Rat Jejunum more than in Ileum

Elaine M. Fisher[1], Xiaoyan Sun[2], Bernadette O. Erokwu[2], and Joseph C. LaManna[2]

Abstract To define some of the specific cellular effects of chronic hypoxia on the small intestine, we measured the concentration of glucose transporter 2 (GLUT2) at two sites, the jejunum and ileum. Wister rats were subjected to 21-day normoxia ($n = 6$) or to continuous 21-day hypobaric hypoxia approximately 0.5 ATM ($n = 5$). Western blot analysis was performed and the abundance of GLUT2 protein was quantified as relative densitometric units and normalized to actin. GLUT2 content was similar in the jejunum and ileum under normoxic (jejunum = 0.65 ± 0.13 mean±SD; ileum = 0.56 ± 0.22 OD; mean difference 0.09, $p = 0.09$) and hypoxic conditions (jejunum = 0.56 ± 0.14 OD mean ± SD; ileum = 0.58 ± 0.16; mean difference -0.01, $p = 0.42$). GLUT2 decreased by 14% of the mean normoxic jejunal levels whereas ileal GLUT2 was slightly increased. A maximum decline in weight of 15% occurred at day 4 followed by a blunted rate of weight gain for rats in the hypoxic group. Thus, sustained exposure to hypobaric hypoxia reduced the availability of GLUT2 for glucose transport in the jejunum. Regulating small intestinal content of glucose transporters may be an important mechanism for tissue adaptation to chronic hypoxia. This finding provides initial insight into hypoxic tolerance of the gut to chronic hypobaric hypoxic exposure.

38.1 Introduction

The gut is an important contributing organ for maintaining whole body glucose homeostasis during normal physiological adaptation. Likewise, it plays a significant role in the protective, restorative, and pathological response to oxidative challenge. At altitudes greater than 4300 meters carbohydrate metabolism

[1]The University of Akron, College of Nursing, Akron, OH.
[2]Case Western Reserve University, Department of Anatomy, School of Medicine, Cleveland, OH. Corresponding author: Elaine M. Fisher, e-mail: efisher@uakron.edu

is impaired [1]. Acute anorexia, nausea and vomiting, weight loss, and malaise. Roberts (1996) and others have described an increase in basal glucose uptake, hyperinsulinemia, alterations in catecholamine concentrations, and increased cortisol during acute exposure to altitude (2–7 days) [2]. Thus, the short-term regulation of glucose may be controlled primarily by changes in insulin and stress activation of the hypothalamic-pituitary-adrenal axis and less under the influence of the metabolic rate [3]. Continued exposure to altitude (7–28 days) however, has been characterized by a return of plasma glucose to baseline (sea level) or lower values and a reduction in the metabolic rate [4]. These changes suggest different regulatory pathways are in effect for short versus long-term control.

While studies have reported on whole body glucose transport, few studies have reported the effect on glucose transport beyond the acute phase at altitude. The body's dependence on glucose as a major energy source at altitude led us to examine the hypoxia induced response on the glucose transport system in the small intestine to sustained hypobaric hypoxia. Approximately two-thirds of glucose absorption in the intestine is by carrier-mediated facilitated diffusion. Glucose transport protein 2 (GLUT2), the most abundant transporter in the small intestine, is a high capacity transporter that exhibits low receptor specificity, hence, versatility in the type of hexose sugar transported (glucose, fructose, galactose) [5]. Histological differences along the small intestine suggest absorption may vary by site (jejunum and ileum); therefore, we partitioned the small intestine to account for potential differences. We measured the concentration of GLUT2 at two major absorptive sites in the small intestine, the jejunum and ileum, to define some of the specific cellular effects of sustained hypobaric hypoxia on glucose transport.

38.2 Methods and Materials

38.2.1 Animal Preparation

Eleven male, Wistar rats were fed standard rat chow with unlimited access to food and water. Rats were weighted on day 1, 2, 4, 12, 19, and 21. Rats in the normoxic group were kept at room atmosphere condition throughout protocol. Animals in the hypoxic group (n = 5) were placed in hypobaric chambers for 21 days at a constant pressure of 0.5 ATM, except for up to one-hour twice weekly as needed to change cages. For comparison, 0.5 ATM (380 mm Hg) is the equivalent of exposure to approximately 10% oxygen at 1 ATM. At the end of 21 days, deep anesthesia was induced with 3-5% isoflorane, followed by immediate decapitation, and tissue retrieval. Institutional approval was obtained prior to initiating experiments.

38.2.2 Tissue Retrieval and Processing

The jejunum was excised approximately 1/3 of the way down the length of the small intestine and at the terminal ileum, 2.5 cm proximal to the cecum. Because rats were used for more than one experiment, they were not fasted and food was found in the intestine. Excrement was freed from the intestine by making a longitudinal cut in the intestine and flushing with 4°C phosphate buffered saline solution (PBS). Tissues were blotted, immediately frozen in liquid nitrogen, and stored at –80°C until analysis.

Total protein content was isolated from the jejunum and ileum. Intestinal membrane proteins were solubilized in RIPA buffer (50 mM Tris HCL pH8, 150 mM NaCl, 1% NP-40, 0.5% sodium deoxycholate, 0.1% SDS, 2 mM EDTA, and 1 mM Na_3VO_4), supplemented with one protease inhibitor cocktail tablet (Roche) per 10 mL RIPA. Tissues were homogenized and rotated for 2 hours in the cold room. Tissue lysates were cold centrifuged at 14,000 rpm for 30 minutes and supernatants were collected. Protein concentrations were determined by Bradford protein assay with bovine serum albumin as the standard (Bio-Rad).

38.2.3 Western Blot Analysis

Samples were diluted with 2X Laemmli buffer and boiled at 100°C for 5 minutes. Seventy-five micrograms of total protein were loaded per lane onto a 10% sodium polyacrylamide gel and electrophoresed. Gel proteins were electroblotted onto a nitrocellulose membrane (75 minutes at 100V). The membrane was blocked for 1 hour with 7.5% nonfat milk and incubated in 7.5% nonfat milk overnight at 4°C with polyclonal rabbit C-terminus GLUT2 antiserum diluted 1:1000 (Santa Cruz Biotechnology Inc., Santa Cruz, CA). The membrane was washed with TBST (TBS+ 0.1% Tweens; 10 minutes × 3) followed by 1 hour incubation at room temperature with a secondary antibody, donkey-anti-rabbit IgG (1:7000). A positive control, human whole cell lysate - glioblastoma was used to verify the GLUT2 band location (sc-2411; Santa Cruz Biotechnology Inc., Santa Cruz, CA). An enhanced chemiluminescence detection system (ECL kit, Amersham) was used to visualize the primary antibody immunoreactive protein bands.

38.2.4 Statistical Methods

The abundance of GLUT2 protein was quantified as relative densitometric units and normalized to actin. Data are reported as mean ± SD. A paired t-test was used to compare sites (jejunum, ileum) at condition (normoxia, hypoxia). The level of significance was defined as $p \leq 0.05$.

38.3 Results

38.3.1 General Observations

Hypoxic rats were initially less active than their normoxic counterparts. On visual examination of the small intestine of hypoxic rats at 21-days, it appeared smaller in lumen diameter with greater vascularity and a bloodier dissection than the small intestine of normoxic rats. Stool was present in the intestine of both normoxic and hypoxic rats indicating the animals were eating.

38.3.2 Effect of Hypobaric Hypoxia on Animal Weight

Initial weights were comparable for normoxic and hypoxic rats (range, 201–209 vs. 204–214 g). A maximum decline of 15% from baseline body weight occurred for hypoxic rats on day 4 followed by a blunted rate of weight gain when compared to the normoxic group (range, 253–311 g vs. 302–350 g) (See Fig. 38.1).

38.3.3 GLUT2 Concentration in the Jejunum and Ileum

GLUT2 presented primarily as a tightly spaced doublet at 62 kDa or on occasion as a single GLUT2 band depending on the antiserum or exposure combination used. For analysis purposes, the entire doublet was included when present. Similar concentrations of GLUT2 were found in the jejunum and ileum

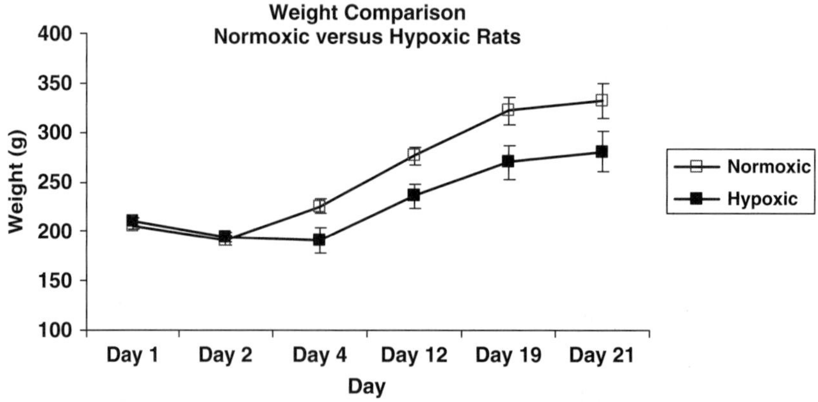

Fig. 38.1 A maximum decline of 15% in the average weight from baseline occurred at Day 4 followed by a parallel but lower increase in weight gain in hypobaric-hypoxic rats by Day 21 (mean ± SD).

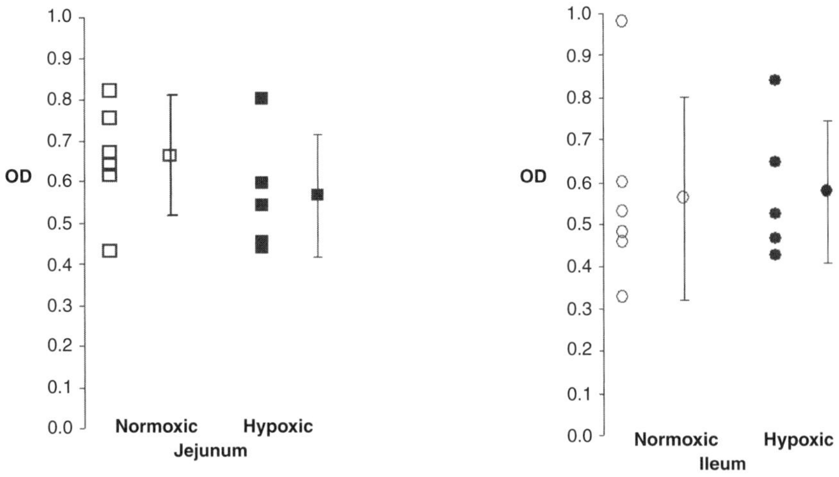

Fig. 38.2 A 14% reduction in the average concentration of GLUT2 transporters was found between normoxic and hypoxic animals.

under both normoxic (0.65 ± 0.13 OD, mean $\pm$ SD and $0.56 \pm .22$; mean difference 0.09, $p = 0.09$) and hypobaric hypoxic conditions ($0.56 + 0.14$ OD, mean $\pm$ SD and $0.58 + 0.16$; mean difference 0.01, $p = 0.42$) (See Fig. 38.2). The jejunum to ileum ratio under normoxic conditions was 1.2 ± 0.3 which fell to 1.0 ± 0.2 during hypoxia.

Figure 38.3 represents a comparison of jejunal GLUT2 and ileal GLUT2 content under the conditions of normoxia and hypobaric hypoxia at 21-days. While not statistically significant, a consistent observation on blots revealed a decline in jejunal GLUT2 with hypoxia. GLUT2 decreased to 14% of mean control jejunal levels. Conversely, a small increase (< 3%) in ileal GLUT2 was detected.

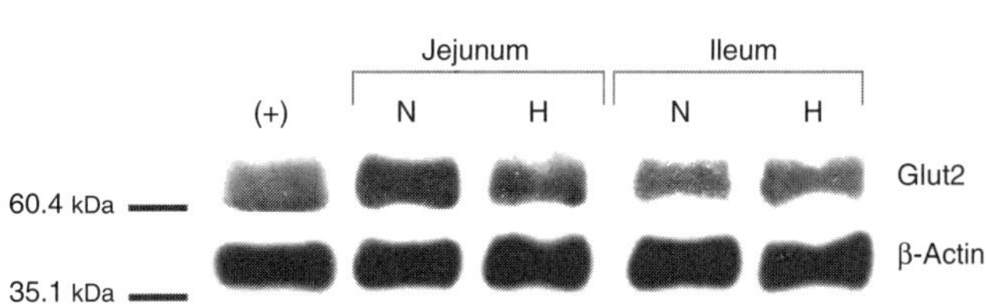

Fig. 38.3 GLUT2 concentration in the jejunum and ileum under normoxic (N) and hypoxic (H) condition, where (+) = positive control, human whole cell lysate, oligoblastoma.

38.4 Discussion

Exposure to a continuous 21 day simulated high altitude of approximately 5300 m resulted in a decline in GLUT2 content in the jejunum while a small increase in ileal GLUT2 was noted. The lack of a substantial difference in ileal GLUT2 may relate to our choice of sample site. To avoid misidentification of tissue between the jejunum and ileum, we sampled from the terminal ileum which may play a lesser role in glucose absorption. The maximal decline in weight loss experienced by rats during acute hypoxia followed by the blunted rate of weight gain when compared to normal controls is consistent with changes reported in the literature.

Glucose homeostasis differs for acute versus prolonged hypoxic exposure at altitude. Acute hypoxic exposure (4300–5500 m) is characterized by increased plasma glucose uptake and hyperinsulinemia, the presence of these two findings is suggestive of insulin resistance [4]. Other studies examining the neurohormonal stress response to acute changes at altitude report transient increases in sympathetic activity and cortisol, with no change in growth hormone. These acute findings suggest mechanisms for the initial changes in glucose are directly regulated by insulin and other stress related hormones, and indirectly influenced by changes in metabolism [6].

Alterations in glucose uptake, insulin release, catecholamines, and cortisol stabilized between 7–10 days of altitude exposure suggesting different mechanisms are responsible for regulating glucose homeostasis for long-term control. With more prolonged exposure to altitude, stabilization of plasma glucose levels at or slightly below sea level values were noted [2]. Previously reported plasma glucose levels at 21-day hypobaric hypoxic condition were 9.4 ± 1 mM, a return toward baseline levels of 9.3 ± 1.4 mM [7]. Plasma glucose is regulated by the liver. The down-regulation we saw in GLUT2 transporters in the jejunum as well as the decline in the jejunal:ileal ratio during hypoxia may be that metabolism is a necessary signal for changes in the transport of glucose across the enterocyte and into the circulation. Diamond and Ferraris hypothesized the number of glucose transporters must be matched to metabolic demand to achieve a maximal effective transport of glucose to the blood. Down-regulation may be adaptive to balance the cost of biosynthesis and maintenance of GLUT2 transporters with changing metabolic demand. Alterations in absorption and transport of glucose as related to metabolism may be explained as: the change in metabolism produces changes in intracellular glucose concentration thereby altering the concentration gradients that determine the facilitated component of transport. While a reduction in metabolic rate seems a likely mechanism, one study examining GLUT transport in heart and skeletal muscle after 28-days of hypobaric hypoxia, reported no change in metabolic enzymes or oxidative capacity (lactate dehydrogenase, pyruvate kinase, hexokinase, citrate synthase, malate dehydrogenase) with hypobaric hypoxia [8].

The longer time period for adaptation may allow for other adjustments to occur in glucose regulation but at present the mechanisms are unknown. For example, activation of apoptotic and/or inflammatory pathways may signal the down-regulation of GLUT2 protein content. Other pathway changes to optimize the use of glucose during chronic hypobaric hypoxia may also occur. Our measurement of GLUT2 using Western blot analysis provided no insight into the efficiency of carrier function in relation to the metabolic state. Therefore, it remains unknown whether adaptation in glucose transport at altitude is in response to the metabolic state or from other triggered pathways.

The down-regulation we noted in GLUT2 to a sustained hypobaric hypoxic stimulus does not hold for all tissue-specific glucose transporters. Dill and colleagues (2001) reported a persistent elevation of GLUT4 in the heart, while in the soleus and plantaris muscle they demonstrated an initial increase in GLUT4 content (7 day) followed by a return to control levels at 28 day hypobaric hypoxia. When cardiac GLUT1 (a second glucose transport protein present in heart tissue) was evaluated, this transporter responded in a manner similar to GLUT2 changes in the small intestine. An increase in GLUT1 transporter occurred at 7 days followed by a decline toward baseline by 28 days. Thus, glucose transporter content is regulated by the duration of exposure to hypobaric hypoxia as well as by a tissue-specific response [8].

In conclusion, chronic exposure to hypoxia reduced the concentration of the glucose transporter GLUT2 in the jejunum with little effect on GLUT2 content in the ileum. Hypobaric hypoxic-induced alterations in basal glucose transport capacity may serve as a protective mechanism against ischemia. Regulating small intestinal content of glucose transporters may be an important mechanism for tissue adaptation to chronic hypoxia. This finding provides initial insight into hypoxic tolerance of the gut to chronic exposure.

Acknowledgment This work was supported by a NIH:NINR Mentored Scientist Award – KO1 NR009787-01 and a Nursing Research Award (The University of Akron, College of Nursing, Akron, OH), both to the first author. Special thanks are extended to Constantinos Tsipis for his assistance with densitometry and figure preparation.

References

1. G.A. Brooks, G.E. Butterfield, R.R. Wolfe, B.M. Groves, R.S. Mazzeo, J.R. Sutton, E.E. Wolfel, and J.T. Reeves, Increased dependence on blood glucose after acclimatization to 4,300 m, J. Appl. Physiol. 70:919 (1991).
2. J.J. Larsen, J.M. Hansen, N.V. Olsen, H. Galbo, and F. Dela, The effect of altitude hypoxia on glucose homeostasis in men, J. Physiol. 504 (Pt 1):241 (1997).
3. K.M. Oltmanns, H. Gehring, S. Rudolf, B. Schultes, S. Rook, U. Schweiger, J. Born, H.L. Fehm, and A. Peters, Hypoxia causes glucose intolerance in humans, Am. J. Respir. Crit. Care Med. 169:1231 (2004).
4. A.C. Roberts, J.T. Reeves, G.E. Butterfield, R.S. Mazzeo, J.R. Sutton, E.E. Wolfel, and G.A. Brooks, Altitude and beta-blockade augment glucose utilization during submaximal exercise, J. Appl. Physiol. 80:605 (1996).

5. C.I. Cheeseman, GLUT2 is the transporter for fructose across the rat intestinal basolateral membrane, Gastroenterology 105:1050 (1993).
6. G.L. Kellett, A. Jamal, J.P. Robertson, and N. Wollen, The acute regulation of glucose absorption, transport and metabolism in rat small intestine by insulin in vivo, Biochem. J. 219:1027 (1984).
7. K. Xu, M.A. Puchowicz, and J.C. LaManna, Renormalization of regional brain blood flow during prolonged, R.P. Dill, S.G. Chadan, C. Li, and W.S. Parkhouse, Aging and glucose transporter plasticity in response to hypobaric hypoxia, Mech. Ageing Dev. 122:533 (2001).
8. R.P. Dill, S.G. Chadan, C.Li, and W.S. Parkhouse, Aging and glucose transporter plasticity in response to hypobaric hypoxia, Mech. Ageing Dev. 122:533 (2001).

Chapter 39
Modeling Oxygen and Carbon Dioxide Transport and Exchange Using a Closed Loop Circulatory System

Brian E. Carlson[1], Joseph C. Anderson[1], Gary M. Raymond[1], Ranjan K. Dash[2], and James B. Bassingthwaighte[1]

Abstract The binding and buffering of O_2 and CO_2 in the blood influence their exchange in lung and tissues and their transport through the circulation. To investigate the binding and buffering effects, a model of blood-tissue gas exchange is used. The model accounts for hemoglobin saturation, the simultaneous binding of O_2, CO_2, H^+, 2,3-DPG to hemoglobin, and temperature effects [1,2]. Invertible Hill-type saturation equations facilitate rapid calculation of respiratory gas redistribution among the plasma, red blood cell and tissue that occur along the concentration gradients in the lung and in the capillary-tissue exchange regions. These equations are well-suited to analysis of transients in tissue metabolism and partial pressures of inhaled gas. The modeling illustrates that because red blood cell velocities in the flowing blood are higher than plasma velocities after a transient there can be prolonged differences between RBC and plasma oxygen partial pressures. The blood-tissue gas exchange model has been incorporated into a higher level model of the circulatory system plus pulmonary mechanics and gas exchange using the RBC and plasma equations to account for pH and CO_2 buffering in the blood.

39.1 Introduction

The exchange of O_2 and CO_2 between the tissue and vasculature depends on adequate delivery and removal of these gases. Oxygen delivery begins with inhalation of ambient air into the airspaces of the lung, transport to the blood from the alveoli, transport through the arterial system, and then exchange between the blood and the peripheral tissue. In a closed circulatory system, venous blood returns to the lungs where CO_2 is expired. Quantifying O_2 and CO_2 transport requires accounting for their solubility in plasma, RBCs and tissue as well as their binding and release from the hemoglobin (Hb) in

[1]Department of Bioengineering, University of Washington, Seattle, Washington, 98195.
[2]Department of Physiology, Medical College of Wisconsin, Milwaukee, Wisconsin, 53226.

K.A. Kang et al. (eds.), *Oxygen Transport to Tissue XXIX*

the RBCs and, in addition, for O_2 only, its binding to myoglobin in tissue. Hemoglobin dissociation curves were developed that described the fraction of O_2 and CO_2 bound to Hb in the steady state as a function of P_{O2}, P_{CO2}, pH, 2,3-DPG and temperature [1]. These expressions were used to describe the steady state transport of O_2 and CO_2 as well as H^+ and HCO_3^- in a blood-tissue exchange model with convective transport and axial diffusion in the capillary along with exchange and metabolism in the surrounding tissue region [2].

The model presented in this study accounts for ventilatory exchange between outside air and lung alveoli, exchange with alveolar capillary blood, convective transport in arteries, the exchange in tissue capillaries and arterioles, and return of venous blood to the lungs. The model describes transport of O_2 and CO_2 to tissue as influenced by respiration rate, composition of inspired gas, H^+ and CO_2 production and O_2 consumption in tissue and buffering in the blood.

A feature of biophysical interest but modest physiological importance is the persistence of disequilibria between plasma and RBC P_{O2} due to the higher velocities of RBC than plasma. This difference in velocity exists in all regions of the vasculature but is at a maximum in the microcirculation. Bloch [3] observed the existence of a layer of plasma close to the vessel wall, which he called the peripheral plasma layer. The average ratio of total layer thickness to vessel inside diameter was 1:4 in the 5-10 μm capillariies in Bloch's study, which agrees with more recent observations of the endothelial surface layer seen by Vink and Duling [4]. Because this layer is close to the capillary wall the velocity of the plasma in that region is slower than the centerline axial velocity of the RBCs. To quantify the relative velocity ratio of RBC to plasma we have looked at indicator dilution studies that document the mean transit time of RBC-tagged versus plasma-tagged indicators by Goresky [5]. Goresky showed that the mean transit time of RBCs was on the order of 2/3 of that of the plasma through the entire hepatic vasculature.

39.2 Description of the Model

39.2.1 Lung-Blood Exchange Region

The lung module is composed of three serial compartments [6]: a low compliance compartment representing the oral/nasal cavity and the cartilaginous airways, a moderately compliant compartment characterizing the collapsible bronchial airways and a high compliance compartment resembling the alveolar space. In the model, the lung can be ventilated by positive pressure or by periodic chest expansion reducing intrapleural pressure, both resulting in bidirectional airflow and inflation and deflation of the lung. Convective flow between compartments is modeled as pressure-driven flow through a resistance. Equations for convective and diffusive transport of oxygen and carbon dioxide between adjacent lung compartments are similar to those used in previous models [6]. Each lung compartment is assumed to be well-mixed. The alveolar compartment exchanges respiratory gases with the plasma region of a blood-tissue exchange unit [2], which contains a region of red cells surrounded by a plasma region. The inhaled partial pressure of each species is a model input.

39.2.2 Blood-Tissue Exchange Region

Figure 39.1 shows a previously described one-dimensional blood-tissue exchange model [2]. The lumen of a permeable vessel is divided into a flowing core of RBCs surrounded by a plasma sleeve. Hemoglobin binding is represented in the RBC region by the invertible hemoglobin dissociation expressions [1]. The exchange vessel endothelial barrier is treated as purely passive and is surrounded by an interstitial fluid layer giving access to parenchymal cells. In the parenchymal cells, where myoglobin buffers O_2, oxygen is consumed and CO_2 is produced in accord with the respiratory quotient: RQ = moles CO_2 per mole O_2 and a specified metabolic rate. Each region is axially-distributed, exhibiting concentration gradients from entrance to exit but well-mixed radially, and is represented by a 1-dimensional partial differential equation. Interregional conductances, defined by the permeability surface area product (PS_x), can be adjusted to accommodate slow diffusional processes. Axial diffusion or dispersion (D_x) smoothes the axial concentration gradients.

The ratio of RBC to plasma velocity in the blood tissue exchange region was calculated using experimental morphometric data on intramyocardial arterioles

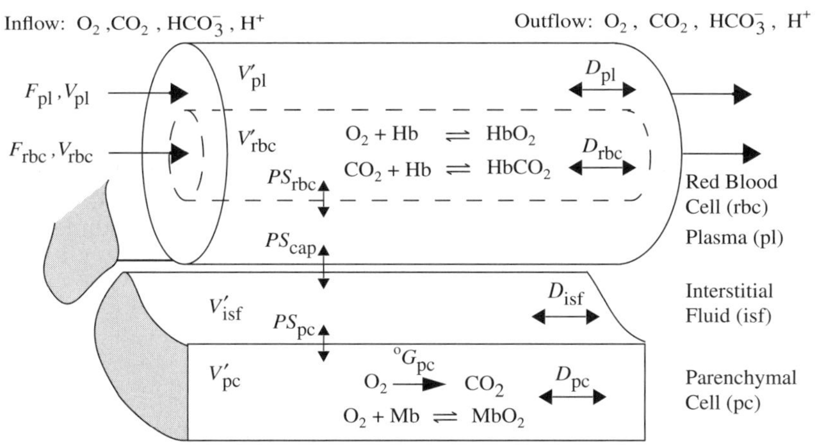

Fig. 39.1 Blood-tissue exchange unit showing red blood cell, plasma, interstitial fluid, and parenchymal cell regions, convection of RBC and plasma, solute transport between regions, the PSs, axial diffusion, binding, and buffering within regions. Bicarbonate buffering occurs in all regions. Each region is axially-distributed and radially well-mixed.

and venules from Kassab et al [7]. and the reduction in hematocrit as a function of vessel diameter documented by Lipowsky [8]. In our model we have represented the blood tissue exchange region as the arterioles and capillaries smaller than 100 μm in diameter because studies by Duling and Berne [9] show significant oxygen loss in the precapillary arterioles. For porcine coronary branching trees of Kassab, flow through the blood tissue exchange region has a relative velocity ratio, v_{RBC}/v_{pl}, of 1.2 to 1.9. Therefore an intermediate value of v_{RBC}/v_{pl}, was taken as 1.5 in lung and tissue exchange regions.

39.2.3 Arterial and Venous Convective Regions

Lung capillary blood is carried through nonexchanging convective conduits (left atrium, left ventricle, aorta, arteries) to the tissue. RBC-plasma solute exchange continues. Red cell-plasma concentration differences persist when their velocities differ. Metabolism is zero, but buffering reactions continue. Values of the ratio of RBC to plasma velocity in the arteries and veins were taken to be 1.1 and 1.05 respectively, though the choice of these numbers is dependent on the range of diameters represented.

39.2.4 Numerical Methods and Simulation Procedures

Numerical methods are those described previously [10] using a Lagrangian sliding fluid element algorithm. Other partial differential equation solvers (MacCormack, TOMS 731) are also available under the JSim simulation system. Parameter values in this model were those used previously [2] with a few exceptions. The flow rate is allowed to range from 5 L/min at rest to 25 L/min during exercise. Also the permeability surface area product for the RBCs (PS_{RBC}) has been recalculated according to experimental evidence [11] to reflect the combination of a small hindrance to permeation of its membrane and the times required for binding and unbinding. The hemoglobin equations incorrectly assume instantaneous equilibration instead of taking several milliseconds. The lung model described here is similar in structure to the model by Lutchen et al. [6].

39.3 Results and Discussion

Two cases are examined in this study. The first concerns respiratory gas transport during normal and elevated ventilation, perfusion, and metabolism. For normal levels, the parameters are respiratory rate of 12 breaths/min, tidal volume of ~500 ml driven by 10 mmHg of inspiratory positive pressure at the mouth for 2 seconds and inspired air with $P_{O2} = 150$ and $P_{CO2} = 0$ mmHg.

Blood flow was set to ~5 L/min and metabolic rate was adjusted so pulmonary end-capillary blood partial pressures of O_2 and CO_2 were ~100 and ~40 mmHg while venous values were 70 and 45 mmHg. The end tidal P_{O2} was 105 and P_{CO2} was 34 mmHg. The second case focuses on the equilibration of O_2 between plasma and RBCs upon entering the arterial region and the effect of the relative velocity difference between plasma and RBCs on equilibration. To investigate the equilibration, the plasma partial pressure of oxygen is increased from 25 to 100 mmHg and the transient effects are observed. To investigate the effect of the relative velocity difference, a pulsed increase in plasma P_{O2} is applied at the upstream end and the difference between the plasma and RBC P_{O2} is quantified along the length of the arterial region.

39.3.1 Respiratory Gas Transport

Figure 39.2 shows that the breathing cycle causes a cyclical variation in the partial pressures of O_2 and CO_2 in bronchiolar and alveolar air and in capillary plasma. Dispersion along the airways dampens the magnitude of the fluctuations. The partial pressure of CO_2 in plasma is almost unaffected by the ventilatory cycle because it is buffered by the large bicarbonate pool in the blood.

To load the red blood cell (RBC) with oxygen, oxygen moves from the well-mixed alveolar space through the plasma and into the RBC as the red blood cell moves along the length of the capillary. The relative speeds of these processes

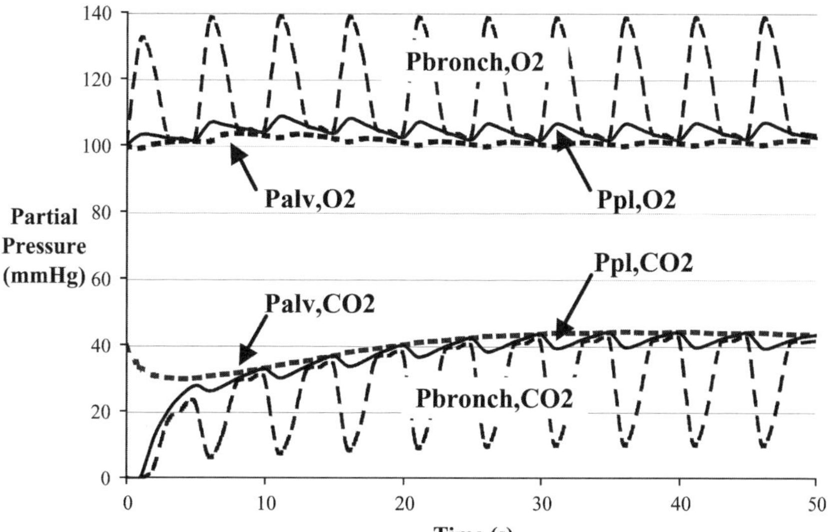

Fig. 39.2 Displays the partial pressures of O_2 and CO_2 in the plasma region of the blood (pl), the alveoli (alv), and the collapsible airways (bronch).

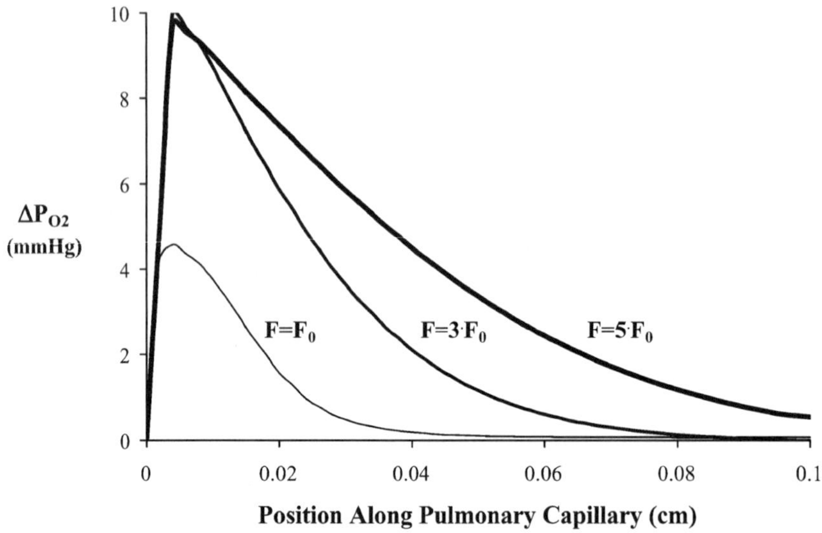

Fig. 39.3 Oxygen partial pressure gradients between the plasma and red blood cell (ΔP_{O2}) at end-exhalation for a normal ventilation rate ($F_{alv,0}$) and blood flow (F_0) and when ventilation, blood flow, and metabolism are increased by 3-times and 5-times the normal rates. $\Delta P_{O2} = P_{O2}$ (plasma) $-$ P_{O2} (RBC).

cause an axial and radial oxygen gradient to be established in the pulmonary capillary. Under normal conditions (given above), a large gradient between the plasma and RBC appears as the blood enters the capillary and disappears after the plasma and RBC have traveled ~40% of the capillary length (Fig. 39.3). However, if the normal ventilation ($F_{alv,0}$) and perfusion (F_0) rate are both increased 3-fold (to offset a corresponding increase in tissue O_2 metabolism), the initial gradient between plasma and RBC increases and the two regions only equilibrate after traversing ~80% of the capillary length. A 5-fold increase in ventilation and perfusion causes the two blood components to never equilibrate while in the pulmonary capillary.

39.3.2 Disequilibrium Between pO_2 in Plasma and RBCs in Arteries

When RBC and plasma velocities are equal then equilibration across the RBC membrane occurs rapidly with a time constant governed by PS_{RBC}/V_{pl}. In contrast, when $v_{RBC}/v_{pl} > 1$ there is a continuing disequilibrium. To show this, the arterial module is isolated from the rest of the closed loop model and beginning with equilibrated P_{O2} at 25mmHg the plasma P_{O2} is increased to 100 mmHg while leaving the RBC P_{O2} unchanged. The relative velocity is given a value of 1.5 to illustrate the relative disequilibrium. In this simulation the concentration front has traveled about 7 cm. before the RBC and plasma concentrations equilibrate. Beyond this, the RBCs near the wavefront of

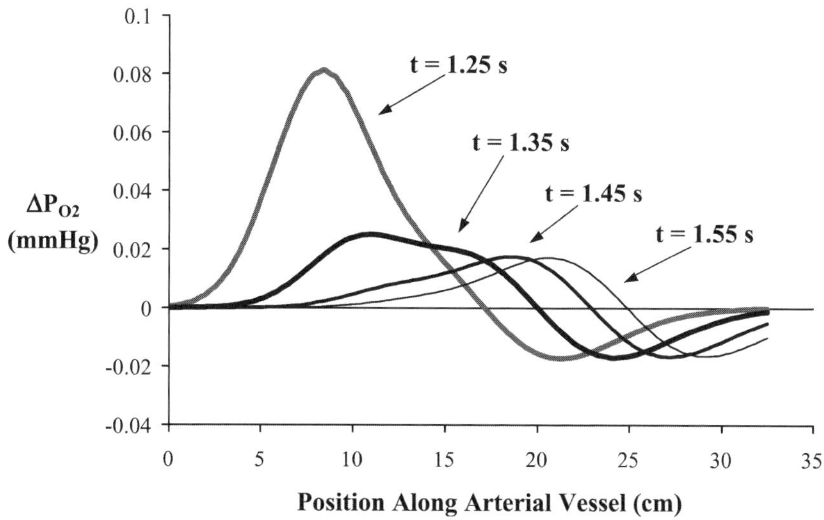

Fig. 39.4 Difference in partial pressure of O_2 between RBC and plasma along vessel length for a Gaussian shaped pulsed increase in plasma P_{O2} at the vessel entry. The four curves represent the pulse at four different times.

heightened plasma P_{O2}, having taken up O_2, advance in the central stream ahead of the plasma front and release O_2 into the plasma where the P_{O2} is still 25 mmHg. This process of RBCs taking up O_2 behind the plasma front and then releasing it to raise the plasma P_{O2} continues. In Fig. 39.4 are shown plots of the ΔP_{O2} across the RBC membrane as a function of position along the aorta at 4 times, a tenth of a second apart after the step increase in plasma P_{O2} at $x = 0$ and $t = 0$. Initially the ΔP_{O2} is $\sim$75 mmHg, but is quickly dissipated as plasma O_2 enters the huge sink of the RBC Hb. Then as the RBCs that have taken up oxygen from the plasma advance ahead of the depleted plasma layer, they have a slightly higher P_{O2} than that in the plasma that had entered the tube before the pulsed increase and therefore lose oxygen to the plasma. The peak in ΔP_{O2} travels at $v_{RBC} = 25$ cm/s. This peak ΔP_{O2} is very small because the carrying capacity for O_2 in plasma is so small compared to that of RBC. The ΔP_{O2} is larger when RBCs are fully loaded as they travel through a region where oxygen is consumed in tissue; it is high when PS_{cap} is high compared to PS_{RBC} and low in the reverse situation. It is relevant to the interpretation of plasma P_{O2} as measured by oxygen-dependent phosphorescent probes [12].

39.4 Conclusions

We have linked together a series of blood-tissue gas exchange models with a model of gas transport in the lung to describe respiratory gas exchange between the lung and tissue via circulating blood. We illustrated that changes in metabolism causing increases in CO_2 production and O_2 consumption can be compensated by simultaneous increases in ventilation and perfusion. O_2 gradients between the plasma and red blood cell can persist along the length of the pulmonary capillary and in the arterial system because RBCs have higher velocities than plasma. The model is ideally suited for investigating questions concerning the integrative effects of pulmonary ventilation, chemical binding kinetics, vascular transport, and tissue metabolism on whole body respiratory gas exchange.

Acknowledgment This research has been supported by NIH/BE-O1973 and HL 073598 and NSF 0506477. Erik Butterworth provided JSim support and assistance in representing this model code in Mathematical Modeling Language (MML). JSim and the model can be downloaded from www.physiome.org.

References

1. R. K. Dash and J. B. Bassingthwaighte, Blood HbO(2) and HbCO(2) dissociation curves at varied O-2, CO2, pH, 2,3-DPG and temperature levels, Ann Biomed Eng 32(12), 1676–1693 (2004).
2. R. K. Dash and J. B. Bassingthwaighte, Simultaneous blood-tissue exchange of oxygen, carbon dioxide, bicarbonate, and hydrogen ion, Ann Biomed Eng 34(7), 1129–1148 (2006).
3. E. H. Bloch, A quantitative study of the hemodynamics in the living microvascular system, Am J Anat 110(2), 125–153 (1962).
4. H. Vink and B. R. Duling, Identification of distinct luminal domains for macromolecules, erythrocytes, and leukocytes within mammalian capillaries, Circ Res 79(3), 581–589 (1996).
5. C. A. Goresky, A linear method for determining liver sinusiodal and extravascular volumes, Am J Physiol 204(4), 626–640 (1963).
6. K. R. Lutchen, F. P. Primiano and G. M. Saidel, A non-linear model combining pulmonary mechanics and gas concentration dynamics, IEEEE Trans Biomed Eng 29(9), 629–641 (1982).
7. G. S. Kassab, C. A. Rider, N. J. Tang and Y. C. B. Fung, Morphometry of pig coronary arterial trees, Am J Physiol Heart Circ Physiol 265(1), H350–H365 (1993).
8. H. H. Lipowsky, S. Usami and S. Chien, Invivo measurements of apparent viscosity and microvessel hematocrit in the mesentery of the cat, Microvasc Res 19(3), 297–319 (1980).
9. B. R. Duling and R. M. Berne, Longitudinal gradients in periarteriolar oxygen tension: A possible mechanism for participation of oxygen in local regulation of blood flow, Circ Res 27(5), 669–678 (1970).
10. J. B. Bassingthwaighte, I. S. J. Chan and C. Y. Wang, Computationally efficient algorithms for convection-permeation-diffusion models for blood-tissue exchange, Ann Biomed Eng 20(6), 687–725 (1992).
11. K. Dalziel and J. R. P. O'Brien, The kinetics of deoxygenation of human haemoglobin, Biochem J 78(236–245 (1961).
12. J. M. Vanderkooi, G. Maniara, T. J. Green and D. F. Wilson, An optical method for measurement of dioxygen concentration based upon quenching of phosphorescence, J Biol Chem 262(12), 5476–5482 (1987).

Chapter 40
Effect of Alternate Energy Substrates on Mammalian Brain Metabolism During Ischemic Events

S.S. Koppaka[1,2], M.A. Puchowicz[2,3], J.C. LaManna[2,3], and J.E. Gatica[1,2]

Abstract Regulation of brain metabolism and cerebral blood flow involves complex control systems with several interacting variables at both cellular and organ levels. Quantitative understanding of the spatially and temporally heterogeneous brain control mechanisms during internal and external stimuli requires the development and validation of a computational (mathematical) model of metabolic processes in brain. This paper describes a computational model of cellular metabolism in blood-perfused brain tissue, which considers the astrocyte-neuron lactate-shuttle (ANLS) hypothesis. The model structure consists of neurons, astrocytes, extra-cellular space, and a surrounding capillary network. Each cell is further compartmentalized into cytosol and mitochondria. Inter-compartment interaction is accounted in the form of passive and carrier-mediated transport. Our model was validated against experimental data reported by Crumrine and LaManna, who studied the effect of ischemia and its recovery on various intra-cellular tissue substrates under standard diet conditions. The effect of ketone bodies on brain metabolism was also examined under ischemic conditions following cardiac resuscitation through our model simulations. The influence of ketone bodies on lactate dynamics on mammalian brain following ischemia is studied incorporating experimental data.

40.1 Introduction

The onset and recovery from cardiac arrest has been associated with loss of neurologic function. Ischemia leads to the loss of intracellular cerebral metabolites and increased lactate in brain. Ischemia and onset of reperfusion have been associated with lactate accumulation in brain. The restoration of

[1]Department of Chemical and Biomedical Engineering, Cleveland State University, Cleveland, OH.
[2]Center for Modeling Integrated Metabolic Systems (MIMS), Cleveland, OH.
[3]Department of Anatomy, School of Medicine, Case Western Reserve University, Cleveland, OH.

neurological function is related to the ability of the brain to recover after an ischemic event. The extent of restoration is dependent upon many variables. Energetics is one such variable that is known to play a major role in the degree of outcome. It is known that anaerobic glycolysis, as a result of an ischemic event, is associated with the accumulation of extra-cellular lactate in brain tissue. Understanding the relationship between glucose metabolism, its compartmentation, and energetics can help discern the mechanism responsible for cellular damage during oxidative stress. Theories have been proposed in an attempt to explain these relationships.

One of the widely accepted theories, the "classical" view is still considered by many to be the underlying mechanism explaining brain metabolism. This theory proposes that glucose is taken up by both neurons and astrocytes (review [1]), where it is completely oxidized in both cell types through the aerobic (tricarboxylic acid cycle) and anaerobic (glycolysis) biochemical pathways. Evidence supporting this suggests that glucose is transported to the neurons by the GLUT transporter system and is readily available to the neurons [2,3]. This is contrary to an alternate view point, referred to as the Astrocyte-Neuron Lactate Shuttle (ANLS), which says that lactate is the primary energy substrate to neuron produced through the metabolic pathways in the astrocyte.

In the early 1990s, it was suggested that lactate produced anaerobically in the astrocyte is shuttled through the extra-cellular space to the neuron as its principal energy substrate (ANLS) [4,5]. They proposed that the lactate transport from the astrocyte to the neuron is based on a demand based regulatory mechanism. This theory was thought to unify the coupling between brain activity and the astrocytic-neuronal metabolism providing an additional perspective in the metabolic modeling of the brain.

Dienel and Hertz [6] observed that the production of lactate increases in the brain under certain conditions and that it is important that lactate be cleared from the cells where it has been produced (due to the redox mechanisms of NAD/NADH). However, they seem to disagree with the theory that lactate is re-accumulated and used as a neuronal fuel. Gjedde et al. [7] stimulated the neuron at different degrees and found that neurons in the baseline condition sustain no net import of pyruvate or lactate *in vivo*. The changes in metabolism, in fact, are linked to the additive increase in the efferent and afferent activity of the brain. A critical review of the ANLS suggests that the possibility of lactate being used as a fuel by brain cannot be completely ruled out owing to the heterogeneity of the cells, however, this scenario is limited to certain conditions such as hypoglycemia [1].

More recently, Aubert et al. [8,9] presented a model, an extension of their previous models. They modeled the brain as comprised of four compartments. The brain was artificially stimulated and they studied the flux of lactate in the sub and extra-cellular compartments. Based on their model, they show that ANLS holds well in the control and excited conditions of the neurons. They explain that the two theories do not contradict each other; rather seem to work concomitantly in explaining the mechanism of metabolism. Using [13] C NMR methodology, Hyder et al. [10] emphasize the importance of the contribution of the astrocyte metabolism of glucose to total glucose metabolism.

It is viewed that brain relies solely on glucose supplied by the periphery for its metabolism and also has little anaerobic capacity with only enough glycogen storage for two minutes of glycolysis during oxygen deprivation [11,12]. Experimental evidence strongly indicates that oxidation of alternate energy substrates such as ketone bodies supplement glucose metabolism in brain during conditions of starvation [12,13]. The utilization of ketone bodies by brain during insufficient glucose supply seems logical since the brain's survival is vital. However, the relative percent of ketone bodies used as an oxidative substrate remains unclear but is thought to be dependent on the duration of ketosis and the transporter capacity. It has been reported that the influx of ketone bodies in the brain depends on the blood concentration of the beta hydroxybutyrate and the activity of the monocarboxylate transporter [14]. Studying ketone body metabolism is an approach to testing other metabolic models since their transporter and compartmentalization are different than glucose, but are likely to be similar to those of lactate.

It has been proposed that the long associated fear of ketosis is exaggerated and state that ketone bodies may provide a better source of energy per brain per unit oxygen [15]. Ketone bodies have been theorized to decrease cell death in neurologic and genetic disorders like Alzheimer's and Parkinson's disease. Ketone bodies have been able to protect neurons in culture and this suggests that the altered energy metabolism in the mitochondria contributes to the patho-physiology of the brain diseases [16]. It has been suggested that ketone bodies are produced by astrocytes and then used together with the lactate produced by the astrocytes to sustain neuronal oxidation [17]. Recently, the effect of ketones on neurons for glutamate toxicity on a rat model was examined [18]. They found that pre-treating the hippocampal cell of a mouse with acetoacetate and β-hydroxybutyrate had a protective effect against glutamate toxicity.

Recently, the effect of chronic hypoxia was investigated, during ketosis induced by diet on the lactate and ketone levels in the tissue [19]. They developed a rat model of ketosis by feeding a high fat diet (no carbohydrate) to study glucose and ketone metabolism in brain. In their chronic hypoxic study, they did not find any elevation in the lactate levels due to the adaptation in rat brain, unlike their study for acute hypoxia [20]. However, the chronic or acute hypoxic conditions were found not to interfere with the induced ketosis. The authors found a sustained increased level of ketones in the plasma and the tissue. It was suggested that ketones alter glucose metabolism possibly through the inhibition of glycolysis or by increased lactate disposal.

Based on the different perspectives on brain metabolism, a comprehensive model of brain that addresses the major bio-chemical pathways is needed. Such a model would facilitate the prediction of trends that are typically challenging to gather experimentally. The basic assumption for this model is that part of the lactate required for sustaining the activity of the neuron is derived from the extra-cellular space supplemented at times by the glycolysis in the neuron.

In this study, the effect of ketone bodies on lactate dynamics following the onset of ischemia induced by cardiac arrest was investigated. The hypothesis of this study was that carbon flux balances would shift towards an increase in flux rates coming from the metabolism of ketone bodies with a proportional decrease in the flux of lactate coming from the astrocyte, as predicted by astrocyte-neuron lactate shuttle (ANLS). We validate our model against the experimental data obtained from the study of Crumrine and LaManna [21], and then examine through our model the profiles of lactate and redox (NAD/NADH). This analysis was thought to be valuable to understanding the clinical significance of ketone bodies toward cerebral energy metabolism during ischemic conditions.

40.2 Model Formulation

Although physiologic modeling of the brain began more than a decade ago, the existing models are either incomplete or neglect the necessary compartmentalization of the various domains within and outside the cellular domain. This could be due to the difficulty in simulating the metabolic dynamics in the brain due to various constraints. With this considered, we developed a new

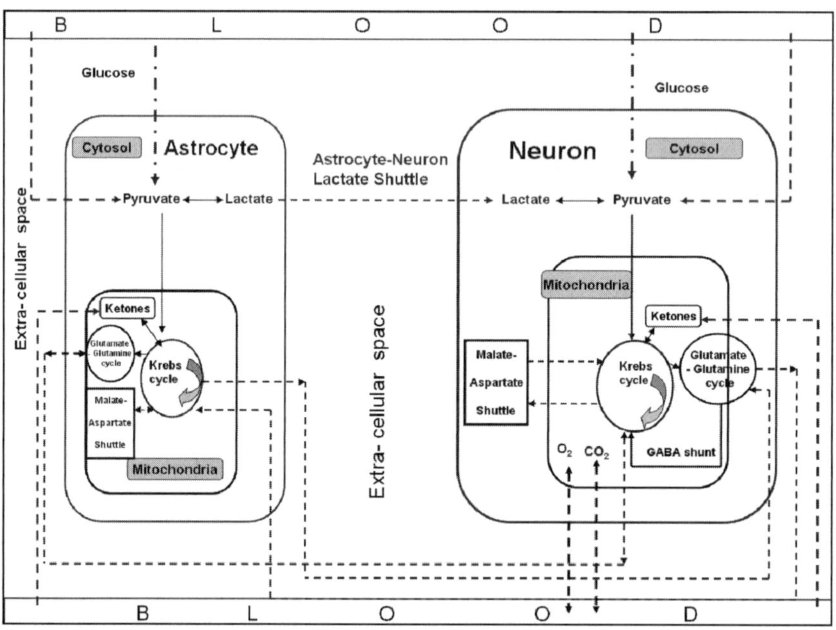

Fig. 40.1 Model diagram of neuro-vascular unit representing various interacting compartments in brain which include: cytosol, mitochondria, extra-cellular space and blood compartments.

multi-domain mathematical model including some of the major pathways of the neuron in the mammalian brain. The approach of this model was to validate it with existing *in vivo* experimental data and then predict physiologic responses which might be difficult to measure experimentally.

Based on the block diagram that shows the components of this model (Fig. 40.1), compartmentalization of the model was formulated with different parameters and concentrations. Flux balances were developed and mass balances of the substrates were formulated as a system of ordinary differential equations representing dynamics of the metabolites. A characteristic neuron cell was modeled into four distinct domains: cytosol, mitochondria, extra-cellular space and the blood compartment. Inter-domain transport exists among the compartments, which could be passive diffusion or carrier-mediated.

40.3 Metabolic Model Components

The model presented in this paper examines the dynamics of 42 key bio-chemical species in brain tissue. The regulators/controllers (NAD^+, NADH, ATP, ADP and Pi) participate in facilitating energy transfer and metabolic regulation. The dynamic mass balance of any species "j" in the blood, transported between the blood and the extra-cellular space can be written as:

$$V_b \frac{dC_{vj}}{dt} = Q(C_{aj} - C_{vj}) - j_{b-xc,\,j} \tag{40.1}$$

where V_b is the blood volume, Q is the blood flow rate, C_{vj} (t) is the venous blood concentration of j, $C_{a\,j}$ (t) the arterial blood concentration, and $J_{b\text{-}x\,c,\,j}$ is the flux from blood to extra-cellular space. The net reaction rate of species "j" in a domain "x", $R_{x\,j}$, can be written as the difference between rate of production of "j", $P_{x\,j}$ and the rate of utilization of "j", $U_{x\,j}$.

$$R_{xj} = P_{xj} - U_{xj} = \sum_{k=1}^{n} \beta_{k,j}\phi_{k,j} - \sum_{k=1}^{m} \beta_{j,k}\phi_{j,k} \tag{40.2}$$

where is the reaction flux of species k forming species j, $\beta_{k,\,j}$ is the corresponding stoichiometric coefficient, n is number of reactions forming j from k, and m is the number of reactions forming k from j. The general form of an equation in our model is:

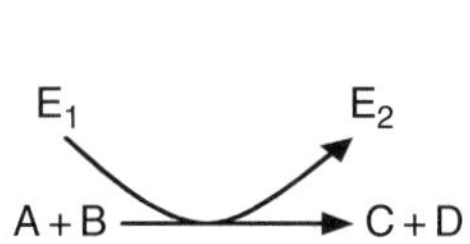

(40.3)

The corresponding reaction flux is assumed to follow a modified Michaelis-Menten form[22] as:

$$\phi_{a-b,c-d} = \frac{V_{a-b,c-d}C_aC_b}{V_{a-b,c-d} + C_aC_b}\left[\frac{PS^{\pm}}{\mu^{\pm} + PS^{\pm}}\right]\left[\frac{RS^{\pm}}{v^{\pm} + RS^{\pm}}\right] \tag{40.4}$$

where $K_{a\text{-}b,\ c\text{-}d}$ and $V_{a\text{-}b,\ c\text{-}d}$ are the Michaelis-Menten coefficients specific to the reaction process. In this model, the two coupled controllers are the phosphorylation state $PS^{+} = C_{ATP}/C_{ADP}$ and the redox state $RS^{+} = C_{NADH}/C_{NAD}$, $PS^{-} = 1/PS^{+}$ and $RS^{-} = 1/RS^{+}$ (for reactions with reverse kinetics). The corresponding controller coefficients $\mu^{\pm}$ and $v^{\pm}$ can vary with each specific reaction process. The derivation and the details of this equation can be found in the cited paper [22].

40.4 Results

In this section, the dynamics of lactate are analyzed during a 30-min ischemia. The qualitative trends of the model results are compared to the trends observed by Crumrine and LaManna [21] who studied lactate dynamics amongst other metabolites during an onset of 30-min ischemia. Figure 40.2 compares our model simulations for tissue lactate trends against the experimental data.

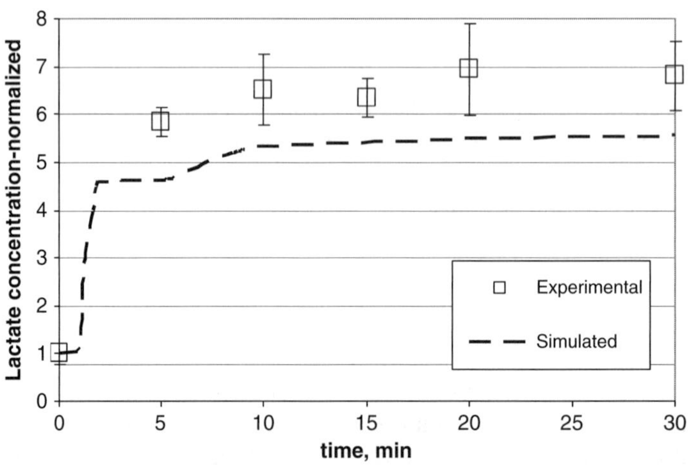

Fig. 40.2 Comparison of the trends of tissue lactate concentration observed in the experimental data (□) gathered [21] (in the cortex of the brain) and model-simulated data (dashed line) generated (in cytosol of the neuron) by the model for ischemia induced over 30 min (no recovery). Plot shows normalized levels of lactate with respect to baseline values. (Normalization refers to ratio of experimental data and model predictions to their corresponding baseline values).

Crumrine and LaManna measured the concentration of tissue lactate in the cortex of the brain tissue which includes lactate in the neuron, astrocyte and the extra-cellular space.

On the other hand, the mathematical model differentiates lactate levels in the cytosol (of the neuron), the extra-cellular space and the plasma. Hence, a qualitative reproduction of the profiles was expected to be a reasonable validation.

A unique feature of the mathematical model is its ability to describe compartmentalization such that the responses of controllers/modulators within specific domains can be predicted and thus different pathways can be analyzed in more detail. Figure 40.3 shows the dynamics of the normalized NAD/NADH ratio in the different intra-cellular compartments. We observe a markedly different dynamic behavior in each compartment. It can be seen that the dynamics of the controllers in the mitochondria have a shorter characteristic time as compared to that of the cytosol.

In Fig. 40.4, the dynamic responses to ischemia are analyzed by following the tissue lactate levels in the standard and ketogenic diets as a function of time. The concentration values are normalized with respect to the tissue lactate concentration baseline value (standard diet condition). It can be seen that, in response to ischemic conditions, the lactate concentration corresponding to the ketotic conditions, exhibits a different dynamics than that observed for standard diet conditions. The lactate levels for ketotic conditions remain below the lactate levels corresponding to the standard diet conditions. This suggests a possible coupling of ketone body metabolism to glycolysis by regulation.

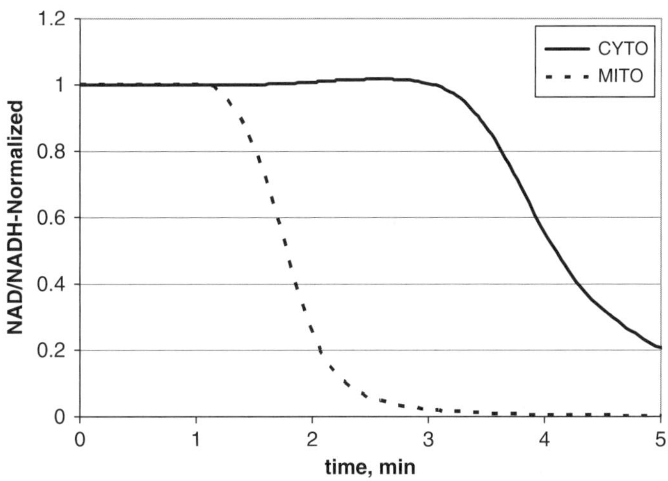

Fig. 40.3 Dynamics of NAD/NADH in the cytosolic and mitochondrial compartments for ischemia induced over 30 min (no recovery) in the neuron. Plot shows normalized dynamic responses up to 5 min (5–30 min truncated).

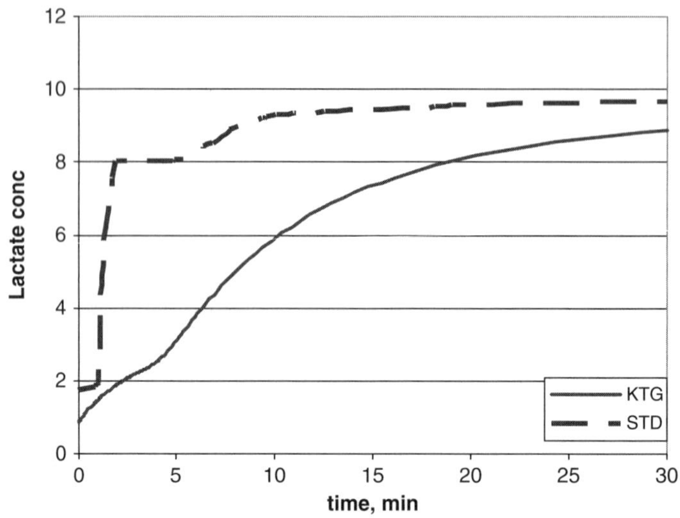

Fig. 40.4 Comparison of the concentrations of tissue lactate in the standard and ketotic conditions for ischemia (concentrations in μ-mol $g^{-1}{}_{ww}$ min^{-1}).

40.5 Discussion and Conclusions

There exist many constraints to studying intact brain metabolism and we realize that an approach to understanding this is through modeling technologies based on in vivo and in vitro data. Most of the current literature on modeling in the brain does not take into account, the possible competition by other fuels at the transport level or the feedback even redox mechanisms. Our model aims at accommodating different aspects of the currently existing theories. This would enable understanding other components which should in theory be testable by the existing models. Hence, this model provides a building block that would serve as a tool towards explaining and testing different theories as well as in incorporating new concepts obtained from newly acquired data. Anyone who wishes to access the model presented in this paper, should contact Dr. Jorge E. Gatica (j.gatica@csuohio.edu) for the latest update of the model.

Using the methodology applied for flux balance analysis, our model predictions for lactate were validated under standard diet conditions against similar experimental conditions from Crumrine and LaManna. The hypothesis of this study was that carbon flux balances would shift towards an increase in flux rates coming from the metabolism of ketone bodies with a proportional decrease in the flux of lactate coming from the astrocyte (as predicted by astrocyte-neuron lactate shuttle). Under the state of ketosis, as a result of feeding a ketogenic diet, our model predictions show a relatively lower accumulation in lactate levels during ischemia than under standard diet conditions. This profile suggests that there is a loose coupling of ketone body metabolism to glucose metabolism in the neuron-astrocyte unit, possibly at the level of glycolysis. Furthermore, ketone body metabolism in brain can be theorized to regulate glycolysis under basal and increased energy demands via a feedback mechanism possibly at the level of citrate.

We anticipate that this work would provide a foundation for further understanding ischemia in the brain under ketotic conditions. Future work would include incorporating blood flow response and metabolite dynamics following ischemia reperfusion.

References

1. C. P. Chih and E. L. Roberts Jr, Energy substrates for neurons during neural activity: a critical review of the astrocyte-neuron lactate shuttle hypothesis, *J Cereb Blood Flow Metab* **23**(11), 1263–1281 (2003).
2. R. L. Leino, D. Z. Gerhart, A. M. van Bueren, A. L. McCall, and L. R. Drewes, Ultrastructural localization of GLUT 1 and GLUT 3 glucose transporters in rat brain, *J Neurosci Res* **49**(5), 617–626 (1997).
3. D. Z. Gerhart, B. E. Enerson, O. Y. Zhdankina, R. L. Leino, and L. R. Drewes, Expression of monocarboxylate transporter MCT1 by brain endothelium and glia in adult and suckling rats, *Am J Physiol* **273**(1 Pt 1), E207–E213 (1997).
4. P. J. Magistretti, L. Pellerin, D. L. Rothman, and R. G. Shulman, Energy on demand, *Science* **283**(5401), 496–497 (1999).
5. L. Pellerin and P. J. Magistretti, Glutamate uptake into astrocytes stimulates aerobic glycolysis: a mechanism coupling neuronal activity to glucose utilization, *Proc Natl Acad Sci U S A* **91**(22), 10625–10629 (1994).
6. G. A. Dienel and L. Hertz, Glucose and lactate metabolism during brain activation, *J Neurosci Res* **66**(5), 824–838 (2001).
7. A. Gjedde, S. Marrett, and M. Vafaee, Oxidative and nonoxidative metabolism of excited neurons and astrocytes, *J Cereb Blood Flow Metab* **22**(1), 1–14 (2002).
8. A. Aubert, R. Costalat, and R. Valabregue, Modelling of the coupling between brain electrical activity and metabolism, *Acta Biotheor* **49**(4), 301–326 (2001).
9. A. Aubert and R. Costalat, Interaction between astrocytes and neurons studied using a mathematical model of compartmentalized energy metabolism, *J Cereb Blood Flow Metab* **25**(11), 1476–1490 (2005).
10. F. Hyder, A. B. Patel, A. Gjedde, D. L. Rothman, K. L. Behar, and R. G. Shulman, Neuronal-glial glucose oxidation and glutamatergic-GABAergic function, *J Cereb Blood Flow Metab* **26**(7), 865–877 (2006).
11. O. H. Lowry and J. V. Passonneau, The relationship between substrates and enzymes of glycolysis in brain, *J Biol Chem* **239,** 31–42 (1964).
12. B. K. Siesjo, Brain Energy Metabolism, (John Wiley & Sons, New York, 1978).
13. O. E. Owen, A. P. Morgan, H. G. Kemp, J. M. Sullivan, M. G. Herrera, and G. F. Cahill, Jr., Brain metabolism during fasting, *J Clin Invest* **46**(10), 1589–1595 (1967).
14. D. Z. Gerhart, B. E. Enerson, O. Y. Zhdankina, R. L. Leino, and L. R. Drewes, Expression of the monocarboxylate transporter MCT2 by rat brain glia, *Glia* **22**(3), 272–281 (1998).
15. R. L. Veech, B. Chance, Y. Kashiwaya, H. A. Lardy, and G. F. Cahill, Jr., Ketone bodies, potential therapeutic uses, *IUBMB Life* **51**(4), 241–247 (2001).
16. Y. Kashiwaya, T. Takeshima, N. Mori, K. Nakashima, K. Clarke, and R. L. Veech, D-beta-hydroxybutyrate protects neurons in models of Alzheimer's and Parkinson's disease, *Proc Natl Acad Sci U S A* **97**(10), 5440–5444 (2000).

17. M. Guzman and C. Blazquez, Is there an astrocyte-neuron ketone body shuttle? *Trends Endocrinol Metab* **12**(4), 169–173 (2001).
18. H. S. Noh, Y. S. Hah, R. Nilufar, J. Han, J. H. Bong, S.S. Kang, G. J. Cho, and W. S. Choi, Acetoacetate protects neuronal cells from oxidative glutamate toxicity, *J Neurosci Res* **83**(4), 702–709 (2006).
19. M. A. Puchowicz, D. S. Emancipator, K. Xu, D. L. Magness, O. I. Ndubuizu, W. D. Lust, and J. C. Lamanna, Adaptation to chronic hypoxia during diet-induced ketosis, *Adv Exp Med Biol* **566** 51–57 (2005).
20. M. A. Puchowicz, Emancipator D.S., K. Xu, and J. C. LaManna, Diet induced ketosis reduces lactate levels in acute hypoxic rat brain, *FASEB* **845.5** (2004).
21. R. C. Crumrine and J. C. LaManna, Regional cerebral metabolites, blood flow, plasma volume, and mean transit time in total cerebral ischemia in the rat, *J Cereb Blood Flow Metab* **11**(2), 272–282 (1991).
22. J. Kim, G. M. Saidel, and M. E. Cabrera, Multi-scale computational model of fuel homeostasis during exercise: effect of hormonal control, *Ann Biomed Eng* **35**(1), 69–90 (2007).

Chapter 41
Cerebral Blood Flow Adaptation to Chronic Hypoxia

Haiying Zhou[1,3], Gerald M. Saidel[1,3], and Joseph C. LaManna[2,3]

Abstract Exposure of rats to mild hypoxia initially increases cerebral blood flow (CBF) as much as two-fold which maintains the arterial oxygen delivery rate. Several days after continued hypoxia, CBF decreases toward its baseline level as the blood oxygen carrying capacity is increased through increased hemoglobin content [1]. Evidently, CBF regulation and the oxygen carrying capacity of blood are correlated. To quantitatively analyze the CBF control mechanisms associated with chronic hypoxia, a mathematical model was developed that describes the concentration dynamics of O_2 and CO_2 transport and metabolic processes in blood and brain tissue. In capillary blood, species transport processes were represented by a one-dimensional convection-dispersion model with diffusion between blood and tissue cells in the cortex and brain stem. Three possible control mechanisms for CBF in response to chronic hypoxia were analyzed: 1) local PO_2 change in cerebral tissue; 2) reduced blood flow associated with elevated blood viscosity from increased Hct; 3) neurogenic input from O_2 receptors in the brain stem. Our hypothesis is that increased PO_2 in the brain stem is the signal for the return of CBF to its baseline condition. This PO_2 increase results from an increase in arterial oxygen carrying capacity and a decrease in local energy metabolism. Model simulations quantify the relative contributions of each of these control mechanisms during 4 days of hypoxic exposure. These simulations are consistent with experimental data that show CBF returns to its baseline even though the cerebral cortical tissue remains hypoxic as indicated by increased levels of the transcription factor Hypoxia Inducible Factor-1 (HIF-1).

[1]Department of Biomedical Engineering, [2]Department of Anatomy, [3]Center for Modeling Integrated Metabolic Systems, Case Western Reserve University, Cleveland, 44106.

K.A. Kang et al. (eds.), *Oxygen Transport to Tissue XXIX*

41.1 Introduction

In mammals (e.g., humans and rats), cerebral blood flow (CBF) increases in response to acute hypoxia [2,3]. The variables that control cerebral circulation are still unclear. Cerebral blood flow in humans does not increase significantly until arterial oxygen partial pressure (PaO_2) decreases below 60 mmHg. According to Brown et al., [4] arterial oxygen content is the major determinant of CBF in humans. In rats, CBF initially increases as much as two-fold in response to hypoxia. As the hypoxic exposure continues, CBF decreases to normoxic baseline in one week and hematocrit increases to restore arterial oxygen concentration (CaO_2) [1,5]. Both the experiments suggest CBF and CaO_2 are correlated and PaO_2 may not have a direct effect on CBF regulation.

The relationships between arterial oxygen content and cerebral blood flow are different under hypoxia and hemodilution conditions. When CaO_2 is decreased, the CBF response is greater in hypoxic animals [6]. This suggests that tissue PO_2 rather than CaO_2 is the predominant factor in controlling CBF. According to Jones et al., [7] CBF is controlled by a mechanism that monitors cerebral O_2 consumption and CaO_2, which implies that tissue PO_2 controls blood flow locally as needed for metabolism.

During hypoxia, CBF and blood viscosity appear correlated, but the relationship ceases when viscosity is high [4]. After accounting for the effects of arterial oxygen content, however, blood viscosity is not a significant factor in CBF regulation. Experiments were performed to separate the effect of arterial oxygen content and hematocrit (Hct) on CBF regulation [8]. At constant PaO_2 and CaO_2, CBF falls after hematocrit is increased by transfusion with red blood cells that contain pure methemoglobin (MHb). Based on this experiment, both CaO_2 and Hct are important in CBF regulation.

In rats, hypoxic excitation of the rostral ventrolateral medulla (RVLM) of the brainstem increases cortical blood flow [9,10]. When the projection effect from RVLM was blocked, the CBF response to normocarbic hypoxia (F_IO_2 = 10%) was decreased by more than 50%, and CBF response to hypercarbia was unchanged. It was implied that neurons of the rostral ventrolateral medulla were oxygen detectors that can globally elevate cerebral blood flow in response to hypoxia. In response to hypoxia, energy metabolism decreases within brain stem regions involved in respiratory and cardiovascular control [11].

Three possible primary mechanisms may be involved in the regulation of the cerebral blood flow response to chronic hypoxia (4 days): 1) local PO_2 change in cerebral tissue (viz., cortical PO_2); 2) reduced blood flow associated with increased Hct that produces higher viscosity; 3) neurogenic input (PO_2) from O_2 receptors in the brain stem. To distinguish and predict the effect of these mechanisms, a mathematical model is needed that can interpret experimental data. To quantitatively analyze the CBF control mechanisms associated with chronic hypoxia, a two-compartment (cortex and brain stem) mathematical model was developed. In capillary blood, species transport processes was represented by a one-dimensional, convection-axial dispersion model with diffusion between blood and tissue cells in the cortex and brain stem. Within the local cortex and brainstem tissues, the concentrations were represented by an average. Model simulations quantify the relative contributions of each of the three control mechanisms during 4 days of hypoxic exposure.

41.2 Methods

41.2.1 Model Description

A model (Fig. 41.1) has been developed to study global CBF regulation during hypoxia. From dynamic mass balances of the cortex (C) and brain stem (S) with distinct metabolic rates, O_2 concentrations are described (Eq. 41.1). The species concentration distribution in the capillary blood from arteriole to venule is represented by a one-dimensional convection-axial dispersion model (Eq. 41.2), in which D is the dispersion coefficient and V_{ib} is the capillary blood volume. The effect of hemoglobin on transport processes is included. This model takes into account the species transport flux (J) between capillary blood and tissue, which depends on the partial pressure differences between the blood and tissue. In the tissue, the species concentration dynamics depend on the mass transport flux between tissue and blood, and metabolic (consumption or production) rate. Since an increase in Hct causes an increase in the effective surface area for O_2 transport [12], the corresponding O_2 and CO_2 transport coefficients are functions of Hct.

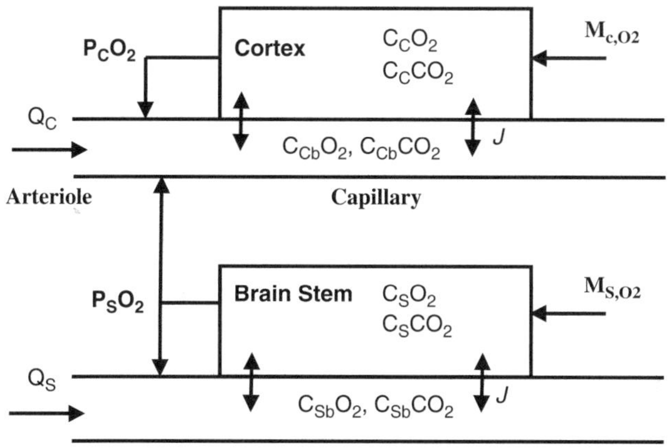

Fig. 41.1 System diagram showing the interaction between cortex and brain stem. $M_{i,O2}$ (i = C,S) : oxygen consumption rate in cortex and brain stem; P_iO_2: O_2 partial pressure in tissue; C_iO_2,C_iCO_2: O_2 and CO_2 concentration in tissue; $C_{ib}O_2$,$C_{ib}CO_2$: O_2 and CO_2 concentration in tissue capillary blood.

$$\frac{dC_iO_2}{dt} = \int_0^{V_{ab}} Jdv - M_{i,O2}; (i = C, S) \tag{41.1}$$

$$\frac{\partial C_{ib}O_2}{\partial t} = -Q_i \frac{\partial C_{ib}O_2}{\partial v} + D \frac{\partial^2 C_{ib}O_2}{\partial v^2} - J \tag{41.2}$$

The expressions (Eqs. 41.3,41.4) for control of cerebral blood flows to cortex (Q_C) and brain stem (Q_S) incorporate the three mechanisms mentioned above. P_CO_2 and P_SO_2 are O_2 partial pressures in the cortex and brain stem. The subscript "0" means baseline value (e.g., Hct_0 is the baseline hematocrit).

$$Q_C = Q_{C0}Hct_0[1 + a(1 - P_CO_2/P_CO_{20}) + b(1 - P_SO_2/P_SO_{20})]/Hct \tag{41.3}$$

$$Q_S = Q_{S0}Hct_0[1 + b(1 - P_SO_2/P_SO_{20})]/Hct \tag{41.4}$$

The general behavior of these controllers is similar. When cortical P_CO_2 and/or the brain stem P_SO_2 decreases, CBF increases; when Hct increases, CBF decreases.

41.2.2 Model Simulations

To simulate the experimental data and predict responses, the model equations were solved numerically. Values of most model parameters were taken from the accepted ranges in previous studies of the rat brain (see Table 41.1). Metabolic rate in the cortex is assumed constant. The O_2 and CO_2 transport coefficients were estimated by matching the model outputs to tissue PO_2 and PCO_2 data under normoxia. The values of parameters in CBF controller were chosen to generate the expected CBF response under hypoxia (see Table 41.1). The control signal from the brain stem is assume to occupy more than 50% [11]. The normoxic steady state of $PaO_2 = 100$ mmHg and $PaCO_2 = 40$ mmHg could be reached by model simulation starting from arbitrary initial conditions. Energy metabolism in brain stem decreases as P_SO_2 decreases in response to hypoxia.

Table 41.1 Parameters in the model

Hct_0	Q_{c0}, Q_{S0} ml/100g/ min	$M_{c,O2}$, $M_{S,O2}$ mlO_2/100g/ min	V_{cb}, V_{sb} ml/ 100g	a $mmHg^{-1}$	b $mmHg^{-1}$	Dl^2/ min	P_CO_{20}, P_SO_{20} mmHg
0.46	120	8.4	3.4	1.0	1.5	5×10^{-4}	15

Cerebral blood flow in response to chronic hypoxia is simulated by changing arterial O_2 and CO_2 partial pressures (PaO_2, $PaCO_2$) and Hct. Initially, PaO_2 decreases quickly from 100 to 40 mmHg, but as the hypoxic exposure continues, PaO_2 slowly increases to 45 mmHg because of hypoxic hyperventilation. Also, $PaCO_2$ decreases quickly from 40 to 30 mmHg. Based on the data of Xu et al., [1]. Hct increases as a mono-exponential function (Eq. 41.5) with time constant $\tau = 127.5\text{h}$

$$Hct = 0.46 + 0.22\left(1 - e^{-\frac{t}{\tau}}\right) \tag{41.5}$$

41.3 Results and Discussion

Model simulation of CBF responses in cortex and brain stem to 4 days of hypoxic exposure are shown in Fig. 41.2. Initially, CBF in cortex increases by 85% (from 120 to 220 ml/100g/min), then decreases slowly to 130 ml/100g/min at the 4th day of hypoxia. Of the initial cortical CBF response, 70% is caused by decreased brain stem PO_2. Initially in the brain stem, CBF increases by 70% and then decreases to a similar level as CBF in cortex at the 4th day of hypoxia.

In response to hypoxia, PO_2 in cortex tissue decreases quickly from 15 to 8 mmHg, then increases slowly to 11 mmHg (Fig. 41.3). In the brain stem, PO_2 decreases quickly to 11 mmHg, then increases slowly to 13 mmHg. The smaller PO_2 change in brain stem is due to the metabolism change during hypoxia. Arterial O_2 concentration (CaO_2) decreases from 9.4 to 7.2 mM at the beginning of hypoxia because PaO_2 decreases. As Hct increases, O_2 concentration increases to 9.7 mM.

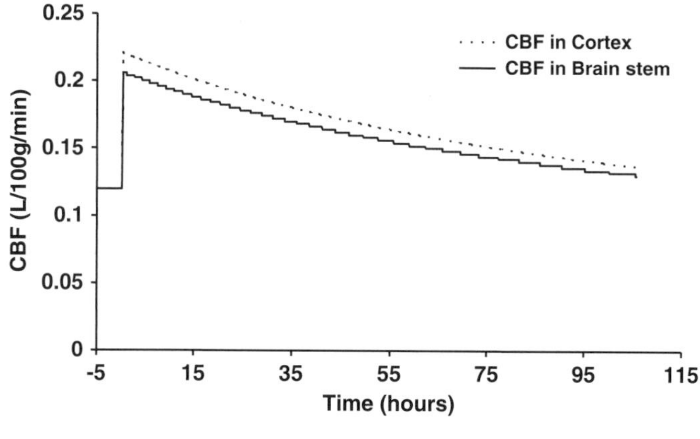

Fig. 41.2 Responses of CBF in cortex and brain stem to 4 days of hypoxia (-5-0 hour: normoxic baseline; 0–100 hour: hypoxia).

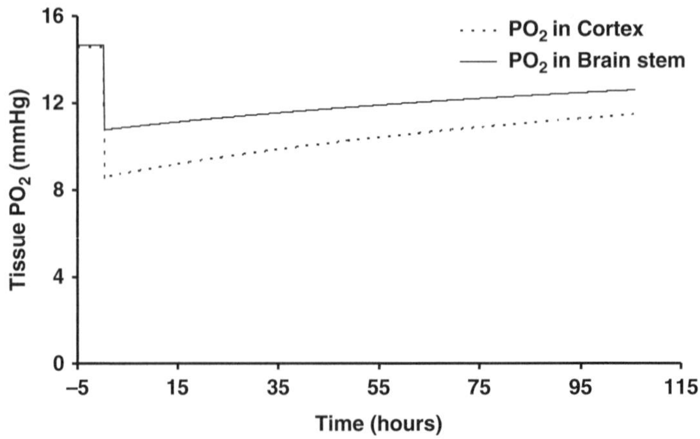

Fig. 41.3 Responses of PO_2 in cortex and brain stem to 4 days of hypoxia.

By examining the values of CBF and CaO_2 during the entire hypoxic exposure, simulations show that oxygen delivery is higher than under normoxic condition. Initially, oxygen delivery increases from the normoxic 0.25 to 0.35 ml O_2/g/min, then slowly declines to 0.3 ml O_2/g/min.

After 4 days of hypoxia, even though CaO_2 has already recovered to the normoxic level, PO_2 in the cortex and brain stem are still less than their normoxic values. This is a consequence of a small transport partial pressure difference between blood and tissue. This simulation is consistent with the experimental measurements by Chávez et al., [13] with HIF-1 as an indicator of tissue hypoxia. HIF-1 initially rose and then fell to about half of the maximum after 7 days, then returned to normoxic level after 21 days. After 4 days of hypoxia, cortical and brain stem PO_2 still can produce an increase CBF. This is offset, however, by the elevated Hct which can decrease CBF by 20% so that the combined effects yield CBF close to the normoxic level.

41.4 Conclusions

The aim of this model is to investigate the cerebral blood flow (CBF) during chronic hypoxia (4 days).The simulation shows that the CBF dynamic response can be explained by the combined effect of tissue PO_2 (cortex and brain stem) and hematocrit (Hct). The initial (minutes) increase of CBF in cortex and brain stem in response to hypoxia is due to decreases in cortical tissue P_CO_2 and brain stem P_SO_2. CBF gradually decreases from the initial increase to the normoxic baseline in response to chronic hypoxia (4 days). This renormalization of CBF is caused by higher Hct: 1) higher Hct causes increase in P_CO_2 and P_SO_2; 2) higher Hct directly decreases CBF (via increased viscosity).

Acknowledgment This work was supported by NIH-NIGMS P50-GM-66309 for Center for Modeling Integrated Metabolic Systems (MIMS) at Case Western Reserve University

References

1. K. Xu, M.A. Puchowicz, and J.C. LaManna. Renormalization of regional brain blood flow during prolonged mild hypoxic exposure in rats. *Brain Res.* 2004; **1027**: 188–191.
2. P.N. Ainslie and M.J. Poulin. Ventilatory, cerebrovascular, and cardiovascular interactions in acute hypoxia: regulation by carbon dioxide. *J. Appl. Physiol.* 2004; **97**: 149–159.
3. J.C. LaManna, L.M. Vendel, and R.M. Farrell. Brain adaptation to chronic hypobaric hypoxia in rats. *J. Appl. Physiol.* 1992; **72**: 2238–2243.
4. M.M. Brown, J.P.H. Wade, and J. Marshall. Fundamental importance of arterial oxygen content in the regulation of cerebral blood flow in man. *Brain* 1985; **108**: 81–93.
5. K. Xu, and J.C. LaManna. Chronic hypoxia and the cerebral circulation. *J. Appl. Physiol.* 2006; **100**: 725–730.
6. M.M. Todd, B. Wu, M. Maktabi, B.J. Hindman, and D.S. Warner. Cerebral blood flow and oxygen delivery during hypoxemia and hemodilution: role of arterial oxygen content. *Am. J. Physiol.* 1994; **267**: H2025–H2031.
7. M.D. Jones, R.J. Traystman, M.A. Simmons, and R. Molteni. Effects of changes in arterial O_2 content on cerebral blood flow in the lamb. *Am. J. Physiol.* 1981; **240**: H209–H215.
8. M.L. Hudak, R.C. Koehler, A.A. Rosenberg, R.J. Traystman, and M.D. Jones. Effect of hematocrit on cerebral blood flow. *Am. J. Physiol.* 1986; **251**: H63–H70.
9. E.V. Golanov, J.R.C. Christensen, and D.J. Reis. Neurons of a limited subthalamic area mediate elevations in cortical cerebral blood flow evoked by hypoxia and excitation of neurons of the rostral ventrolateral medulla. *J. Neurosci.* 2001; **21**: 4032–4041.
10. E.V. Golanov, D.A. Ruggiero, and D.J. Reis. A brainstem area mediating cerebrovascular and EEG responses to hypoxic excitation of rostral ventrolateral medulla in rat. *J. Physiol.* 2000; **529**: 413–429.
11. J.C. LaManna, M.A. Haxhiu, K.L. Kutina-Nelaon, S. Pundik, B. Erokwu, E.R. Yeh, W.D. Lust, and N.S. Cherniack. Decreased energy metabolism in brain stem during central respiratory depression in response to hypoxia. *J. Appl. Physiol.* 1996; **81**: 1772–1777.
12. K. Groebe, and G. Thews. Basic mechanisms of diffusive and diffusion-related oxygen transport in biological systems: a review, In: *Oxygen transport to tissue XIV*, W. Erdman and D. F. Bruley, ed., Plenum Press, New York. 1992; pp. 21–33.
13. J.C. Chávez, F. Agani, P. Pichiule, and J.C. LaManna. Expression of hypoxia-inducible factor-1α in the brain of rats during chronic hypoxia. *J. Appl. Physiol.* 2000; **89**: 1937–1942.

Chapter 42
Mitochondrial Dysfunction in Aging Rat Brain Following Transient Global Ischemia

Kui Xu, Michelle A. Puchowicz, Xiaoyan Sun, and Joseph C. LaManna[1]

Abstract Aged rat brain is more sensitive to reperfusion injury induced by cardiac arrest and resuscitation. The mitochondrial respiratory chain, the major source of free radicals during reperfusion, is likely to be the target of lipid peroxidation. Previous work has shown a higher mortality and lower hippocampal neuronal survival in older rats. 4-hydroxy-2-nonenal (HNE), a major product of lipid peroxidation, was found to be elevated in cortex and brainstem after resuscitation. In this study we investigated the acute changes of mitochondrial function in aging rat brain following cardiac arrest and resuscitation; the effect of an antioxidant, *alpha*-phenyl-*tert*-butyl-nitrone (PBN) was also tested. Fischer 344 rats, 6 and 24-month old, were subjected to cardiac arrest (7–10 minutes) and allowed to recover 1 hour after resuscitation. Mitochondria of cortex and brainstem were isolated and assayed for respiratory function. Compared to their respective non-arrested control group, 1h untreated groups (both 6 month and 24 month) had similar state 3 (ADP-stimulated) but higher state 4 (resting state) respiratory rates. The respiratory control ratio (state 3/state 4) of cortex in the 1h untreated group was 26% lower than the non-arrested control group; similar results were found in brainstem. The decreased mitochondrial respiratory function was improved by PBN treatment. HNE–modified mitochondrial proteins were elevated 1h after resuscitation, with an evident change in the aged. Treatment with PBN reduced the elevated HNE production in mitochondria of cortex. The data suggest (i) there is increased sensitivity to lipid peroxidation with aging, (ii) mitochondrial respiratory function related to coupled oxidation decreases following cardiac arrest and resuscitation, and (iii) treatment with antioxidant, such as PBN, reduces the oxidative damage following cardiac arrest and resuscitation.

[1]Kui Xu, Michelle A. Puchowicz, Xiaoyan Sun, and Joseph C. LaManna, Department of Anatomy, Case Western Reserve University, 10900 Euclid Avenue, Cleveland, OH 44106, USA.

42.1 Introduction

Transient global brain ischemia induced by cardiac arrest and resuscitation results in reperfusion injury associated with oxidative stress [1]. We have previously shown that aged rats are more sensitive to reperfusion injury induced by cardiac arrest and resuscitation. Higher mortality and lower hippocampal neuronal survival were found in the older rats [2]. 4-hydroxy-2-nonenal (HNE), a major product of lipid peroxidation, was found to be elevated in cortex and brainstem after resuscitation [3]. An antioxidant, *alpha*-phenyl-*tert*-butyl-nitrone (PBN) [4], improved the outcome following cardiac arrest and resuscitation [2]. The cellular mechanism that contributes to the age-related changes in brain aerobic capacity remains to be discerned. Mitochondria are both a major source and target of free radicals during insult of oxidative stress. Modification of mitochondrial proteins by lipid peroxidation products HNE, have been described to decrease oxidative capacity [5]. Oxidative damage is therefore likely to play a role in the decline of mitochondrial function [6]. With aging, increased susceptibility to oxidative stress is known to lead to declined mitochondrial function, especially following an ischemic-reperfusion insult [5]. In this study we investigated the effects of aging on mitochondrial function in brain following cardiac arrest and resuscitation. Mitochondrial oxidative capacity was assessed by measuring respiratory rates (state 3/state 4) from freshly isolated mitochondria. Western blot analysis was used to detect HNE-modified mitochondrial proteins. We hypothesized that an antioxidant treatment with PBN in the aged rats would result in improved mitochondrial function following cardiac arrest and resuscitation by attenuation of lipid peroxidation.

42.2 Methods and Materials

42.2.1 Animal Preparation

Male Fischer 344 rats (6- and 24-month-old) were purchased and allowed to acclimate in the animal facility at Case Western Reserve University for one week before being utilized. Surgical procedures for each experiment were as follows: [7] anesthesia was induced by isoflurane (2.5% isoflurane, 70% N_2O in O_2) and maintained with 1–2% isoflurane, 70% N_2O in O_2 through a nasal cone. Cannulae were placed in: (i) Ventral tail artery using polyethylene tubing (PE-50, 0.023" i.d., 0.038" o.d.) for the purpose of monitoring of systemic arterial blood pressure and to obtain samples for blood gas, plasma glucose and lactate determinations (ii) External jugular vein into the right atrium using a Silastic catheter (0.025" i.d., 0.047" o.d.) for administration of drug. After surgery, the rats were allowed to recover for at least 1 hour while restrained in plastic cages. Throughout the experiment, the body temperature was maintained at 37°C by an infrared heat lamp (250W, 45 cm above the body) regulated by feedback from a rectal probe.

42.2.2 Induction of Total Cerebral Ischemia in Rat

Reversible total cerebral ischemia was achieved using a cardiac arrest and resuscitation model [7]. Cardiac arrest was induced in the conscious rat by rapid sequential intra-atrial injection of d-tubocurare (0.3mg) and ice-cold KCl solution (0.5 M; 0.12 ml/100g of body weight). Resuscitation was initiated 5 minutes after arrest following orotreacheal intubation with a 14-gauge catheter attached to a rodent ventilator (100% O_2, tidal volume: 10cc/kg, respiratory rate: 80 breaths/min). Simultaneous chest compressions and the infusion of normal saline (0.5 ml/min) were given until a spontaneous heartbeat returned. Epinephrine (4–10g) was given intravenously to establish a mean blood pressure greater than 80% of the pre-arrest value, at which point the animal was considered to be resuscitated (7–10 min ischemia). Ventilation was then adjusted to ~30% O_2 and 70% N_2O, depending on the normal range of blood gas, until spontaneous respiration was regained. For the PBN-treated rats, PBN (100 mg/kg) was infused intravenously immediately after resuscitation for 60 minutes. The untreated rats were given normal saline for the same period of time. Non-arrested rats went through the same surgical procedures except cardiac arrest. Non-arrested controls and resuscitated rats (1 h post-resuscitation) were decapitated and brains were removed for further process.

42.2.3 Isolation of Brain Mitochondria

Brain mitochondria were isolated using a method previously described with slight modifications [8]. In brief, tissue of cortex (whole layer, bilateral, 0.7–1.0g) and brainstem (~0.3g) were dissected (see Fig. 42.1 below) and rinsed

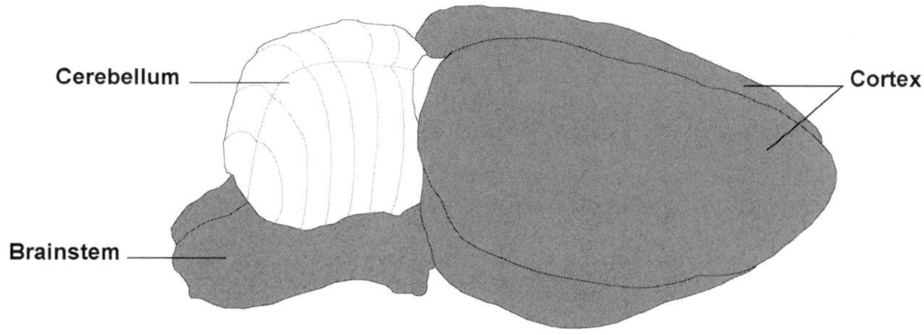

Fig. 42.1 Dissection of cortex and brainstem (gray-shaded area) of a rat brain.

in ice-cold isolation buffer (200 mM Mannitol, 70 mM Sucrose, and 5.0 mM MOPS, pH 7.4). The tissue was blotted dry, freed of visible blood vessels, then weighed and minced thoroughly. The tissue was suspended in isolation buffer (10 ml/g tissue) containing defatted bovine albumin (BSA, 0.2%) and EDTA (0.2 mM) and then treated with the protease Subtilisin A (5mg/g), for 30 second with light shaking. The suspension was then homogenized with a Teflon pestle (4 strokes). The homogenate was then centrifuged at 4 °C. The resulting mitochondrial pellet was washed (x2) with isolation buffer, centrifuged and resuspended to a final protein concentration of approximately 25 mg/ml and 10 mg/ml, cortex and brainstem, respectively.

42.2.4 Measurement of Mitochondrial Respiratory Rates

Oxidative rates were assessed by measuring oxygen consumption using a polarographic system consisting of a Clark-type electrode in the presence of the substrates glutamate plus malate [8]. The NADH-linked oxidative rates (state 3: ADP-stimulated; state 4: resting state, ADP-limited) were then calculated (natom oxygen/min/mg protein). The respiratory control ratio (RCR) was determined (state 3 / state 4). ADP-to-oxygen (ADP/O) ratios (nmol ADP per nanoatom O) were calculated as previously described.

42.2.5 Detection of HNE-Modified Mitochondrial Protein

Western blot analysis was used to detect HNE-modified mitochondrial protein. Samples of mitochondria were prepared by addition of isolation buffer containing 1 mM dithiothreitol and 1 mM phenylmethylsulfonyl. Samples (100 μg of protein) were electrophoresed on 10% SDS-polyacrilamide gels. The proteins on the gels were transferred to nitrocellulose membranes then incubated with 5% skim milk blocking buffer for 1 hour (room temperature). HNE modified proteins were detected by incubating the membranes with a 1:500 dilution of polyclonal anti-HNE antibody (Calbiochem) overnight (4°C) followed by incubation with horseradish peroxidase-conjungated anti-rabbit IgG (1:5000) for 1 hour (Jackson ImmunoResearch). The primary antibody immunoreactive protein bands were visualized using enhanced chemiluminescence detection system (ECL kit, Amersham).

42.2.6 Statistical Methods

All values were represented as mean ± S.D. Statistical analyses were performed using SPSS v13.0 for Windows. Group comparisons are made by one-way analysis of variance (ANOVA) using t-test. Significance was considered at the level of $p < 0.05$.

42.3 Results

42.3.1 Mitochondrial Respiratory Function

To determine the effect of aging on the overall mitochondrial oxidative capacity following cardiac arrest and resuscitation, we measured polarographically, oxygen consumption in the presence of the substrates of glutamate plus malate. As seen in Table 42.1, there were no significant differences in state 3 oxidative rates between 6-month and 24 -month-old rat brain mitochondria (cortex and brainstem) under all conditions. However, in brainstem, the state 3 oxidative rates in the 24-month-old rat brains were decreased in both conditions (22% and 17%, non-arrested and 1 h recovery groups, respectively), compared to the 6-month-old. The state 4 rates in cortex and brainstem were significantly higher at 1 h recovery compared to their respective non-arrested controls (6 and 24-month-old). In both the cortex and brainstem, there appeared to be no aging effect on the state 4 respiratory rates (non-arrested and 1 h recovery), acutely. In both age groups, the respiratory control ratios in cortex and brainstem were significantly lower at 1 h recovery compared to non-arrested controls. Since the state 3 oxidative rates were similar, the lower respiratory control ratios were as a result of higher state 4 rates. There were no differences in ADP/O in any conditions or between age groups (Table 42.1).

Figure 42.2 shows the RCR in the 24-month group (cortex and brainstem), non-arrested, 1 h recovery-untreated and 1 h recovery-PBN treated. The data show a decrease of 26% in cortex and 28% in brainstem, compared to the values of non-arrested controls. PBN treatment resulted in similar RCR values to the non-arrested baseline.

42.3.2 HNE Detection in Isolated Mitochondria of Brain

HNE-modified proteins were detected by Western blot analysis in isolated mitochondria from of cortex in 6 and 24-month-old rats. HNE adduct formation was observed within the molecular weight range of 65 to 110 kDa, with intense bands at 1h recovery in both age groups compared to the non-arrested controls (Fig. 42.3). The increase of HNE production was more evident in the aged group. The elevated levels of HNE adducts were reduced with PBN treatment in the 24-month group.

Table 42.1 Respiratory properties of rat brain mitochondria. Values are mean ± standard deviation

	Age (mos)	Condition	n	State 3 (nO/mg/min)	State 4 (nO/mg/min)	RCR	ADP/O
Cortex	6	Non-arrested	5	338.4 ± 34.5	29.0 ± 3.0	11.7 ± 1.0	2.1 ± 0.3
	6	1 h recovery	3	313.1 ± 16.8	34.9 ± 2*	9.0 ± 0.1*	2.4 ± 0.3
	24	Non-arrested	4	317.0 ± 34.4	27.5 ± 2.7	11.5 ± 0.8	2.2 ± 0.3
	24	1 h recovery	3	287.8 ± 51.2	34.2 ± 1.3*	8.5 ± 1.0*	2.2 ± 0.1
	24	PBN-treated	3	302.5 ± 71.9	28.3 ± 8.8	10.9 ± 1.2	2.5 ± 0.4
Brainstem	6	Non-arrested	5	319.7 ± 40.0	27.4 ± 3.9	11.9 ± 2.2	2.3 ± 0.3
	6	1 h recovery	3	297.7 ± 31.7	35.2 ± 1.2*	8.5 ± 0.8*	2.3 ± 0.4
	24	Non-arrested	4	248.3 ± 39.0	25.5 ± 2.7	9.7 ± 1.1	2.2 ± 0.1
	24	1 h recovery	3	246.8 ± 33.8	35.1 ± 1.3*	7.0 ± 0.8*§	2.3 ± 0.1
	24	PBN-treated	3	278.5 ± 66.4	28.5 ± 4.9	9.7 ± 0.7	2.3 ± 0.2

RCR, respiratory control ratio; ADP/O, ADP-to-oxygen ratio. * indicates significant difference (t-test, $p<0.05$) from the pre-arrest value in the same age group. §indicates significant difference (t-test, $p<0.05$) between the untreated and PBN-treated groups.

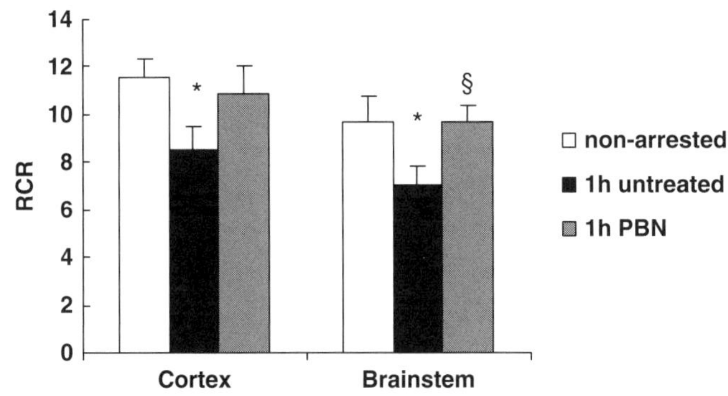

Fig. 42.2 Respiratory control ratio (RCR) in mitochondria isolated from cortex and brainstem of 24-month- old rats, non-arrested control (n = 4), untreated (n = 3) and *alpha*-phenyl-*tert*-butyl-nitrone (PBN)-treated rats (n = 3). Values are mean ± S.D., * indicates significant difference (t-test, p<0.05) from the values of non-arrested controls. § indicates significant difference between (t-test, p<0.05) the untreated and PBN-treated group.

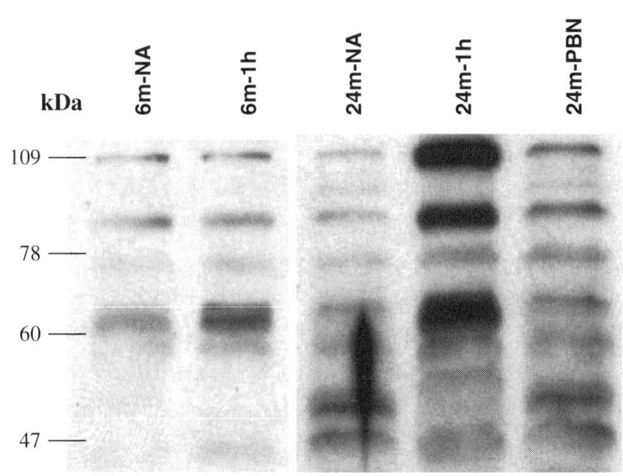

Fig. 42.3 Western blot analyses showing HNE adducts formation in th cortex of 6 and 24-month-old rats. NA: non-arrested controls; 6m-NA: 6-month non-arrested rat; 6m-1h: 6-month untreated 1 h recovery rat; 24m-NA: 24-month non-arrested rat; 24m-1h: 24-month 1 untreated 1h recovery rat; 24m-PBN: *alpha*-phenyl-*tert*-butyl-nitrone (PBN)-treated 24-month-old rat at 1 h recovery.

42.4 Discussion

This study focused on the early recovery phase from cardiac arrest. Our data showed that at 1 h recovery following cardiac arrest and resuscitation in both young and old rat brain RCR are decreased. This was evident by increased state

4 rates and not decreased state 3 rates. These data indicate that ischemic reperfusion injury acutely affects mitochondrial oxidative function through uncoupling. Furthermore, the HNE–modified mitochondrial proteins were elevated 1 h after resuscitation and were more apparent in the aged rats compared to the younger rats, suggesting that mitochondria in aged brain are more susceptible to damage as a result of lipid peroxidation. The data also show that with PBN treatment, there was improved mitochondrial respiratory function and reduced HNE modified mitochondrial proteins, possibly through free radical scavenging properties of PBN. The importance of this work is to provide information which may aid in potential therapeutic strategies aimed at early phase treatment in brain towards oxidative stress induced damage following cardiac arrest and resuscitation.

Acknowledgment This work was supported by NIH grants NS 46074 and GM 066309. We would like to especially thank Constantinos Tsipis for his assistance in preparation of the manuscript.

References

1. J. Lehotsky, R. Murin, A. Strapkova, A. Urikova, Z. Tatarkova, and P. Kaplan, Time course of ischemia/reperfusion-induced oxidative modification of neural proteins in rat forebrain, *Gen. Physiol. Biophys.* **23**(4), 401–415 (2004).
2. K. Xu, X. Sun, M. A. Puchowicz, and J. C. LaManna, Increased sensitivity to transient global ischemia in aging rat brain, *Adv. Exp. Med. Biol.* 599, 199–206 (2007).
3. J. C. LaManna, N. L. Neubauer, and J. C. Chavez, In: *Matuation Phenomenon in Cerebral Ischemia IV*, edited by Bazen NG, Ito U, Maecheselli VL, Kuroiwa T, and Klatzo I (Springer-Verlag, Berlin Heidelberg, 2001), pp. 223–227.
4. J. W. Phillis and C. Clough-Helfman, Protection from cerebral ischemic injury in gerbils with the spin trap agent N-tert-butyl-alpha-phenylnitrone (PBN), *Neurosci. Lett.* **116**(3), 315–319 (1990).
5. D. T. Lucas and L. I. Szweda, Cardiac reperfusion injury: aging, lipid peroxidation, and mitochondrial dysfunction, *Proc. Natl. Acad. Sci. U. S. A.* **95**(2), 510–514 (1998).
6. M. F. Anderson and N. R. Sims, Mitochondrial respiratory function and cell death in focal cerebral ischemia, *J. Neurochem.* **73**(3), 1189–1199 (1999).
7. K. Xu, M. A. Puchowicz, W. D. Lust, and J. C. Lamanna, Adenosine treatment delays postischemic hippocampal CA1 loss after cardiac arrest and resuscitation in rats, *Brain Res.* **1071**(1), 208–217 (2006).
8. J. Kerner, P. J. Turkaly, P. E. Minkler, and C. L. Hoppel, Aging skeletal muscle mitochondria in the rat: decreased uncoupling protein-3 content, *Am. J. Physiol Endocrinol. Metab.* **281**(5), E1054–E1062 (2001).

Part X
Others

Chapter 43
Measurement of Cerebral Tissue Oxygenation in Young Healthy Volunteers During Acetazolamide Provocation: A Transcranial Doppler and Near-Infrared Spectroscopy Investigation

Ilias Tachtsidis[1], Martin Tisdall[2], David T. Delpy[1], Martin Smith[1], and Clare E. Elwell[1]

Abstract Recent advances in near-infrared spectroscopy (NIRS) allow measurements of absolute tissue oxygen saturation (TOI) using spatially resolved spectroscopy (SRS), while enabling better depth sensitivity. However concerns remain regarding the relative contribution of the extracranial circulation to the cerebral NIRS TOI signal. In this study we investigated this during a period of selective rise in cerebral blood flow (CBF) produced by the administration of acetazolamide (ACZ) in 10 healthy volunteers. A two channel spectrometer (NIRO 300, Hamamatsu Photonics KK) was used to measure absolute cerebral TOI over the frontal cortex using the SRS technique using an optode spacing of 5 cm and 1.5 cm for channel 1 and 2 respectively. After ACZ administration we were able to observe a significant increase in the velocity of middle cerebral artery (V_{mca}, measured with the transcranial Doppler (TCD)) which was accompanied by an increase in TOI as monitored by the NIRO 300 with an optode spacing of 5 cm but not with an optode spacing of 1.5 cm. Furthermore a direct relationship was seen between the V_{mca} and the TOI measured at 5 cm optode spacing. This work suggests that using this commercial NIRS instrument with an optode spacing of 5 cm one is able to detect the intracranial changes.

Keywords: Near-infrared spectroscopy · Brain oxygenation · Brain blood flow · Acetazolamide

43.1 Introduction

Near-infrared spectroscopy (NIRS) has been widely used to evaluate changes in cerebral oxygenation and blood volume non-invasively. The technique involves near-infrared emitting and detecting optical probes placed on the

[1]Medical Physics and Bioengineering, University College London, Malet Place Engineering Building, Gower Street, London WCIE 6BT.
[2]Department of Neuroanaesthesia & Neurocritical Care, The National Hospital for Neurology & Neurosurgery, London.
Corresponding author: Illias Tachtsidis, e-mail: iliastac@medphys.ucl.ac.uk

scalp, transmitting light through cerebral as well as extracerebral tissues, such as skin, cranial bone, and cerebrospinal fluid (CSF). It is obvious that the position and volume of the interrogated tissue will depend on the position and spacing of the source and the detection fibres and results from mathematical modelling of light transport in tissue suggest that the average penetration depth of the detected near-infrared (NIR) light in the tissue increases with probe separation [1–3]. As a result, it is believed that oxygenation changes in deeper tissues of the brain can be detected with large optode spacing [4,5].

Recent advances in NIRS technology allow measurements of absolute tissue oxygen saturation (TOI) using spatially resolved spectroscopy (SRS) [6], while enabling better depth sensitivity [7]. However concerns remain regarding the relative contribution of the extracranial circulation to the cerebral NIRS TOI signal. The aim of this study was to investigate whether cerebral tissue oxygenation monitored using spatially resolved NIRS reflects changes in intracranial haemodynamics. This was done by monitoring the NIR derived cerebral tissue oxygenation signal using two different inter-optode spacings during a period of selective rise in cerebral blood flow produced by the administration of acetazolamide (ACZ).

43.2 Method

43.2.1 Subjects

After local ethics committee approval and written informed consent, 10 healthy volunteers (7 men and 3 women; age 27 to 33 years; mean 29.6 years) took part in this study.

43.2.2 Instrumentation

Figure 43.1 shows a schematic of the experimental set up. Middle cerebral artery flow velocity (V_{mca}) was measured in the basal right middle cerebral artery using a transcranial Doppler ultrasonography instrument (Pioneer Nicolet Biomedical Inc). After artery identification a permanently fixed 2-MHz probe was used. The envelope velocity was collected at 50Hz sampling rate and the mean V_{mca} was calculated every second using a trapezoidal integration function (MatLab Mathworks Inc).

A two channel continuous wave near-infrared spectrometer (NIRS), with a sampling rate of 6 Hz (NIRO 300, Hamamatsu Photonics KK) was used to measure absolute cerebral TOI over the frontal cortex using the SRS technique. The optodes were placed on the forehead (taking care to avoid the midline sinuses) and were secured in position by using an elastic bandage. The optodes

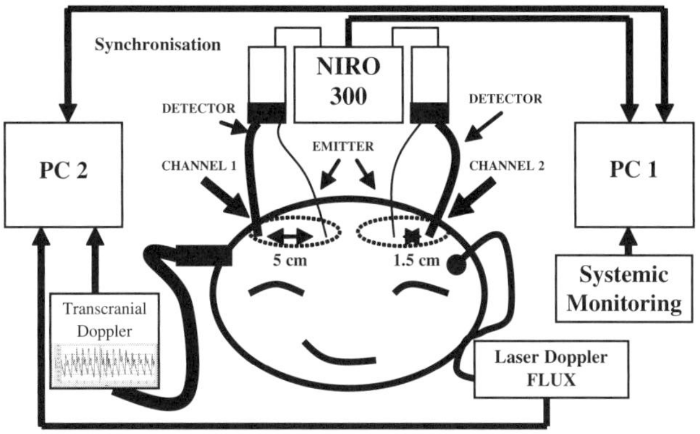

Fig. 43.1 Schematic representation of the experimental set up showing also the placement of the probes on the forehead.

for Channel 1 were placed over the right frontal lobe with an optode spacing of 5 cm; this measurement of TOI was termed 'Head TOI'. The optodes for Channel 2 were placed over the left frontal lobe with an optode spacing of 1.5 cm. This measurement of TOI was termed 'Superficial TOI'. Optical attenuators were used where necessary to optimise the signal to noise ratio.

Skin blood perfusion in the forehead was measured with a laser Doppler system (Moor Instruments). The laser Doppler probe was placed on the left forehead at the height of the eyebrows. The laser Doppler flux signal (the product of skin blood concentration and erythrocyte velocity) was collected continuously with a sampling rate of 50 Hz.

End tidal CO_2 ($EtCO_2$) was measured continuously (1Hz sampling rate) with nasal prongs (HP Merlin). Mean blood pressure (MBP) was monitored using the Portapres® system, which uses small finger cuffs to continuously and non-invasively measure the blood pressure waveform.

43.2.3 Procedure

Five minutes resting baseline data were recorded. All volunteers were then given 1 g of ACZ. ACZ was injected intravenously over a 2 minute period with the subject resting in a semi-recumbent position. Monitoring continued for 20 minutes post ACZ administration.

43.2.4 Analysis

The data from all instruments was collected and resampled to 1 point per minute (0.016 Hz) using a cubic interpolation function (MatLab Mathworks Inc.).

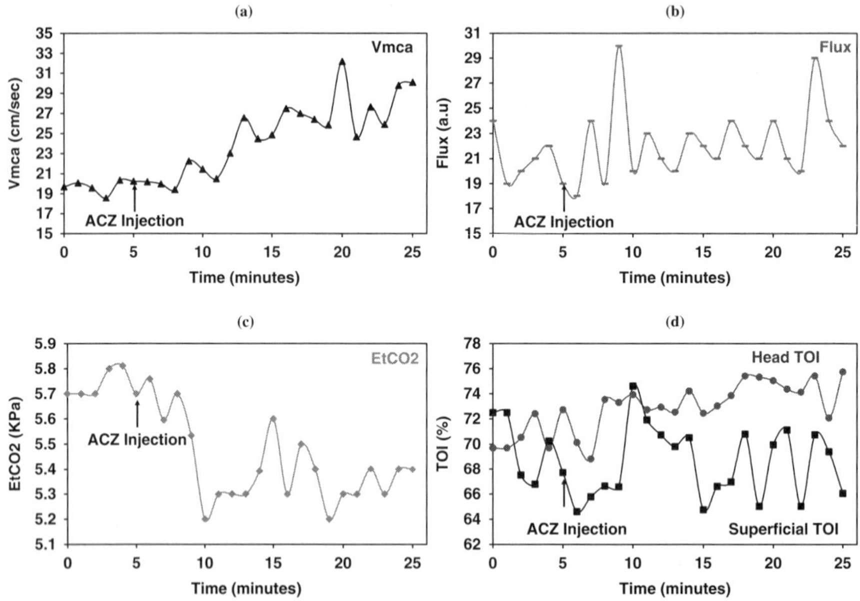

Fig. 43.2 Typical data from one volunteer. (a) $EtCO_2$ (b) Skin blood perfusion (c) Middle cerebral artery flow velocity (d) NIRS Tissue oxygenation index.

This optimised data handling without losing the temporal resolution of the ACZ response. Data was then normalised to the start of the ACZ injection. Figure 43.2 shows typical data from one volunteer during the 25 minute study.

Group data is presented as mean ± standard deviations. All p values were calculated for two-tailed tests of significance, and differences were considered statistically significant from baseline at $p<0.05$. Correlations between variables were analysed using the Pearson coefficient.

43.3 Results

Group summary data is shown in Fig. 43.3. Mean V_{mca} before the ACZ injection was 48.6 ± 28.1 cm/s; the maximum increase in velocity was 21.8 ± 10.3 cm/s which was equal to a mean increase of $51 \pm 17\%$ ($p<0.01$ from baseline). The mean V_{mca} at the end of the study was 63.3 ± 30.6 cm/s. The mean Head TOI before the ACZ injection was $70.1 \pm 3.5\%$; the maximum increase was $1.8 \pm 2.4\%$ ($p<0.05$ from baseline). The $EtCO_2$ at baseline was 5.1 ± 0.6 kPa; the maximum decrease was 0.8 ± 0.5 kPa ($p<0.01$ from baseline) which was reached 20 minutes after the injection.

No statistical differences were seen in the laser Doppler perfusion signal, the mean blood pressure and the mean Superficial TOI.

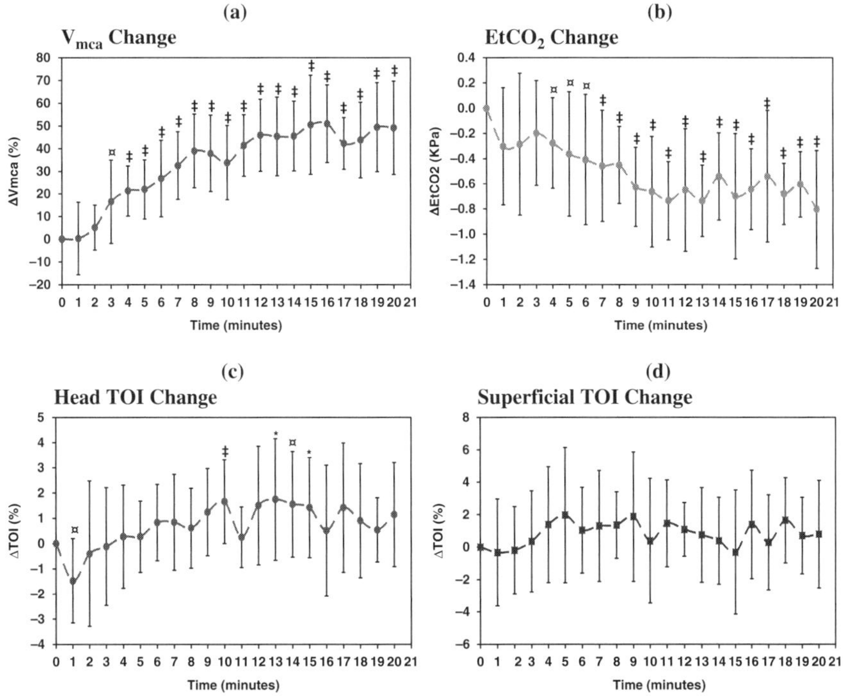

Fig. 43.3 Group summary data (a) Percentage changes in V_{mca} (b) Changes in $EtCO_2$ (c) Changes in Head TOI (d) Changes in the Superficial TOI. (t-test from baseline; *$p<0.05$ ¤ $p<0.03$ ‡ $p<0.01$).

The relationship between the group mean percentage changes in V_{mca} and the mean cerebral Head TOI changes show a significant association ($r = 0.77$, $p<0.01$), no correlation was seen with the Superficial TOI. These results are shown in Fig. 43.4.

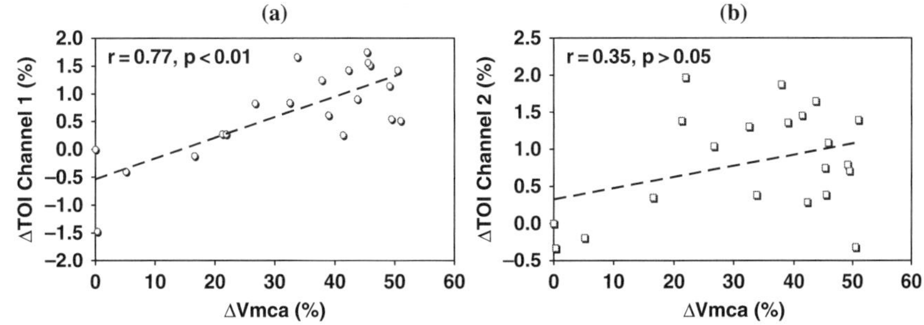

Fig. 43.4 Correlation analysis of the group summary data (a) Changes in Head TOI with percentage changes in V_{mca} (b) Changes in Superficial TOI with percentage changes in V_{mca}.

43.4 Discussion

In this study a significant rise in the V_{mca} and Head TOI were observed after ACZ injection (see Table 43.1). These results are similar with a previous study from Kaminogo et al.[8] who found a mean increase in V_{mca} of 44% and in regional cerebral oxygenation of 5.4% using the INVOS-3100 spectrometer (Somanetics Corp).

Skin blood flow and the Superficial TOI did not show any significant rise after the ACZ injection. These results agree with the study of Kohri et al. [9] who observed a significant increase in the head tissue oxygenation (optode distance 4 cm) of 2.4% but no increase in the superficial tissue oxygenation (optode distance 2 cm) using a combination of a spatially- and time-resolved NIRS system.

The changes in the Head TOI were not as large as the changes seen in the V_{mca} signal. This may be because the NIRS TOI signal represents a combination of arterial, capillary and venous oxygen saturation and volume data (1). Cranial NIRS interrogates a multi-compartmental system of arteries, arterioles, capillaries, venules and veins; the NIRS TOI, is the average ratio of oxygenated to total tissue haemoglobin concentrations in all these compartments. In order to investigate the origins of the Head TOI signal it is useful to simplify the above multi-compartmental system and it is usual to consider just two compartments, one arterial and one venous, with a typical arterial:venous volume ratio Va:Vv of 25–75% in a healthy adult brain.[10] By considering the definitions of arterial oxygen saturation ($SaO_2 = [HbO_2]_{art}/[HbT]_{art}$) and venous oxygen saturation ($SvO_2 = [HbO_2]_{ven}/[HbT]_{ven}$), where $[HbO_2]$ is oxygenated haemoglobin and $[HbT]$ is the total haemoglobin, we have:

$$\mathrm{TOI} = \mathrm{SaO_2} \cdot \frac{\mathrm{V_a}}{\mathrm{V_a} + \mathrm{V_v}} + \mathrm{SvO_2} \cdot \frac{\mathrm{V_v}}{\mathrm{V_a} + \mathrm{V_v}} \tag{43.1}$$

One now can replace SvO_2 in (1) by considering the definition of the Fick equation[11] shown below:

$$\mathrm{SvO_2} = \mathrm{SaO_2} - \frac{\mathrm{CMRO_2}}{\mathrm{k} \cdot \mathrm{CBF} \cdot \left[\mathrm{Hb} \cdot 10^{-2}\right]} \tag{43.2}$$

where $CMRO_2$ is the oxygen consumption (in ml of Oxygen/min), k is the oxygen combining power of Hb (~1.306 ml of Oxygen/g of Hb) and CBF is

Table 43.1 Maximum mean changes for all the volunteers

	Maximum Changes (Mean ±SD)
ΔV_{mca} (%)	51±17.1
$\Delta EtCO_2$ (KPa)	−0.8±0.5
Head ΔTOI (%)	1.8±2.4

cerebral blood flow (in ml/min) and *Hb* is the haemoglobin (in g of Hb/dL of blood). Therefore:

$$TOI = SaO_2 - \left(\frac{V_v}{V_a + V_v}\right) \cdot \left(\frac{CMRO_2}{k \cdot CBF \cdot \left[Hb \cdot 10^{-2}\right]}\right) \times 100\% \qquad (43.3)$$

Equation 43.3 demonstrates the direct relationship of the NIRO 300 TOI signal with the arterial/venous volume ratio, oxygen consumption and the indirect relationship with CBF.

ACZ is a selective inhibitor of carbonic anhydrase, which reversibly catalyses the conversion of $CO_2+H_2O \leftrightarrow H_2CO_3$. It therefore causes an increase in the H^+, HCO_3^- and CO_2 concentrations in the extracellular fluid of the brain, which are assumed to be the stimuli for the increase in CBF [12]. Bearing in mind (3) and with the knowledge that ACZ markedly increases CBF without any changes in oxygen consumption [13] or arterial oxygen saturation, it is possible that an increase in the venous volume ratio will attenuate the increase in TOI resulting from the rise in CBF. Furthermore each of the cerebral haemodynamics compartments has different saturations and volumes which may respond differently to ACZ causing a possible change in the Va:Vv.

43.5 Conclusions

After ACZ administration we were able to observe a significant increase in CBF (measured with the transcranial Doppler) which was accompanied by a small increase in absolute tissue oxygenation as monitored by the NIRO 300 with an optode spacing of 5 cm but not with an optode spacing of 1.5 cm. Furthermore a direct relationship was seen between the changes in V_{mca} and the Head TOI. This work suggests that using this commercial NIRS instrument with an optode spacing of 5 cm one is able to detect the intracranial changes. Further experimental work is needed to determine the relationship between TOI, CBF, CBV and $CMRO_2$.

Acknowledgment This work was funded by the MRC/EPSRC MIAS-IRC, the Clinical Research & Development Committee of RF&UCMS/UCLH Charitable Trustees (IT) and the Wellcome Trust (MT).

References

1. G.H. Weiss, R. Nossal, R.F. Bonner, "Statistics of penetration depth of photons re-emitted from irradiate tissue" J. Mod. Opt. **36**, 349–359 (1989).
2. C. Cui, C. Kumar, B. Chance, "Experimental study of migration depth for the photons measured at sample surface" Proc SPIE **1431**, 180–191 (1991).

3. E. Okada, M. Firbank, D.T. Delpy, "The effect of overlying tissue on the spatial sensitivity profile of near-infrared spectroscopy" Phys. Med. Biol. **40**, 2093–2108 (1995).
4. D.N.F. Harris, F.M. Cowans, D.A. Wertheim, S. Hamid, "NIRS in adults – effects of increasing optode separation" Adv. Exp. Med. Biol. **345**, 837–840 (1994).
5. E. Okada, D.T. Delpy, "Near-infrared light propagation in an adult head model. I. Modelling of low-level scattering in the cerebrospinal fluid layer" Applied Optics **42**(16), 2906–2914 (2003).
6. S. Suzuki, S. Takasaki, T. Ozaki, Y. Kobayashi, "A tissue oxygenation monitor using NIR spatially resolved spectroscopy" Proc. SPIE **3597**, 582–592 (1999).
7. P.G. Al Rawi, P. Smielewski, P.J. Kirkpatrick, "Evaluation of a near-infrared spectrometer (NIRO 300) for the detection of intracranial oxygenation changes in the adult head" Stroke **32**, 2492–2500 (2001).
8. M. Kaminogo, A. Ichikura, S. Shibata, T. Toba, M. Yonekura, "Effect of acetazolamide on regional cerebral oxygen saturation and regional cerebral blood flow" Stroke **26**, 2358–2360 (1995).
9. S. Kohri, Y. Hoshi, M. Tamura, C. Kato, Y. Kuge, N. Tamaki, "Quantitative evaluation of the relative contribution ratio of cerebral tissue to near-infrared signals in the adult human head: a preliminary study" Physiol. Meas. **23**, 301–312 (2002).
10. H. An, W. Lin, "Cerebral venous and arterial blood volumes can be estimated separately in humans using magnetic resonance imaging," Magn. Reson. Med. **48**, 583–588 (2002).
11. G.A. Dienel, "Energy generation in the central nervous system," Cerebral blood flow and metabolism, Edvinsson L. and Krause D.N., eds., (Lippincott Williams & Wilkins, Philadelphia, 2001), 140–161.
12. N.A. Lassen, "Is central chemoreceptor sensitive to intracellular rather than extracellular Ph?" Clin. Physiol. **10**(4), 311–319 (1990).
13. S. Vorstrup, L. Henriksen, O.B. Paulson, "Effect of acetazolamide on cerebral blood flow and cerebral metabolic rate of oxygen" J. Clin. Invest. **74**, 1634–1639 (1984).

Chapter 44
Measurement of Frontal Lobe Functional Activation and Related Systemic Effects: A Near-Infrared Spectroscopy Investigation

Ilias Tachtsidis, Terence S. Leung, Laurence Devoto, David T. Delpy, and Clare E. Elwell[1]

Abstract Near-infrared spectroscopy (NIRS) has been used to measure changes in cerebral oxy- and deoxy- haemoglobin ($\Delta[HbO_2]$, $\Delta[HHb]$) in response to functional activation. It has been previously reported that during functional activation of the motor cortex heart rate increases. The aim of this study was to investigate systemic changes during functional activation of the frontal cortex. The responses to anagram presentations with varying difficulty (4-Letters and 7-Letters) over a 6 minute period were recorded. A Hamamatsu NIRO 200 NIRS system recorded $\Delta[HbO_2]$ and $\Delta[HHb]$ using the modified Beer Lambert law (MBL) and tissue oxygenation index (TOI) employing spatial resolved spectroscopy (SRS) over the left and right frontal hemisphere. Mean blood pressure (MBP) and heart rate (HR) were measured continuously. Nine young healthy volunteers (mean age 23) were included in the analysis. Significant task related changes were observed in both the NIRS and systemic signals during the anagram solving with increases in $[HbO_2]$ and [HHb] accompanied by changes in MBP and HR. The $[HbO_2]$ and [HHb] signals measured over the frontal region were found to have a varying association with the MBP signal across different volunteers. The effect of these systemic changes on measured NIRS signals must be considered

Keywords: Near-infrared spectroscopy, Frontal lobe activation, Anagrams

44.1 Introduction

Near infrared spectroscopy (NIRS) has been widely used to investigate haemodynamic changes which occur in response to functional activation of specific regions of the cerebral cortex [1]. With conventional continuous wave NIRS

[1]Medical Physics and Bioengineering, University College London, Malet Place Engineering Building, Gower Street, London WC1E 6BT.
Corresponding author: Ilias Tachtsidis, e-mail: iliastac@medphys.ucl.ac.uk

systems it is not possible to determine exactly where the changes in attenuation have taken place within the illuminated tissue A general assumption is usually made that the changes seen in oxyhaemoglobin (HbO_2) and deoxyhaemoglobin (HHb) which are coincident with the period of stimulation originate from the cortical layers. We have previously reported that significant changes in heart rate occur during a finger tapping protocol for activation of the motor cortex in adults [2]. NIRS is increasingly being used to monitor the haemodynamic response to cognitive tasks by making measurements over the frontal and prefrontal regions [3–5]. It is possible that some mental tasks used in these studies may elicit a systemic response which may affect the measured NIRS signals. The aim of this study was to investigate the systemic changes during functional activation of the frontal cortex by measuring heart rate and mean blood pressure during anagram solving in adult volunteers.

44.2 Method

44.2.1 Subjects

Nine healthy male volunteers all right handed with English as their first language (age 20–25 years; mean 22.9 years) took part in this study.

44.2.2 Instrumentation

A continuous wave near-infrared spectrometer with a sampling rate of 6 Hz (NIRO 200, Hamamatsu Photonics KK) was used to measure changes in tissue oxygenation index (TOI) using spatially resolved spectroscopy and [HbO_2]) and [HHb] using the modified Beer-Lambert law. The optodes from the dual channel system were placed on the left and right forehead respectively (taking care to avoid the midline sinuses) and were shielded from ambient light by using an elastic bandage and a black cloth. An optode spacing of 4cm was used and the differential pathlength factor (DPF) applied was 6.26 [6]. A Portapres® system (TNO Institute of Applied Physics) was used to continuously and non-invasively measure mean blood pressure (MBP) and heart rate (HR) from the finger.

44.2.3 Procedure

All the volunteers were positioned in a sitting position. After 2 minutes of baseline rest measurements activation started with a minute period of solving 4-Letter anagrams (15 anagrams, 4 seconds per anagram) which was followed by a minute period of solving 7-Letter anagrams (6 anagrams, 10 seconds per anagram). Each period was repeated a total of three times, with the study ending after a 2 minute rest period (total study time 10 minutes). In this study solving an anagram was defined as producing one coherent word using only the letters from another (e.g. golf–flog; disease–seaside).

44.2.4 Analysis

The NIRS haemoglobin signals were first detrended to remove the slow drift, then all the signals including MBP and HR, were low pass filtered at 0.08 Hz to minimise the effects of other signal components. The filtering was carried out by a 5th order low pass Butterworth digital filter in forward backward directions to avoid introducing a phase delay (MatLab Mathworks Inc). The filtered signals from each volunteer were ensemble averaged over the repetition cycles (per volunteer two rest periods, three 4-Letter periods and three 7-Letter periods). Changes in total haemoglobin concentration ([HbT]) were calculated from the sum of changes in [HbO_2] and [HHb].

A 'Student t-test' was used to assess the significance of the responses (the threshold of significance was set at $p<0.05$ from baseline). Correlations between variables were analysed with the Pearson correlation model.

44.3 Results

44.3.1 Activation Results

Figure 44.1 shows the grand average of the NIRS, MBP and HR data from all nine volunteers during the entire ten minute test. The response to stimulation was calculated as the difference between the average of 10 seconds worth of baseline rest data, and the average of 10 seconds of data 20 seconds after the onset of the 4-Letter anagram solving period and the 7-Letter anagram solving period respectively. These changes are shown in Table 44.1. There was a significant change in [HbO_2], [HHb] and [HbT] between rest and the 4-Letter anagram solving period and between rest and the 7-Letter anagram solving period. There was no significant difference in the NIRS signals between the 4-Letter anagram solving period and the 7-Letter anagram solving period. The systemic signals (MBP and HR) also showed a significant difference between rest and the 4-Letter anagram solving period and between rest and the 7-Letter anagram solving period.

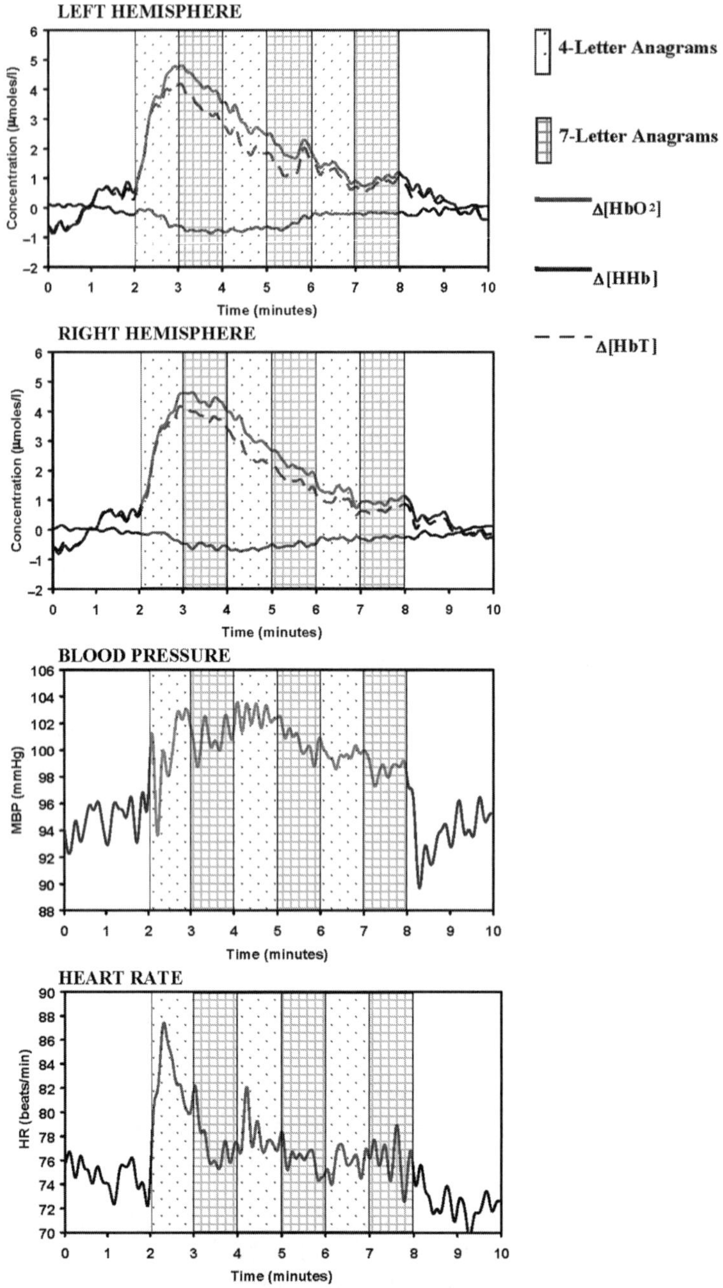

Fig. 44.1 Grand averaged responses from all nine subjects of NIRS haemoglobin signals and systemic measurements for the 10 minute study period.

Table 44.1 Response of NIRS signals over the left and right brain frontal regions and MBP and HR during 4- and 7- Letter anagram solving. Data from nine volunteers is presented as means±SD. (t-test *p<0.01; †p<0.03; ‡p<0.05)

	4-Letters - Rest		7-Letters - Rest	
	LH	RH	LH	RH
Δ[HbO_2] (μmoles/1)	2.48 ±2.42*	2.41±1.72*	2.19±2.48*	2.30±1.76*
Δ[HHb] (μmoles/1)	−0.28±0.46‡	−0.38±0.33*	−0.46±0.48*	−0.44±0.30*
TOI(%)	−1.03±3.20	−0.29±1.22	0.03±2.01	0.25±1.30
Δ[HbT](μmoles/1)	2.19±2.69†	2.03±1.92*	1.73±2.76‡	1.86±1.92*
MBP (mmHg)	4.6±4.1*		4.8±3.2*	
HR (beats/min)	7.7±5.2*		3.5±3.0*	

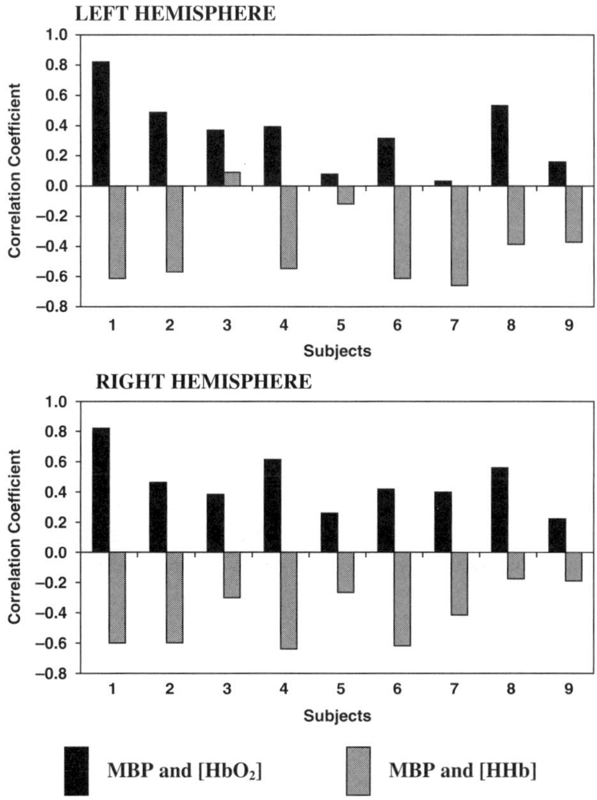

Fig. 44.2 Individual correlation coefficients between MBP and [HbO_2] and MBP and [HHb] for each subject.

44.3.2 Inter-Subject Correlation

The [HbO_2] and [HHb] signals measured over the frontal region were found to have a varying association with the MBP signal across different volunteers. In order to investigate this we calculated the correlation coefficient between the filtered [HbO_2] and MBP and [HHb] and MBP for both hemispheres in all subjects. These results are shown in Fig. 44.2.

44.4 Discussion

In this study we demonstrated significant changes in NIRS variables ([HbO_2], [HHb] and [HbT]) measured over both the right and left frontal region between rest and a 4-letter anagram solving period and between rest and a 7-letter anagram solving period. Furthermore, in the group data, we observed a significant increase from rest in both MBP and HR during periods when the subjects were solving the 4 and 7 letter anagrams. We have found that the haemoglobin changes measured by NIRS during frontal lobe functional activation were in some volunteers significantly correlated with the systemic changes in MBP and HR.

Given that the anagram task involves both language and spatial processing it is reasonable that the response is not lateralised. A previous study by Chance et al.[7] describes a robust prefrontal oxygenation signal in response to anagram solving which also appears to be bilateral. To minimise the likelihood of movement artifact or non stimulation related changes, we chose in this study not to ask the subject to provide the answers to the presented anagrams and we were therefore unable to score the subjects' performance on the task and determine whether the systemic changes were related to this level of performance.

To our knowledge this is the first report of simultaneous measurements of MBP and NIRS variables during a functional activation task of the frontal cortex. In a previous study we reported an increase in heart rate during a finger tapping task in adult volunteers [2]. Obrig et al [8]. measured arterial blood pressure and heart rate in three subjects during visual stimulation (annular checkerboard alternating at 8 Hz) and found a coherence between arterial blood pressure and [HbO_2] at frequencies coinciding with the heart rate and spontaneous low frequency oscillations (centred around 0.1 Hz), but made no specific comment about activation related changes in the systemic variables.

Nearly all studies of task-specific activation using functional neuroimaging rely on the existence of a close coupling between regional changes in brain metabolism and regional cerebral blood flow, sometimes referred to as activation-flow coupling or neurovascular coupling. Regional haemodynamic changes are used as a surrogate marker for changes in regional brain function that occur due to changes in metabolism during excitatory or inhibitory neurotransmission, both of which are energy consuming processes. The relatively high correlation coefficient found in some subjects in this study between [HbO_2] and MBP and [HHb] and MBP suggests a centrally mediated mechanism that might play a role in the overall functional haemodynamic changes seen in the brain in these individuals during stimulation. Caution therefore should be taken when analysing the cerebrovascular response of the activated brain due to the unknown haemodynamic contribution from the systemic alterations occurring during stimulation.

Acknowledgment Thanks to the MIAS-IRC and the Medical Physics group of the Institute of Physics for travel support (IT) and the Clinical Research & Development Committee of RF&UCMS/UCLH Charitable Trustees (IT).

References

1. H. Obrig, A. Villringer "Beyond the visible-imaging the human brain with light" J. Cereb. Blood Flow Metab. **23**, 1–18 (2003).
2. T. Leung, C.E. Elwell, J. Henty, D.T. Delpy "Simultaneous measurement of cerebral tissue oxygenation over the adult frontal and motor cortex during rest and functional activation" Adv. Exp. Med & Biol **530**, 385–389 (2002).
3. J. Fallgater, W. Strik "Frontal brain activation during the Wiscosin Card Sorting Test assessed with two channel near-infrared spectroscopy" Eur. Arch. Psychiatry Clin. Neurosci. **248**, 245–249 (1998).
4. M.J. Hermann, A.C. Ehlis, A.J. Fallgater "Prefrontal activation through task requirements of emotional induction measured with NIRS" Biol. Psychol. **64**, 255–263 (2003).
5. Y. Hoshi, B.H. Tsou, V.A. Billock, M. Tanosaki, Y. Iguchi, M. Shimada, T. Shinba, Y. Yamada, I. Oda "Spatiotemporal characteristics of hemodynamic changes in the human lateral prefrontal cortex during working memory tasks" Neuroimage **20**, 1493–1504 (2003).
6. A. Duncan, J. Meek, M. Clemence, C.E. Elwell, L. Tyszczuk, M. Cope, D.T. Delpy "Optical pathlength measurements on adult head, calf and forearm and the head of the newborn infant using phase resolved optical spectroscopy" Phys. Med. Biol. **40**, 295–304 (1995).
7. B. Chance, S. Nioka, S. Sadi, C. Li "Oxygenation and blood concentration changes in human subjects prefrontal activation by anagram solutions" Adv. Exp. Med. & Biol. **510**,397–401 (2003).
8. H. Obrig, M. Neufang, R. Wenzel, M. Kohl, J. Steinbrink, K. Einhaupl, A. Villringer "Spontaneous low frequency oscillations of cerebral hemodynamics and metabolism in human adults" Neuroimage **12**, 623–639 (2000).

Author Index

Subject Index